Lecture Notes in Mathematics 2191

More information about this series at http://www.springer.com/series/304

Gene Abrams • Pere Ara • Mercedes Siles Molina

Leavitt Path Algebras

Gene Abrams
Department of Mathematics
University of Colorado
Colorado Springs, USA

Pere Ara
Departament de Matemàtiques
Universitat Autònoma de Barcelona
Barcelona, Spain

Mercedes Siles Molina
Departamento de Álgebra, Geometría y
Topología
Universidad de Málaga
Málaga, Spain

ISSN 0075-8434 ISSN 1617-9692 (electronic)
Lecture Notes in Mathematics
ISBN 978-1-4471-7343-4 ISBN 978-1-4471-7344-1 (eBook)
DOI 10.1007/978-1-4471-7344-1

Library of Congress Control Number: 2017946554

Mathematics Subject Classification (2010): 16Sxx, 16D25, 16D40, 16D50, 16E20, 16E40, 16E50, 16N60, 16P20, 16P40, 46L05

Printed on acid-free paper

This Springer imprint is published by Springer Nature
The registered company is Springer-Verlag London Ltd.
The registered company address is: The Campus, 4 Crinan Street, London, N1 9XW, United Kingdom

A Juan Luis. Gracias por tu generosidad y por hacer tuyos mis éxitos.

A Fina. Gràcies per la teva generositat i per fer teus els meus èxits.

To Mickey. Unbounded thanks for all you do.

Preface

The great challenge in writing a book about a topic of ongoing mathematical research interest lies in determining *who* and *what*. *Who* are thc readers for whom the book is intended? *What* pieces of the research should be included?

The topic of **Leavitt path algebras** presents both of these challenges, in the extreme. Indeed, much of the beauty inherent in this topic stems from the fact that it may be approached from many different directions, and on many different levels.

The topic encompasses classical ring theory at its finest. While at first glance these Leavitt path algebras may seem somewhat exotic, in fact many standard, well-understood algebras arise in this context: matrix rings and Laurent polynomial rings, to name just two. Many of the fundamental, classical ring-theoretic concepts have been and continue to be explored here, including the ideal structure, $\mathbb{Z}$-grading, and structure of finitely generated projective modules, to name just a few.

The topic continues a long tradition of associating an algebra with an appropriate combinatorial structure (here, a directed graph), the subsequent goal being to establish relationships between the algebra and the associated structures. In this particular setting, the topic allows for (and is enhanced by) visual, pictorial representation via directed graphs. Many readers are no doubt familiar with the by-now classical way of associating an algebra over a field with a directed graph, the standard *path algebra*. The construction of the *Leavitt path algebra* provides another such connection. The path algebra and Leavitt path algebra constructions are indeed related, via algebras of quotients. However, one may understand Leavitt path algebras without any prior knowledge of the path algebra construction.

The topic has significant, deep connections with other branches of mathematics. For instance, many of the initial results in Leavitt path algebras were guided and motivated by results previously known about their analytic cousins, the *graph C^*-algebras*. The study of Leavitt path algebras quickly matured to adolescence (when it became clear that the algebraic results are not implied by the C^* results), and almost immediately thereafter to adulthood (when in fact some C^* results, including some *new* C^* results, were shown to follow from the algebraic results). A number of long-standing questions in algebra have recently been resolved using Leavitt path algebras as a tool, thus further establishing the maturity of the subject.

The topic continues a deep tradition evident in many branches of mathematics in which K-theory plays an important role. Indeed, in retrospect, one can view Leavitt path algebras as precisely those algebras constructed to produce specified K-theoretic data in a universal way, data arising naturally from directed graphs. Much of the current work in the field is focused on better understanding just how large a role the K-theoretic data plays in determining the structure of these algebras.

Our goal in writing this book, the *Why?* of this book, simultaneously addresses both the *Who?* and *What?* questions. We provide here a self-contained presentation of the topic of Leavitt path algebras, a presentation which will allow readers having different backgrounds and different topical interests to understand and appreciate these structures. In particular, graduate students having only a first year course in ring theory should find most of the material in this book quite accessible. Similarly, researchers who don't self-identify as algebraists (e.g., people working in C^*-algebras or symbolic dynamics) will be able to understand how these Leavitt path algebras stem from, or apply to, their own research interests. While most of the results contained here have appeared elsewhere in the literature, a few of the central results appear here for the first time. The style will be relatively informal. We will often provide historical motivation and overview, both to increase the reader's understanding of the subject and to play up the connections with other areas of mathematics. Although space considerations clearly require us to exclude some otherwise interesting and important topics, we provide an extensive bibliography for those readers who seek additional information about various topics which arise herein.

More candidly, our real *Why?* for writing this book is to share what we know about Leavitt path algebras in such a way that others might become prepared, and subsequently inspired, to join in the game.

Colorado Springs, USA — Gene Abrams
Barcelona, Spain — Pere Ara
Málaga, Spain — Mercedes Siles Molina

Acknowledgments

Throughout the years during which this book was being written, dozens of individuals offered topic suggestions, pointed out typos, and helped to improve the final version. We are deeply appreciative of their contributions.

While much of the time and energy which went into writing this book was expended while each of the authors was in residence at her/his own university, the three authors were able on a number of occasions to spend time visiting each other's home institutions to help more effectively move the project toward completion. We collectively thank the Universitat Autònoma de Barcelona, the Universidad de Málaga, and the University of Colorado Colorado Springs for their warm hospitality during these visits.

The first author was partially supported by a Simons Foundation Collaboration Grant, #20894.

The second author was partially supported by the Spanish MEC and Fondos FEDER through projects MTM2008-06201-C02-01, MTM2011-28992-C02-01, and MTM2014-53644-P and by the Generalitat de Catalunya through project 2009SGR 1389.

The third author was partially supported by the Spanish MEC and Fondos FEDER through projects MTM2010-15223, MTM2013-41208-P, and MTM2016-76327-C3-1-P and by the Junta de Andalucía and Fondos FEDER, jointly, through projects FQM-336 and FQM-02467.

Contents

Chapter 1
The Basics of Leavitt Path Algebras: Motivations, Definitions and Examples

In this initial chapter of the book we introduce the *Leavitt path algebra* $L_K(E)$ which arises from a directed graph E and a field K. We begin in Sect. 1.1 by reviewing a class of algebras defined and investigated in the early 1960s by W.G. Leavitt, the now-so-called *Leavitt algebra* $L_K(1, n)$ corresponding to any positive integer n and a field K. The importance of these algebras is that they are the universal examples of algebras which fail to have the *Invariant Basis Number* property; to wit, if $R = L_K(1, n)$, then the free left R-modules R and R^n are isomorphic. Once the definition of $L_K(E)$ is given for any graph E, we will recover $L_K(1, n)$ as $L_K(R_n)$, where R_n is the graph having one vertex and n loops at that vertex.

With the general definition of a Leavitt path algebra presented in Sect. 1.2 in hand, we give in Sect. 1.3 the three fundamental examples of Leavitt path algebras: the Leavitt algebras; full matrix rings over K; and the Laurent polynomial algebra $K[x, x^{-1}]$. These three types of Leavitt path algebras will provide the motivation and intuition for many of the general results in the subject.

The subject did not arise in a vacuum. Indeed, there are intimate connections between Leavitt path algebras and a powerful monoid-realization result of Bergman. In addition, there are strong and historically significant connections between Leavitt path algebras and *graph* C^**-algebras*. We describe both of the connections in Sect. 1.4.

As we will see, there are natural modifications to the definition of a Leavitt path algebra which provide the data to construct a (seemingly) more general class of algebras, the *relative Cohn path algebras* $C_K^X(E)$ corresponding to a graph E, a subset X of the vertices of E, and a field K. Although the class of relative Cohn path algebras contains as specific examples the class of Leavitt path algebras, we will see in Sect. 1.5 that every relative Cohn path algebra $C_K^X(E)$ is in fact isomorphic to the Leavitt path algebra $L_K(E(X))$ for some germane graph $E(X)$.

Although the motivating examples of Leavitt path algebras arise from finite graphs, the definition of $L_K(E)$ allows for the construction even when E is infinite. Indeed, much of the interesting work and many of the applications-related results

G. Abrams et al., *Leavitt Path Algebras*, Lecture Notes in Mathematics 2191,
DOI 10.1007/978-1-4471-7344-1_1

about Leavitt path algebras arise in the situation where E is infinite. We show in Sect. 1.6 that, perhaps surprisingly, every Leavitt path algebra may be viewed as a direct limit (in an appropriate category) of Leavitt path algebras associated to finite graphs.

We conclude the chapter by presenting in Sect. 1.7 a brief historical overview of the subject.

1.1 A Motivating Construction: The Leavitt Algebras

A student's first exposure to the theory of rings more than likely involves a study of various "basic examples", typically including fields, $\mathbb{Z}$, matrix rings over fields, and polynomial rings with coefficients in a field. It is not hard to show that each of these rings R has the *Invariant Basis Number (IBN) property*:

IBN: If m and m' are positive integers with the property that the free left modules R^m and $R^{m'}$ are isomorphic, then $m = m'$.

Less formally, a ring has the IBN property (more succinctly: *is IBN*) if any two bases (i.e., linearly independent spanning sets) of any finitely generated free left R-module have the same number of elements. It turns out that many general classes of rings have this property (e.g., noetherian rings and commutative rings), classes of rings which include all of the basic examples with which the student first made acquaintance. (Typically, the student would have encountered the fact that the field of real numbers has the IBN property in an undergraduate course on linear algebra.)

Unfortunately, since all of the examples the student first encounters have the IBN property, the student more than likely is left with the wrong impression, as there are many important classes of rings which are not IBN. Perhaps the most common such example is the ring $B = \text{End}_K(V)$, where V is an infinite-dimensional vector space over a field K. Then B is not IBN (with a vengeance!): it is not hard to show that the free left B-modules B^m and $B^{m'}$ are isomorphic for *all* positive integers m, m'.

Definition 1.1.1 Suppose R is not IBN. Let $m \in \mathbb{N}$ be minimal with the property that $R^m \cong R^{m'}$ as left R-modules for some $m' > m$. For this m, let n denote the minimal such m'. In this case we say that R has *module type* (m, n).

So, for example, $B = \text{End}_K(V)$ has module type $(1, 2)$. We note that in the definition of module type it is easy to show that the same m, n arise if one considers free right R-modules, rather than left.

As we shall see, there is a perhaps surprising amount of structure inherent in non-IBN rings. To start with, in the groundbreaking article [112], Leavitt proves the following fundamental result.

Theorem 1.1.2 *For each pair of positive integers $n > m$ and field K there exists a unital K-algebra $L_K(m,n)$, unique up to K-algebra isomorphism, such that:*

(i) *$L_K(m,n)$ has module type (m,n), and*
(ii) *for each unital K-algebra A having module type (m,n) there exists a unit-preserving K-algebra homomorphism $\phi : L_K(m,n) \to A$ which satisfies certain (natural) compatibility conditions.*

Our motivational focus here is on non-IBN rings of module type $(1,n)$ for some $n > 1$. In particular, such a ring then has the property that there exist isomorphisms of free modules

$$\phi \in \mathrm{Hom}_R(R^1, R^n) \text{ and } \psi \in \mathrm{Hom}_R(R^n, R^1), \quad \text{for which } \psi \circ \phi = \iota_R \text{ and } \phi \circ \psi = \iota_{R^n},$$

where ι denotes the identity map on the appropriate module. Using the usual interpretation of homomorphisms between free modules as matrix multiplications (a description which the student encounters for the real numbers in an undergraduate linear algebra course, and which is easily shown to be valid for any unital ring), we see that such isomorphisms exist if and only if there exist $1 \times n$ and $n \times 1$ R-vectors

$$\begin{pmatrix} x_1 \ x_2 \cdots x_n \end{pmatrix} \text{ and } \begin{pmatrix} y_1 \\ y_2 \\ \vdots \\ y_n \end{pmatrix},$$

$$\text{for which } \begin{pmatrix} x_1 \ x_2 \cdots x_n \end{pmatrix} \cdot \begin{pmatrix} y_1 \\ y_2 \\ \vdots \\ y_n \end{pmatrix} = (1_R) \text{ and } \begin{pmatrix} y_1 \\ y_2 \\ \vdots \\ y_n \end{pmatrix} \cdot \begin{pmatrix} x_1 \ x_2 \cdots x_n \end{pmatrix} = \begin{pmatrix} 1_R & 0 & \cdots & 0 \\ 0 & 1_R & \cdots & 0 \\ \vdots & & \ddots & \vdots \\ 0 & 0 & \cdots & 1_R \end{pmatrix}.$$

Rephrased,

$$_R R^1 \cong {}_R R^n \text{ for some } n > 1$$

if and only if there exist $2n$ elements $x_1, \ldots, x_n, y_1, \ldots, y_n$ of R for which

$$\sum_{i=1}^{n} x_i y_i = 1_R \text{ and } y_i x_j = \delta_{ij} 1_R \text{ (for all } 1 \leq i, j \leq n). \tag{1.1}$$

The relations displayed in (1.1) provide the key idea in constructing the Leavitt algebras, and will play a central role in motivating the subsequent more general construction of Leavitt path algebras. For example, in the ring $B = \mathrm{End}_K(V)$ having module type $(1,2)$, it is straightforward to describe a set x_1, x_2, y_1, y_2 of $2 \cdot 2 = 4$ elements of B which behave in this way.

Indeed, given $n > 1$, it is relatively easy to construct an algebra A which contains $2n$ elements behaving as do those in (1.1). Specifically, let K be any field, let

$$S = K\langle X_1, \ldots, X_n, Y_1, \ldots, Y_n \rangle$$

be the free associative K-algebra in $2n$ non-commuting variables, let I denote the ideal of S generated by the relations

$$I = \langle \sum_{i=1}^{n} X_i Y_i - 1,\ Y_i X_j - \delta_{ij} 1 \mid 1 \leq i, j \leq n \rangle,$$

and let

$$A = S/I.$$

Then the set $\{x_i = \overline{X_i}, y_j = \overline{Y_j} \mid 1 \leq i, j \leq n\}$ behaves in the desired way (by construction), so that $A^1 \cong A^n$ as left A-modules.

At this point one must be careful: although we have just constructed a K-algebra A for which $A^1 \cong A^n$, we cannot conclude that the module type of A is $(1, n)$ until we can guarantee the minimality of n. (For instance, it is not immediately clear that the algebra $A = S/I$ is necessarily nonzero.) But this is precisely what Leavitt establishes in [112]. Indeed, the K-algebra $L_K(1, n)$ of Theorem 1.1.2 is exactly the algebra $A = S/I$ constructed here. We formalize this in the following.

Definition 1.1.3 Let K be any field, and $n > 1$ any integer. Then the *Leavitt K-algebra of type* $(1, n)$, denoted $L_K(1, n)$, is the K-algebra

$$K\langle X_1, \ldots, X_n, Y_1, \ldots, Y_n \rangle \ / \ \langle \sum_{i=1}^{n} X_i Y_i - 1,\ Y_i X_j - \delta_{ij} 1 \mid 1 \leq i, j \leq n \rangle.$$

Notationally, it is often more convenient to view $R = L_K(1, n)$ as the free associative K-algebra on the $2n$ variables $x_1, \ldots, x_n, y_1, \ldots, y_n$, subject to the relations $\sum_{i=1}^{n} x_i y_i = 1_R$ and $y_i x_j = \delta_{ij} 1_R$ $(1 \leq i, j \leq n)$. Specifically, $L_K(1, n)$ is the universal K-algebra of type $(1, n)$.

We summarize our discussion thus far. Although non-IBN rings might seem exotic at first sight, they in fact occur naturally. Non-IBN rings having module type $(1, n)$ can be constructed with relative ease. The key ingredient to produce such rings is the existence of elements $x_1, \ldots, x_n, y_1, \ldots, y_n$ for which the relations displayed in (1.1) are satisfied.

For those readers curious about the previous "surprising amount of structure" comment, we conclude this section with the following morsel of supporting evidence, established by Leavitt in [113].

Theorem 1.1.4 *For every integer $n \geq 2$, and for any field K, $L_K(1, n)$ is a simple K-algebra.*

This remarkable result will in fact follow as a corollary of the more general results presented in Chap. 2.

1.2 Leavitt Path Algebras

With the construction of the Leavitt algebras $L_K(1,n)$ as a motivational backdrop, we are nearly in position to present the central idea of this book, the Leavitt path algebras. We start by fixing some basic notation and definitions.

Notation 1.2.1 If K is a field, then by $K^\times$ we denote the nonzero elements of K, i.e., the invertible elements. $\mathbb{Z}$ denotes the set of integers; $\mathbb{Z}^+ = \{0,1,2,\ldots\}$; $\mathbb{N} = \{1,2,3,\ldots\}$.

Unless otherwise indicated, an *R-module* will mean a *left* R-module. In the sequel we will write our left-module homomorphisms on the side opposite the scalars; in particular, the composition fg of left R-module homomorphisms means 'first f, then g'. In all other situations (e.g., for ring homomorphisms, or lattice maps), composition of functions will be written so that $f \circ g$ means 'first g, then f'.

Definitions 1.2.2 A *(directed) graph* $E = (E^0, E^1, r, s)$ consists of two sets E^0, E^1 and two functions $r, s : E^1 \to E^0$. The elements of E^0 are called *vertices* and the elements of E^1 *edges*. We place no restriction on the cardinalities of E^0 and E^1, nor on properties of the functions r and s. Throughout, the word "graph" will always mean "directed graph".

If $s^{-1}(v)$ is a finite set for every $v \in E^0$, then the graph is called *row-finite*. A vertex v for which $s^{-1}(v) = \emptyset$ is called a *sink*, while a vertex v for which $r^{-1}(v) = \emptyset$ is called a *source*. In other words, v is a sink (resp., source) if v is not the source (resp., range) of any edge of E. A vertex which is both a source and a sink is called *isolated*. A vertex v such that $|s^{-1}(v)|$ is infinite is called an *infinite emitter*. If v is either a sink or an infinite emitter, we call v a *singular vertex*; otherwise, v is called a *regular vertex*. The expressions Sink(E), Source(E), Reg(E), and Inf(E) will be used to denote, respectively, the sets of sinks, sources, regular vertices, and infinite emitters of E.

A *path* μ in a graph E is a sequence of edges $\mu = e_1, e_2, \ldots, e_n$ such that $r(e_i) = s(e_{i+1})$ for $i = 1, \ldots, n-1$. In this case, $s(\mu) = s(e_1)$ is the *source* of μ, $r(\mu) = r(e_n)$ is the *range* of μ, and $n = \ell(\mu)$ (or $n = |\mu|$) is the *length* of μ. We typically denote μ by using the more efficient notation $e_1e_2\cdots e_n$. We view the vertices of E as paths of length 0; to streamline notation, we will sometimes extend the functions s and r to E^0 by defining $s(v) = r(v) = v$ for $v \in E^0$. If $\mu = e_1e_2\cdots e_n$ is a path then we denote by μ^0 the set of its vertices, that is, $\mu^0 = \{s(e_1), r(e_i) \mid 1 \le i \le n\}$. For $n \ge 2$ we define E^n to be the set of paths in E of length n, and define Path(E) $= \bigcup_{n \ge 0} E^n$, the set of all paths in E.

Here are the main objects of our desire.

Definition 1.2.3 (Leavitt Path Algebras) Let E be an arbitrary (directed) graph and K any field. We define a set $(E^1)^*$ consisting of symbols of the form $\{e^* \mid e \in E^1\}$. The *Leavitt path algebra of E with coefficients in K*, denoted $L_K(E)$, is the free associative K-algebra generated by the set $E^0 \cup E^1 \cup (E^1)^*$, subject to the following relations:

(V) $vv' = \delta_{v,v'} v$ for all $v, v' \in E^0$,
(E1) $s(e)e = er(e) = e$ for all $e \in E^1$,
(E2) $r(e)e^* = e^* s(e) = e^*$ for all $e \in E^1$,
(CK1) $e^* e' = \delta_{e,e'} r(e)$ for all $e, e' \in E^1$, and
(CK2) $v = \sum_{\{e \in E^1 | s(e) = v\}} ee^*$ for every regular vertex $v \in E^0$.

Phrased another way, $L_K(E)$ is the free associative K-algebra on the symbols $E^0 \cup E^1 \cup (E^1)^*$, modulo the ideal generated by the five types of relations indicated in the previous list.

Remark 1.2.4 There is a connection between the classical notion of *path algebras* and the notion of Leavitt path algebras, which we describe here. As a brief reminder, if K is a field and $G = (G^0, G^1)$ is a directed graph then the *path K-algebra of G*, denoted KG, is defined as the free associative K-algebra generated as an algebra by the set $G^0 \cup G^1$, with relations given by (V) and (E1) of Definition 1.2.3. Equivalently, KG is the K-algebra having Path(G) as basis, and in which multiplication is defined by the K-linear extension of path concatenation (i.e., $p \cdot q = pq$ if $r(p) = s(q)$, 0 otherwise).

Given a graph E, we define the *extended graph of E* (also sometimes called the *double graph of E*) as the new graph $\widehat{E} = (E^0, E^1 \cup (E^1)^*, r', s')$, where $(E^1)^* = \{e^* \mid e \in E^1\}$, and the functions r' and s' are defined as

$$r'|_{E^1} = r,\ s'|_{E^1} = s,\ r'(e^*) = s(e), \text{ and } s'(e^*) = r(e) \text{ for all } e \in E^1.$$

(In other words, each edge e^* in $(E^1)^*$ has orientation the reverse of that of its counterpart $e \in E^1$.) Then $L_K(E)$ is the quotient of the path K-algebra $K\widehat{E}$ by the ideal of $K\widehat{E}$ generated by relations given in (CK1) and (CK2) of Definition 1.2.3.

Remark 1.2.5 (The Universal Property of $L_K(E)$) Suppose E is a graph, and A is a K-algebra which contains a set of pairwise orthogonal idempotents $\{a_v \mid v \in E^0\}$, and two sets $\{a_e \mid e \in E^1\}$, $\{b_e \mid e \in E^1\}$ for which

(i) $a_{s(e)} a_e = a_e a_{r(e)} = a_e$ and $a_{r(e)} b_e = b_e a_{s(e)} = b_e$ for all $e \in E^1$,
(ii) $b_f a_e = \delta_{e,f} a_{r(e)}$ for all $e, f \in E^1$, and
(iii) $a_v = \sum_{\{e \in E^1 | s(e) = v\}} a_e b_e$ for every regular vertex $v \in E^0$.

We call such a family an *E-family* in A. By the relations defining the Leavitt path algebra, there exists a unique K-algebra homomorphism $\varphi : L_K(E) \to A$ such that $\varphi(v) = a_v$, $\varphi(e) = a_e$, and $\varphi(e^*) = b_e$ for all $v \in E^0$ and $e \in E^1$. We will often refer to this as the *Universal Property* of $L_K(E)$.

Notation 1.2.6 We sometimes refer to the edges in the graph E as the *real edges*, and the additional edges of $\widehat{E}$ (i.e., the elements of $(E^1)^*$) as the *ghost edges*. If $\mu = e_1 e_2 \cdots e_n$ is a path in E, then the element $e_n^* \cdots e_2^* e_1^*$ of $L_K(E)$ is denoted by μ^*.

Remark 1.2.7 Less formally (but no less accurately), one may view the Leavitt path algebra $L_K(E)$ as follows. Consider the standard path algebra $K\widehat{E}$ of the extended graph. Then impose on $K\widehat{E}$ the following relations:

(i) If e is an edge of E, we replace any expression of the form e^*e in $K\widehat{E}$ by the vertex $r(e)$.
(ii) If e and f are distinct edges in E, then we define $e^*f = 0$ in $K\widehat{E}$.
(iii) If v is a regular vertex, then the sum over *all* terms of the form ee^* for which $s(e) = v$ is replaced by v in $K\widehat{E}$.

The resulting algebra is precisely $L_K(E)$.

In the standard pictorial description of a directed graph E, we use the notation $\bullet^v \xrightarrow{(n)} \bullet^w$ to indicate that there are n distinct edges e_i in E for which $s(e_i) = v$ and $r(e_i) = w$; the value of n may be finite or infinite.

Example 1.2.8 An example will no doubt help clarify the definition of a Leavitt path algebra. Let E be the graph pictorially described by

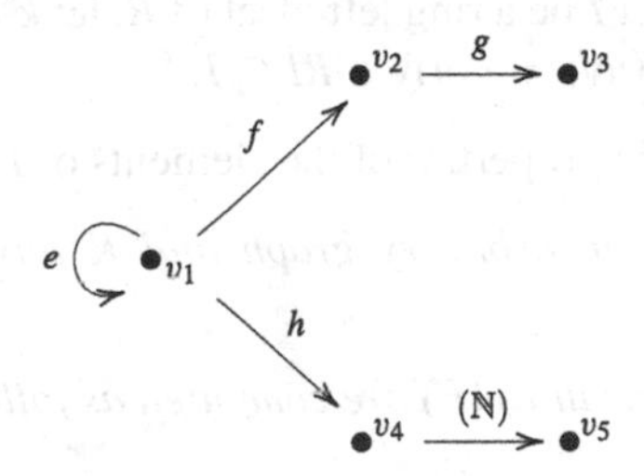

Here are some representative computations in $L_K(E)$ (for any field K).

$$v_1 f = f = f v_2 \text{ by (E1), while } v_2 f^* = f^* = f^* v_1 \text{ by (E2)}$$

$$f^*f = v_2, \text{ while } f^*h = f^*e = 0 \text{ both by (CK1)}$$

$$v_1 = ee^* + ff^* + hh^* \text{ by (CK2)}$$

$$gg^* = v_2 \text{ by (CK2) (the sum contains only one term)}$$

We observe that there is no (CK2) relation at v_4 (as $v_4 \in \text{Inf}(E)$); neither is there a (CK2) relation at the sinks v_3 and v_5.

Remark 1.2.9 We note that the construction of the Leavitt path algebra for a graph E over a field K can be extended in the obvious way to the construction of the Leavitt path *ring* for a graph E over an arbitrary unital ring R. (See for example [148], where the author studies Leavitt path algebras with coefficients in a commutative ring.)

The existence of a multiplicative identity in $L_K(E)$ depends on whether or not E^0 is finite (see Lemma 1.2.12 below). But even in non-unital situations, there is still much structure to be exploited.

Definition 1.2.10 An associative ring R is said to have a *set of local units* F if F is a set of idempotents in R having the property that, for each finite subset $r_1, \dots, r_n$ of R, there exists an $f \in F$ for which $fr_if = r_i$ for all $1 \leq i \leq n$. Rephrased, a set of idempotents $F \subseteq R$ is a set of local units for R if each finite subset of R is contained in a (unital) subring of the form fRf for some $f \in F$.

An associative ring R is said to have *enough idempotents* if there exists a set of nonzero orthogonal idempotents E in R for which the set F of finite sums of distinct elements of E is a set of local units for R. Note that, when this happens, ${}_RR = \oplus_{e \in E} Re$ as left R-modules.

For a ring with local units, an abelian group M is a left R-*module* if there is a (standard) module action of R on M, but with the added proviso that $RM = M$. (This is the appropriate generalization of the requirement that $1_R \cdot m = m$ for all m in a left module M over a unital ring R.)

For a field K, a ring R with local units is said to be a K-*algebra* if R is a K-vector space (with scalar action $\cdot$), and $(k \cdot r)s = k \cdot (rs)$ for all $k \in K$, $r, s \in R$.

Remark 1.2.11 In any K-algebra R with local units, every (one-sided, resp., two-sided) ring ideal of R is a (one-sided, resp., two-sided) K-algebra ideal of R. This is easy to see: for instance, let I be a ring left ideal of R, let $k \in K$ and $y \in I$. Let $u \in R$ with $y = uy$. Then $ky = k(uy) = (ku)y \in RI \subseteq I$.

We now give some basic properties of the elements of $L_K(E)$.

Lemma 1.2.12 *Let E be an arbitrary graph and K any field. Let $\gamma, \lambda, \mu, \rho$ be elements of* $\mathrm{Path}(E)$.

(i) *Products of monomials in $L_K(E)$ are computed as follows:*

$$(\gamma\lambda^*)(\mu\rho^*) = \begin{cases} \gamma\kappa\rho^* & \text{if} \quad \mu = \lambda\kappa \text{ for some } \kappa \in \mathrm{Path}(E) \\ \gamma\sigma^*\rho^* & \text{if} \quad \lambda = \mu\sigma \text{ for some } \sigma \in \mathrm{Path}(E) \\ 0 & \text{otherwise.} \end{cases}$$

In particular, if $\ell(\lambda) = \ell(\mu)$, then $\lambda^*\mu \neq 0$ if and only if $\lambda = \mu$, in which case $\lambda^*\mu = r(\lambda)$.

(ii) *The K-action on the algebra $L_K(E)$ is trivial; that is,*

$$(k\gamma\lambda^*)(k'\mu\rho^*) = kk'(\gamma\lambda^*\mu\rho^*)$$

for $k, k' \in K$.

(iii) *The algebra $L_K(E)$ is spanned as a K-vector space by the set of monomials of the form*

$$\{\gamma\lambda^* \mid \gamma, \lambda \in \mathrm{Path}(E) \text{ for which } r(\gamma) = r(\lambda)\}.$$

In other words, every nonzero element x of $L_K(E)$ may be expressed as

$$x = \sum_{i=1}^{n} k_i \gamma_i \lambda_i^*,$$

where $k_i \in K^\times$, and $\gamma_i, \lambda_i \in \mathrm{Path}(E)$ with $r(\gamma_i) = r(\lambda_i)$ for each $1 \leq i \leq n$. We note that, except for trivial cases, this representation is not unique; i.e., the displayed monomials do not form a basis of $L_K(E)$.

(iv) *The algebra $L_K(E)$ is unital if and only if E^0 is finite. In this case,*

$$1_{L_K(E)} = \sum_{v \in E^0} v.$$

(v) *For each $\alpha \in L_K(E)$ there exists a finite set of distinct vertices $V(\alpha)$ for which $\alpha = f\alpha f$, where $f = \sum_{v \in V(\alpha)} v$. Moreover, the algebra $L_K(E)$ is a ring with enough idempotents (consisting of the vertices E^0), and thus a ring with local units (consisting of sums of distinct elements of E^0).*

Proof (i) By (CK1), any expression of the form e^*f in $L_K(E)$ reduces either to 0 or to the vertex $r(e)$, from which the statement follows by a straightforward computation.

(ii) follows directly from the definition of $L_K(E)$ as the free K-algebra on various generators.

(iii) follows easily from (i).

(iv) For E^0 finite, the indicated element acts as the identity by the representation of elements of $L_K(E)$ given in (iii). If E^0 is infinite, then there is no element of $L_K(E)$ which acts as an identity on each element of the set $\{v \mid v \in E^0\}$.

(v) By the orthogonality given in Definition 1.2.3(V), it is clear that any sum of distinct vertices in $L_K(E)$ yields an idempotent. Now let $\alpha = \sum_{i=1}^{m} k_i \gamma_i \lambda_i^*$ be an arbitrary element of $L_K(E)$, and let $V(\alpha)$ denote the (finite) set of vertices which appear either as $s(\gamma_i)$ or as $s(\lambda_i)$ for some $1 \leq i \leq m$. If we define $f = \sum_{v \in V(\alpha)} v$, then an easy computation yields that $\alpha = f\alpha f$. The additional statements follow in the same manner. □

Definitions 1.2.13 We say that a graph E is *connected* if $\widehat{E}$ is a connected graph in the usual sense, that is, if given any two vertices $u, v \in E^0$ there exist $h_1, h_2, \ldots, h_m \in E^1 \cup (E^1)^*$ such that $\eta = h_1 h_2 \cdots h_m$ is a path in $\widehat{E}$ such that $s(\eta) = u$ and $r(\eta) = v$. The *connected components* of a graph E are the graphs $\{E_i\}_{i \in \Lambda}$ such that E is the disjoint union $E = \sqcup_{i \in \Lambda} E_i$, where every E_i is connected.

We close the section by recording the following observation, which is easily verified utilizing the Universal Property of $L_K(E)$ 1.2.5.

Proposition 1.2.14 *Let E be an arbitrary graph and K any field. Suppose $E = \sqcup_{i \in \Lambda} E_i$ is a decomposition of E into its connected components. Then $L_K(E) \cong \oplus_{i \in \Lambda} L_K(E_i)$.*

1.3 The Three Fundamental Examples of Leavitt Path Algebras

Part of the beauty of the Leavitt path algebras is that they include many well-known, but seemingly disparate, classes of algebras. To make these connections clear, we introduce some notation which will be used throughout.

Notation 1.3.1 We let R_n denote the *rose with n petals* graph having one vertex and n loops:

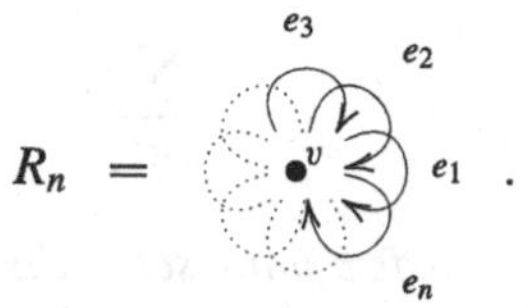

In particular, a special role in the theory is played by the graph R_1:

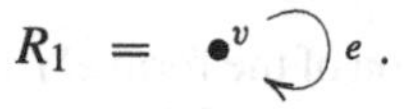

For any $n \in \mathbb{N}$ we let A_n denote the *oriented n-line graph* having n vertices and $n-1$ edges:

The examples presented in the following three propositions may be viewed as the three primary colors of Leavitt path algebras. Making good now on an earlier promise, we validate our claim that the Leavitt algebras $L_K(1, n)$ are truly motivating examples for the more general notion of a Leavitt path algebra.

Proposition 1.3.2 *Let $n \geq 2$ be any positive integer, and K any field. Let $L_K(1, n)$ be the Leavitt K-algebra of type $(1,\ n)$ presented in Definition 1.1.3, and let R_n be the rose with n petals. Then*

$$L_K(1, n) \cong L_K(R_n).$$

Proof That these two algebras are isomorphic follows directly from the definition of $L_K(1, n)$ as a quotient of the free associative algebra on $2n$ variables, modulo the relations given in display (1.1). Specifically, we map $x_i \mapsto e_i$ and $y_i \mapsto e_i^*$. Then the relations given in (1.1) are precisely the relations provided by the (CK1) and (CK2) relations of Definition 1.2.3. □

The rose with one petal produces a more-familiar (although less-exotic) algebra. Prior to the description of $L_K(R_1)$, the following remark is very much in order.

Remark 1.3.3 If E is a graph and $e \in E^1$, then the element ee^* of $L_K(E)$ is always an idempotent, since using (CK1) we have $(ee^*)(ee^*) = e(e^*e)e^* = er(e)e^* = ee^*$. However, ee^* does *not* equal $s(e)$ unless e is the only edge emitted by $s(e)$ (since in that case the (CK2) relation reduces to the equation $s(e) = ee^*$).

For any field K, the *Laurent polynomial K-algebra* is the associative K-algebra generated by the two symbols x and y, with relations $xy = yx = 1$. For obvious reasons this algebra is denoted by $K[x, x^{-1}]$. The elements of $K[x, x^{-1}]$ may be written as $\sum_{i=m}^{n} k_i x^i$ (where $k_i \in K$ and $m \leq n \in \mathbb{Z}$); note in particular that the exponents are allowed to include negative integers. Viewed another way, $K[x, x^{-1}]$ is the group algebra of $\mathbb{Z}$ over K.

Proposition 1.3.4 *Let K be any field. Then*

$$K[x, x^{-1}] \cong L_K(R_1).$$

Proof By the (CK1) relation and Lemma 1.2.12(iv) we have $x^*x = v = 1$ in $L_K(R_1)$. But since v emits only the edge x, Remark 1.3.3 yields $xx^* = v = 1$ in $L_K(R_1)$ as well, and the result now follows. □

The third of the three primary colors of Leavitt path algebras moves us from the less-exotic $K[x, x^{-1}]$ to the almost-mundane matrix algebras $M_n(K)$.

Proposition 1.3.5 *Let K be any field, and $n \geq 1$ any positive integer. Then*

$$M_n(K) \cong L_K(A_n).$$

Proof Let $\{f_{i,j} \mid 1 \leq i, j \leq n\}$ denote the standard matrix units in $M_n(K)$. We define the map $\varphi : L_K(A_n) \to M_n(K)$ by setting $\varphi(v_i) = f_{i,i}$, $\varphi(e_i) = f_{i,i+1}$, and $\varphi(e_i^*) = f_{i+1,i}$. Using Remark 1.3.3, it is then easy to check that φ is an isomorphism of K-algebras, as desired. □

The title of this section notwithstanding, we provide a fourth example of a well-known classical algebra which arises as a specific example of a Leavitt path algebra.

Example 1.3.6 The *Toeplitz graph* is the graph

$$E_T = \quad {}^{e}\circlearrowright \bullet^u \xrightarrow{f} \bullet^v \quad .$$

Let K be any field. We denote by $\mathscr{T}_K$ the *algebraic Toeplitz K-algebra*

$$\mathscr{T}_K = L_K(E_T).$$

Proposition 1.3.7 *For any field K, the Leavitt path algebra $L_K(E_T)$ is isomorphic to the free associative K-algebra $K\langle x, y\rangle$, modulo the single relation $xy = 1$. Rephrased, the algebraic Toeplitz K-algebra $\mathcal{T}_K$ is the K-algebra $K\langle U, V\rangle$ investigated by Jacobson in* [98].

Proof We begin by noting that in $L_K(E_T)$ we have the relations $ee^* + ff^* = u$ and $u + v = 1$. We consider the elements $X = e^* + f^*$ and $Y = e + f$ of $L_K(E_T)$. Then by (CK1) we have $XY = u + v = 1$, while $YX = ee^* + ff^* = u \neq 1$ by (CK1) and (CK2). The subalgebra of $\mathcal{T}_K = L_K(E_T)$ generated by X and Y then contains $1 - u = v$, which in turn gives that this subalgebra contains $e = Yu$, $f = Yv$, $e^* = uX$, and $f^* = vX$. These observations establish that the map $\varphi : K\langle U, V\rangle \to L_K(E_T)$ given by the extension of $\varphi(U) = e^* + f^*, \varphi(V) = e + f$ is a surjective K-algebra homomorphism. The injectivity of φ will follow from results in Sect. 1.5; see specifically Example 1.5.20. □

1.4 Connections and Motivations: The Algebras of Bergman, and Graph C^*-Algebras

In presenting a description of the Leavitt algebras $L_K(1, n)$ in the very first section of this book, our intent was to provide some sort of "natural" motivation for the relations which define the more general Leavitt path algebras. In this section we present two additional avenues which lead in a natural way to the description of Leavitt path algebras. The first such avenue takes us through a description of the finitely generated projective modules over a ring, while the second provides an expedition through the world of C^*-algebras. These two topics will be explored much more extensively, and in more generality, in Chaps. 3 and 5, respectively.

Definition 1.4.1 Let R be any unital ring. We denote by $\mathcal{V}(R)$ the semigroup whose elements are the isomorphism classes of the finitely generated projective left R-modules, with operation given by $[P] + [Q] = [P \oplus Q]$.

Clearly $\mathcal{V}(R)$ is a commutative monoid for any ring R, with zero element $[\{0\}]$. In addition, it is apparent that $\mathcal{V}(R)$ has the property that

$$x + y = [\{0\}] \text{ in } \mathcal{V}(R) \text{ if and only if } x = y = [\{0\}]. \tag{1.2}$$

Since R is assumed here to be unital (we will relax this requirement later), then each finitely generated projective left R-module is isomorphic to a direct summand of R^n for some integer n, so it is similarly apparent that the element $I = [R]$ of $\mathcal{V}(R)$ has the property that

$$\forall\, x \in \mathcal{V}(R)\ \exists\, y \in \mathcal{V}(R) \text{ and } n \in \mathbb{N} \text{ for which } x + y = nI. \tag{1.3}$$

In a groundbreaking construction conceived and executed by Bergman in [51], it is shown that, in this context, anything that *can* happen in fact *does* happen. That is, if S is any finitely generated commutative monoid having the (necessary) properties described in displays (1.2) and (1.3), and K is any field, then there exists an explicitly constructed unital K-algebra R for which $\mathcal{V}(R) \cong S$. Moreover, this K-algebra is universal in the sense that, for any unital K-algebra T having $\mathcal{V}(T) \cong S$, there exists a nonzero homomorphism $\varphi : R \to T$ which induces the identity on S.

We now define, for any graph E, an associated semigroup M_E; with the previous three sections in mind, the relations which describe M_E should seem familiar.

Definition 1.4.2 Let E be an arbitrary graph. We denote by M_E the free abelian monoid on a set of generators $\{a_v \mid v \in E^0\}$, modulo relations given by

$$a_v = \sum_{\{e \in E^1 \mid s(e)=v\}} a_{r(e)} \tag{1.4}$$

for each $v \in \mathrm{Reg}(E)$.

So to any graph E we can associate the semigroup M_E, and to any graph E and field K we can associate the semigroup $\mathcal{V}(L_K(E))$. We will prove the following in Chap. 3; this result shows that these two semigroups are intimately related.

Theorem 1.4.3 *Let E be any row-finite graph and K any field. Then, using the presentation of the monoid M_E given in Definition 1.4.2, $L_K(E)$ is precisely the universal K-algebra corresponding to the monoid M_E as constructed by Bergman in [51, Theorem 6.2]. In particular,*

$$\mathcal{V}(L_K(E)) \cong M_E.$$

The upshot of this discussion is that, with the Leavitt algebras $L_K(1, n)$ having been presented as our first motivational offering, there is now a second motivating description of the Leavitt path algebras (arising from row-finite graphs): they are precisely the universal K-algebras which arise in [51, Theorem 6.2] for monoids of the form M_E. This is no small conclusion, in the sense that for general commutative monoids which satisfy displayed conditions (1.2) and (1.3), it is rare that one can so explicitly describe the corresponding universal K-algebras.

In fact, the Leavitt algebras $L_K(1, n)$ play a basic role in Bergman's analysis. Specifically, let $\mathbb{Z}_{n-1}$ be the standard cyclic group of order $n - 1$, and let S be the semigroup $\mathbb{Z}_{n-1} \cup \{z\}$ where $z + g = g = g + z$ for all $g \in S$. Then S is a commutative monoid satisfying (1.2) and (1.3) above, and $L_K(1, n)$ is the universal K-algebra corresponding to S. We will investigate this construction much more deeply in Chap. 3.

And now for something completely different. While the next few paragraphs (and various subsequent portions of this book) discuss the notion of a C^*-algebra, readers may choose to skip these portions while still gaining an in-focus picture of Leavitt path algebras. In any event, it behooves us to remark that C^*-algebras are always algebras in the usual ring-theoretic sense over the field of complex numbers $\mathbb{C}$.

Definitions 1.4.4 Let E be an arbitrary graph. (In the following context it is typically assumed that the sets E^0 and E^1 are at most countable, but we need not make those assumptions here.) A *Cuntz–Krieger E-family* in a C^*-algebra B consists of a set of mutually orthogonal projections $\{p_v \mid v \in E^0\}$ and a set of partial isometries $\{s_e \mid e \in E^1\}$ satisfying

$$s_e^* s_e = p_{r(e)} \text{ for } e \in E^1, \; p_v = \sum_{\{e \mid s(e)=v\}} s_e s_e^* \text{ whenever } v \in \mathrm{Reg}(E), \text{ and } s_e s_e^* \le p_{s(e)} \text{ for } e \in E^1.$$

It is shown in [105] that there is a C^*-algebra $C^*(E)$, called the *graph C^*-algebra of E*, generated by a universal Cuntz–Krieger E-family $\{s_e, p_v\}$; in other words, for every Cuntz–Krieger E-family $\{t_e, q_v\}$ in a C^*-algebra B, there is a homomorphism $\pi = \pi_{t,q} : C^*(E) \to B$ such that $\pi(s_e) = t_e$ and $\pi(p_v) = q_v$ for all $e \in E^1, v \in E^0$.

The relations presented in Definitions 1.4.4 clearly smack of those which generate the Leavitt path algebras, so it is probably not surprising that there is a strong connection between the structures $L_{\mathbb{C}}(E)$ and $C^*(E)$. In fact, we will show in Chap. 5 that $L_{\mathbb{C}}(E)$ embeds as a $\mathbb{C}$-algebra inside $C^*(E)$ in a natural way, and that $C^*(E)$ may be realized as the completion of $L_{\mathbb{C}}(E)$ in an appropriate topology.

The main point to be made here is that the Leavitt path $\mathbb{C}$-algebra $L_{\mathbb{C}}(E)$ can be realized and motivated as an algebraic foundation upon which $C^*(E)$ can be built. We will note often throughout the later chapters that while there are striking (indeed, *compellingly mysterious*) similarities amongst some of the results pertaining to the two structures $L_{\mathbb{C}}(E)$ and $C^*(E)$, there are other situations in which perhaps-anticipated parallels between these structures are indeed different. Further, while the Leavitt path $\mathbb{C}$-algebra $L_{\mathbb{C}}(E)$ is then naturally motivated by the $\mathbb{C}$-algebra $C^*(E)$ in this way, we shall see that the structural properties of $L_{\mathbb{C}}(E)$ typically pass to identical structural properties of $L_K(E)$ for any field K.

As of the writing of this book, there is no vehicle which allows one to easily establish results on the algebra side as direct consequences of results on the analytic side, or vice versa.

1.5 The Cohn Path Algebras and Connections to Leavitt Path Algebras

In the previous section we focused on two different constructions, both of which naturally led to the construction of Leavitt path algebras: the "realization algebras" of Bergman, and the graph C^*-algebras. In this section we present a third construction, the *relative Cohn path algebras* $C_K^X(E)$, and in particular the *Cohn path algebras* $C_K(E)$, which also can be used to produce Leavitt path algebras.

The relative Cohn path algebras will serve two main purposes here. First, it will be trivial to show that every Leavitt path algebra is a quotient of a relative Cohn path algebra by an appropriately defined ideal. As will become apparent, the vector space structure of a Cohn path algebra is straightforward (e.g., a basis of $C_K(E)$ is easy to describe). This structure in turn will allow us to almost seamlessly obtain various results about Leavitt path algebras simply by appealing to quotient-preserving properties. Second, the relative Cohn path algebras will allow us to further showcase the ubiquity of the Leavitt path algebras. Specifically, for any graph E we will show that each relative Cohn path algebra $C_K^X(E)$ (including $C_K(E)$ itself) is isomorphic to the Leavitt path algebra $L_K(F)$ of some graph F.

The motivational information given in the previous section was presented almost as an advertising teaser ("stay tuned for further details!", the hard work to be confronted in subsequent chapters). In contrast, our description and use of the relative Cohn path algebras will require us to get our hands dirty right away. We start with the most important of these.

Definition 1.5.1 Let E be an arbitrary graph and K any field. We define a set $(E^1)^*$ consisting of symbols of the form $\{e^* \mid e \in E^1\}$. The *Cohn path algebra of E with coefficients in K*, denoted by $C_K(E)$, is the free associative K-algebra generated by the set $E^0 \cup E^1 \cup (E^1)^*$, subject to the relations given in (V), (E1), (E2), and (CK1) of Definition 1.2.3.

In other words, $C_K(E)$ is the algebra generated by the same symbols as those which generate $L_K(E)$, but on which we do *not* impose the (CK2) relation. Since by (CK1) we have $e^*f = \delta_{e,f}r(e)$ in $C_K(E)$ for $e,f \in E^1$ (and the lack of the (CK2) relation in $C_K(E)$ notwithstanding), it is easy to show that there is still some information to be had about expressions of the form ee^* in $C_K(E)$: namely, that the family $\{ee^* \mid e \in E^1\}$ is a set of orthogonal idempotents in $C_K(E)$. What we do not impose in $C_K(E)$ is any relationship between this family and the set of vertices E^0 in $C_K(E)$.

Remark 1.5.2 In a manner similar to the explanation given in Remark 1.2.4, another way of looking at Cohn path algebras is the following: $C_K(E)$ is the quotient of the path K-algebra over the extended graph $K\widehat{E}$ by the ideal of $K\widehat{E}$ generated by the relations given in (CK1).

In [64], P.M. Cohn introduced and studied the collection of K-algebras $\{U_{1,n} \mid n \in \mathbb{N}\}$ (for any field K); these have come to be known as the *Cohn algebras*, and as such we now use the notation $C_K(1,n)$ for these. It is clear that for each $n \in \mathbb{N}$ we have $C_K(R_n) \cong C_K(1,n)$. Thus the algebras $C_K(1,n) \cong C_K(R_n)$ stand in relation to the more general Cohn path algebras in precisely the same way that the Leavitt algebras $L_K(1,n) \cong L_K(R_n)$ stand in relation to the more general Leavitt path algebras.

Remark 1.5.3 As with Leavitt path algebras, we can define analogously the *Cohn path ring* $C_R(E)$ for any unital ring R and graph E.

Example 1.5.4 The algebra investigated by Jacobson which was presented in Proposition 1.3.7 is the quintessential example of a Cohn path algebra. Specifically, the free associative K-algebra $K\langle U,V\rangle$ modulo the single relation $UV = 1$ is precisely the Cohn path algebra $C_K(R_1)$, where R_1 is as usual the graph with one vertex and one loop.

The following result follows directly from the definition of the indicated algebras.

Proposition 1.5.5 *Let E be an arbitrary graph and K any field. Let I be the ideal of the Cohn path algebra $C_K(E)$ generated by the set*

$$\{v - \sum_{e \in s^{-1}(v)} ee^* \mid v \in \mathrm{Reg}(E)\}.$$

Then

$$L_K(E) \cong C_K(E)/I$$

as K-algebras.

Unlike the situation in the Leavitt path algebras, inside the Cohn path algebras every element can be expressed *in a unique way* as a linear combination of the terms $\lambda\nu^*$, with λ and ν paths in E for which $r(\lambda) = r(\nu)$.

Proposition 1.5.6 *Let E be an arbitrary graph and K any field. Then*

$$\mathscr{B} = \{\lambda\nu^* \mid \lambda, \nu \in \mathrm{Path}(E), r(\lambda) = r(\nu)\}$$

is a K-basis of $C_K(E)$.

Proof Let A be the K-vector space with basis $\mathscr{B}$. We define a bilinear product on A by the formula

$$(\lambda_1\nu_1^*)(\lambda_2\nu_2^*) = \begin{cases} \lambda_1\lambda_2'\nu_2^* & \text{if} \quad \lambda_2 = \nu_1\lambda_2' \text{ for some } \lambda_2' \in \mathrm{Path}(E) \\ \lambda_1(\nu_1')^*\nu_2^* & \text{if} \quad \nu_1 = \lambda_2\nu_1' \text{ for some } \nu_1' \in \mathrm{Path}(E) \\ 0 & \text{otherwise.} \end{cases}$$

To see that this gives the structure of an associative K-algebra on A we only need to check that $x = y$, where $x = (\lambda_1\nu_1^*)((\lambda_2\nu_2^*)(\lambda_3\nu_3^*))$ and $y = ((\lambda_1\nu_1^*)(\lambda_2\nu_2^*))(\lambda_3\nu_3^*)$. A tedious computation shows that

$$x = y = \begin{cases} \lambda_1\lambda_2'\lambda_3'\nu_3^* & \text{if} \quad \lambda_3 = \nu_2\lambda_3' \quad \text{and} \quad \lambda_2 = \nu_1\lambda_2' \\ \lambda_1\lambda_3'\nu_3^* & \text{if} \quad \lambda_3 = \nu_2\lambda_3''\lambda_3' \quad \text{and} \quad \nu_1 = \lambda_2\lambda_3'' \\ \lambda_1(\nu_1')^*\nu_3^* & \text{if} \quad \lambda_3 = \nu_2\lambda_3' \quad \text{and} \quad \nu_1 = \lambda_2\lambda_3'\nu_1' \\ \lambda_1\lambda_2'(\nu_2')^*\nu_3^* & \text{if} \quad \nu_2 = \lambda_3\nu_2' \quad \text{and} \quad \lambda_2 = \nu_1\lambda_2' \\ \lambda_1(\nu_1')^*(\nu_2')^*\nu_3^* & \text{if} \quad \nu_2 = \lambda_3\nu_2' \quad \text{and} \quad \nu_1 = \lambda_2\nu_1' \\ 0 & \text{otherwise,} \end{cases}$$

as desired. This clearly yields the result. □

Corollary 1.5.7 *Let E be an arbitrary graph and K any field. The restriction of the canonical projection $K\widehat{E} \to C_K(E)$ is injective on the subspace generated by the paths in E and the paths in E^*. In particular, the maps $KE \to C_K(E)$ and $KE^* \to C_K(E)$ are injective.*

Now we construct certain natural quotient algebras of Cohn path algebras. For $v \in \mathrm{Reg}(E)$, consider the following element q_v of $C_K(E)$:

$$q_v = v - \sum_{e \in s^{-1}(v)} ee^*.$$

Proposition 1.5.8 *The elements q_v are idempotents of $C_K(E)$. Moreover, $q_v C_K(E) q_w = \delta_{v,w} q_v K$ for each pair $v, w \in \mathrm{Reg}(E)$.*

Proof A simple computation shows that $\{q_v \mid v \in \mathrm{Reg}(E)\}$ is a family of pairwise orthogonal idempotents in $C_K(E)$. Now let $v \in E^0$ and $f \in E^1$. If $f \notin s^{-1}(v)$ then $e^*f = 0$ for all $e \in s^{-1}(v)$. On the other hand, if $f \in s^{-1}(v)$ then $ee^*f = 0$ for $e \neq f$, while $ff^*f = f$. Thus we see that $\sum_{e \in s^{-1}(v)} ee^*f = vf$, and in a similar way that $\sum_{e \in s^{-1}(v)} f^*ee^* = f^*v$, for all $f \in E^1$. So

$$f^* q_v = 0 = q_v f \tag{1.5}$$

for all $f \in E^1$ and $v \in \mathrm{Reg}(E)$. This yields that $q_v C_K(E) q_w = K q_v q_w = \delta_{v,w} q_v K$, as desired. □

Definition 1.5.9 Let E be an arbitrary graph and K any field. Let X be any subset of $\mathrm{Reg}(E)$. We denote by I^X the K-algebra ideal of $C_K(E)$ generated by the idempotents $\{q_v \mid v \in X\}$. The *Cohn path algebra of E relative to X*, denoted $C_K^X(E)$, is defined to be the quotient K-algebra

$$C_K(E)/I^X.$$

Clearly this notion of the relative Cohn path algebra links the Cohn and Leavitt path algebra constructions, as we see immediately that

$$C_K(E) = C^{\emptyset}(E) \ \text{ and } \ L_K(E) = C_K^{\mathrm{Reg}(E)}(E).$$

Generalizing the Universal Property for Leavitt path algebras 1.2.5, we have the following.

Remark 1.5.10 Suppose E is a graph, X is a subset of $\mathrm{Reg}(E)$, and A is a K-algebra which contains a set of pairwise orthogonal idempotents $\{a_v \mid v \in E^0\}$, and two sets $\{a_e \mid e \in E^1\}$, $\{b_e \mid e \in E^1\}$ for which

(i) $a_{s(e)} a_e = a_e a_{r(e)} = a_e$ and $a_{r(e)} b_e = b_e a_{s(e)} = b_e$ for all $e \in E^1$,
(ii) $b_f a_e = \delta_{e,f} a_{r(e)}$ for all $e, f \in E^1$, and
(iii) $a_v = \sum_{\{e \in E^1 \mid s(e) = v\}} a_e b_e$ for every vertex $v \in X$.

By the relations defining the relative Cohn path algebra, there exists a unique K-algebra homomorphism $\varphi : C_K^X(E) \to A$ such that $\varphi(v) = a_v$, $\varphi(e) = a_e$, and $\varphi(e^*) = b_e$ for all $v \in E^0$ and $e \in E^1$. We will often refer to this as the *Universal Property* of $C_K^X(E)$.

Proposition 1.5.11 *Let E be an arbitrary graph and K any field. Let X be a subset of* $\mathrm{Reg}(E)$. *Then a K-basis of I^X is given by the family $\lambda q_v \mu^*$, where $v \in X$ and $\lambda, \mu \in \mathrm{Path}(E)$ with $r(\lambda) = r(\mu) = v$. For $v \in X$ let $\{e_1^v, \ldots, e_{n_v}^v\}$ be an enumeration of the elements of $s^{-1}(v)$. Then a K-basis of $C_K^X(E)$ is given by the family*

$$\mathscr{B}'' = \mathscr{B} \setminus \{\lambda e_{n_v}^v (e_{n_v}^v)^* \nu^* \mid r(\lambda) = r(\nu) = v\},$$

where $\mathscr{B} = \{\lambda\nu^* \mid r(\lambda) = r(\nu)\}$ is the canonical basis of $C_K(E)$ given in Proposition 1.5.6.

Proof By the displayed Eq. (1.5), we have that the elements $\lambda q_v \mu^*$, for $v \in X$ and $\lambda, \mu \in \mathrm{Path}(E)$ with $r(\lambda) = v = r(\mu)$, generate I^X. To show that they are linearly independent, assume that there is an equation

$$\sum k_{\gamma,\mu} \gamma q_v \mu^* = 0$$

in $C_K(E)$, with $k_{\gamma,\mu} \in K$. Expressing the left-hand side as a linear combination of monomials $\lambda\nu^*$, and using the linear independence of these monomials (Proposition 1.5.6), we immediately get $k_{\gamma,\mu} = 0$ for all γ, μ.

Let $\mathscr{B}'$ be the basis of I^X just constructed. To show the second part of the proposition, it is enough to prove that $\mathscr{B}' \cup \mathscr{B}''$ is a basis of $C_K(E)$. Clearly every element $\lambda\nu^*$ of the basis $\mathscr{B}$ of $C_K(E)$ can be written as a linear combination of the elements in $\mathscr{B}' \cup \mathscr{B}''$. On the other hand, any nonzero linear combination of elements in $\mathscr{B}'$ must involve (with a nonzero coefficient) a monomial of the form $\lambda e_{n_v}^v (e_{n_v}^v)^* \nu^*$, and so it cannot be a linear combination of elements in $\mathscr{B}''$. This shows that $\mathscr{B}' \cup \mathscr{B}''$ is a basis of $C_K(E)$. □

As $L_K(E) = C_K^{\mathrm{Reg}(E)}(E)$, Proposition 1.5.11 immediately yields the following.

Corollary 1.5.12 *Let E be an arbitrary graph and K any field. Let $\mathscr{B} = \{\lambda\nu^* \mid r(\lambda) = r(\nu)\}$ be the canonical basis of $C_K(E)$ given in Proposition 1.5.6. For each vertex $v \in \mathrm{Reg}(E)$, let $\{e_1^v, \ldots, e_{n_v}^v\}$ be an enumeration of the elements of $s^{-1}(v)$. Then a basis of $L_K(E)$ is given by the family*

$$\mathscr{B}'' = \mathscr{B} \setminus \{\lambda e_{n_v}^v (e_{n_v}^v)^* \nu^* \mid r(\lambda) = r(\nu) = v \in \mathrm{Reg}(E)\}.$$

Proposition 1.5.11 easily yields the following three consequences as well.

Corollary 1.5.13 *Let E be an arbitrary graph and K any field. The restriction of the canonical projection $K\widehat{E} \to L_K(E)$ is injective on the subspace generated by the paths in E and the paths in E^*. In particular, the maps $KE \to L_K(E)$ and $KE^* \to L_K(E)$ are injective.*

Corollary 1.5.14 *Let R and S be unital rings, with R commutative, and suppose there exists a unital ring homomorphism $R \to Z(S)$ (where $Z(S)$ denotes the center of S). Let E be an arbitrary graph, and suppose $X \subseteq \mathrm{Reg}(E)$. Then there are ring* *isomorphisms*

$$C_R^X(E) \otimes_R S \cong C_S^X(E) \cong S \otimes_R C_R^X(E).$$

In particular,

$$L_R(E) \otimes_R S \cong L_S(E) \cong S \otimes_R L_R(E).$$

Proof We see that the computations made in Propositions 1.5.6 and 1.5.11 are independent of the coefficient ring, so that we have, for instance, $C_R^X(E) \otimes_R S = (\bigoplus_{\mathfrak{b} \in \mathscr{B}''} \mathfrak{b}R) \otimes_R S \cong \bigoplus_{\mathfrak{b} \in \mathscr{B}''} \mathfrak{b}S = C_S^X(E)$. □

Corollary 1.5.15 *Let E be an arbitrary graph and K any field. Then any set of distinct elements of* Path(E) *is linearly independent in the Cohn path algebra $C_K(E)$, as well as in the Leavitt path algebra $L_K(E)$.*

One of the nice things about Cohn path algebras is that they turn out, perhaps unexpectedly, to be Leavitt path algebras. In fact, we will show that any relative Cohn path algebra $C_K^X(E)$ is isomorphic to the Leavitt path algebra of a graph $E(X)$ which is obtained by adding various new vertices and edges to E.

Definition 1.5.16 Let E be an arbitrary graph and K any field. Let X be a subset of $\mathrm{Reg}(E)$, and define $Y := \mathrm{Reg}(E) \setminus X$. Let $Y' = \{v' \mid v \in Y\}$ be a disjoint copy of Y. For $v \in Y$ and for each edge $e \in r_E^{-1}(v)$, we consider a new symbol e'. We define the graph $E(X)$ as follows:

$$E(X)^0 = E^0 \sqcup Y' \quad \text{and} \quad E(X)^1 = E^1 \sqcup \{e' \mid r_E(e) \in Y\}.$$

For $e \in E^1$ we define $r_{E(X)}(e) = r_E(e)$ and $s_{E(X)}(e) = s_E(e)$, and define $s_{E(X)}(e') = s_E(e)$ and $r_{E(X)}(e') = r_E(e)'$ for the new symbols e'.

Less formally, the graph $E(X)$ is built from E and X by adding a new vertex to E corresponding to each element of $Y = \mathrm{Reg}(E) \setminus X$, and then including new edges to each of these new vertices as appropriate. Observe in particular that each of the new vertices $v' \in Y'$ is a sink in $E(X)$, so that $\mathrm{Reg}(E) = \mathrm{Reg}(E(X))$. If $X = \mathrm{Reg}(E)$, then $E = E(X)$.

Example 1.5.17 Let E be the following graph:

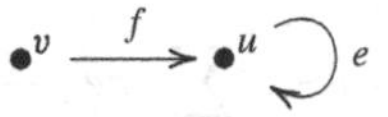

$$\bullet^v \xrightarrow{\ f\ } \bullet^u \circlearrowleft e$$

Take $X = \emptyset$, so that $Y = \mathrm{Reg}(E) = \{u, v\}$. Then the graph $E(X)$ is the following:

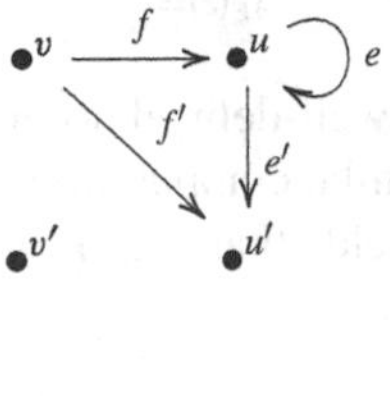

$$\begin{array}{ccc} \bullet^v & \xrightarrow{\ f\ } & \bullet^u \circlearrowleft e \\ & \searrow^{f'} & \downarrow^{e'} \\ \bullet^{v'} & & \bullet^{u'} \end{array}$$

For any ring R, if f and g are idempotents of R then it is standard in the literature to write $f \leq g$ if $fg = gf = f$. (We note, however, that this notation is not consistent with the notation $v \leq w$ where $v, w \in E^0$ and v, w are viewed as idempotent elements of $L_K(E)$. This notation will be presented in Definition 2.0.4 below; however, used in context, this should not cause confusion.)

As noted previously, every Leavitt path algebra arises (easily) as a relative Cohn path algebra, to wit, $L_K(E) = C_K^{\mathrm{Reg}(E)}(E)$. Perhaps more surprising is the following (very useful) result, which shows the converse.

Theorem 1.5.18 *Let E be an arbitrary graph and K any field. Let X be any subset of* Reg(E)*, and let $E(X)$ be the graph constructed in Definition 1.5.16. Then*

$$C_K^X(E) \cong L_K(E(X)).$$

Proof We define a K-algebra homomorphism $\phi: C_K^X(E) \to L_K(E(X))$ as follows. Write $Y = \mathrm{Reg}(E) \setminus X$. For a vertex v of E define $\phi(v) = v + v'$ if $v \in Y$, and $\phi(v) = v$ otherwise. Moreover, for $e \in E^1$, define $\phi(e) = e$ if $r_E(e) \notin Y$ and $\phi(e) = e + e'$ if $r_E(e) \in Y$, and define $\phi(e^*) = \phi(e)^*$. Clearly relation (V) is preserved by ϕ. To show that relation (E1) is preserved by ϕ, we consider first the case where $r_E(e) \notin Y$. Then $\phi(e) = e$, $\phi(r_E(e)) = r_E(e)$ and $s_{E(X)}(e) = s_E(e) \leq \phi(s_E(e))$, so

$$\phi(s_E(e))\phi(e) = s_E(e)e = e = er_E(e) = \phi(e)\phi(r_E(e)).$$

If $v := r_E(e) \in Y$ then $\phi(e) = e + e'$ and $\phi(v) = v + v'$, and $s_{E(X)}(e) = s_{E(X)}(e') \leq \phi(s_E(e))$, so that

$$\phi(s_E(e))\phi(e) = s_E(e)(e + e') = e + e' = \phi(e) = (e + e')(v + v') = \phi(e)\phi(r_E(e)),$$

as desired. Relations (E2) follow by applying $*$ to the above. Now we consider relation (CK1). If $e \neq f$ then clearly $\phi(e)^*\phi(f) = 0$. If $r_E(e) \notin Y$ then $\phi(e)^*\phi(e) = e^*e = r_E(e) = \phi(r_E(e))$. If $r_E(e) \in Y$ then

$$\phi(e)^*\phi(e) = (e^* + (e')^*)(e + e') = r_E(e) + r_E(e)^* = \phi(r_E(e)).$$

We must check that the (CK2) relation holds for the vertices in X. If $v \in X$ then $\phi(v) = v$ and $s_{E(X)}^{-1}(v) = s_E^{-1}(v) \sqcup \{e' \mid s_E(e) = v \text{ and } r_E(e) \in Y\}$, so that

$$\begin{aligned}\phi(v) - \sum_{e \in s_E^{-1}(v)} \phi(e)\phi(e)^* &= v - \sum_{r_E(e) \notin Y} ee^* + \sum_{r_E(e) \in Y} (e + e')(e^* + (e')^*) \\ &= v - \sum_{s_E(e)=v} ee^* - \sum_{s_E(e)=v, r_E(e) \in Y} e'(e')^* = 0.\end{aligned}$$

So we have shown that ϕ is a well-defined homomorphism.

Assume that $v \in Y$. Then a similar computation to the one presented above, using this time that $\phi(v) = v + v'$, yields that $\phi(q_v) = v'$, where q_v is defined prior to

Proposition 1.5.8. It follows that $v, v' \in \operatorname{Im}(\phi)$. Now we have, for $e \in E^1$ such that $r_E(e) = v \in Y$, that $\phi(e)v = (e + e')v = e$ and $\phi(e)v' = e'$, so that $e, e' \in \operatorname{Im}(\phi)$. It follows that ϕ is surjective.

Now we build the inverse homomorphism $\psi: L_K(E(X)) \to C_K^X(E)$. This is dictated by the above computations, so that we necessarily must set $\psi(v) = v$ if $v \notin Y$, and $\psi(v) = v - q_v$, $\psi(v') = q_v$ if $v \in Y$. For $e \in E^1$, set $\psi(e) = e$ if $r_E(e) \notin Y$, and set $\psi(e) = e(v - q_v)$, $\psi(e') = eq_v$ if $r_E(e) = v \in Y$. It is straightforward to show that all the defining relations of $L_K(E(X))$ are preserved by ψ, so that we get a well-defined homomorphism from $L_K(E(X))$ to $C_K^X(E)$. We check here the preservation of the (CK2) relation, and leave the others to the reader. Since $\operatorname{Reg}(E(X)) = \operatorname{Reg}(E)$ we need to consider only the regular vertices of E. Let $v \in \operatorname{Reg}(E)$. Relation (CK2) in $L_K(E(X))$ may be presented as

$$v = \sum_{s_E(e)=v, r_E(e)\notin Y} ee^* + \sum_{s_E(e)=v, r_E(e)\in Y} ee^* + \sum_{s_E(e)=v, r_E(e)\in Y} e'(e')^*.$$

If $v \in X$ then

$$\sum_{s_E(e)=v, r_E(e)\notin Y} \psi(e)\psi(e)^* + \sum_{s_E(e)=v, r_E(e)\in Y} \psi(e)\psi(e)^* + \sum_{s_E(e)=v, r_E(e)\in Y} \psi(e')\psi(e')^*$$

$$= \sum_{s_E(e)=v, r_E(e)\notin Y} ee^* + \sum_{s_E(e)=v, r_E(e)\in Y} e(r_E(e) - q_{r_E(e)})e^* + \sum_{s_E(e)=v, r_E(e)\in Y} eq_{r_E(e)}e^*$$

$$= \sum_{s_E(e)=v} ee^* = v = \psi(v).$$

On the other hand, if $v \in Y$ then the same computation as above gives

$$\sum_{s_E(e)=v, r_E(e)\notin Y} \psi(e)\psi(e)^* + \sum_{s_E(e)=v, r_E(e)\in Y} \psi(e)\psi(e)^* + \sum_{s_E(e)=v, r_E(e)\in Y} \psi(e')\psi(e')^* = v - q_v = \psi(v),$$

as desired.

It is now straightforward to show that both compositions $\psi \circ \phi$ and $\phi \circ \psi$ give the identity on the generators of the corresponding algebras, thus these maps are the identity on their respective domains. It follows that ϕ is an isomorphism. □

Here are two specific consequences of Theorem 1.5.18.

Example 1.5.19 Consider the graphs

$$E = \bullet^v \xrightarrow{f} \bullet^u \circlearrowright e \quad \text{and} \quad F = \bullet^v \xrightarrow{f} \bullet^u \circlearrowright e, \quad \bullet^v \xrightarrow{f'} \bullet^{u'}, \quad \bullet^u \xrightarrow{e'} \bullet^{u'}, \quad \bullet^{v'}$$

Then $C_K(E) \cong L_K(F)$ since $C_K(E) = C_K^{\emptyset}(E)$ (this is true for any graph E), and, as observed in Example 1.5.17, $F = E(X)$ for $X = \emptyset$.

As with the Leavitt path algebras, the "rose with n petals" graphs R_n $(n \geq 1)$ plays an important role in the context of Cohn path algebras as well. We demonstrate now what the graph $R_n(X)$ looks like for $X = \emptyset$. This in particular will demonstrate how the Toeplitz algebra arises naturally from the Cohn path algebra point of view.

Example 1.5.20 If

$R_n =$ 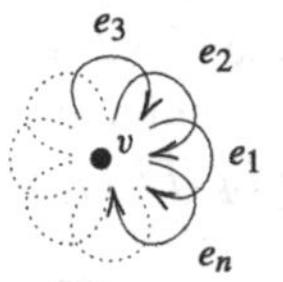

and $X = \emptyset$, then it is easy to show that

$R_n(X) =$ 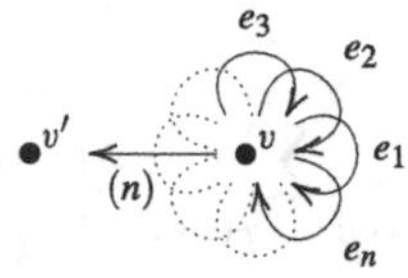 .

In particular, for $E = R_1 = \bullet^v \circlearrowright e$, we get $R_1(X) = \bullet^{v'} \longleftarrow \bullet^v \circlearrowright e = E_T$, the graph of Example 1.3.6. So Proposition 1.3.7 together with Theorem 1.5.18 give K-isomorphisms

$$K\langle U, V \mid UV = 1\rangle \cong C_K(R_1) \cong L_K(E_T) = \mathcal{T}_K.$$

We finish the section by making some easily checked, eventually useful observations about the relationship between the graphs E and $E(X)$ for any $X \subseteq \text{Reg}(E)$.

Proposition 1.5.21 *Let E be any graph, and X any subset of* $\text{Reg}(E)$. *Let Y denote* $\text{Reg}(E) \setminus X$.

(i) *E is acyclic if and only if $E(X)$ is acyclic.*
(ii) *E is finite if and only if $E(X)$ is finite.*
(iii) *E is row-finite if and only if $E(X)$ is row-finite.*
(iv) *The sinks of $E(X)$ are precisely the sinks of E together with the vertices $\{v' | v \in Y\}$.*
(v) *If v is a source in E, then v is also a source in $E(X)$. If moreover $v \in Y$, then v' is an isolated vertex in $E(X)$. Any isolated vertex of E is also isolated in $E(X)$.*

1.6 Direct Limits in the Context of Leavitt Path Algebras

The Leavitt path algebras of finite graphs not only play an historically important role in the theory, they also quite often provide key information regarding the structure of Leavitt path algebras corresponding to arbitrary graphs. We show in this section how the Leavitt path algebra $L_K(E)$ of any graph E may be viewed as the direct limit of certain subalgebras of $L_K(E)$, where each of these subalgebras is isomorphic to the Leavitt path algebra of some finite graph.

We start by offering the following cautionary note. It may be tempting to think that if F is a subgraph of E, then, using the obvious identification, we should have that $L_K(F)$ is a subalgebra of $L_K(E)$. However, this is not true in general, as a moment's reflection reveals that the (CK2) relation at a vertex v viewed in $L_K(F)$ need not be compatible with the (CK2) relation at that same vertex v when viewed as an element of $L_K(E)$. For example, the obvious graph embedding of R_2 into R_3 does not extend to an algebra homomorphism from $L_K(R_2)$ to $L_K(R_3)$. However, in certain situations a subgraph F embeds in E in a way compatible with the (CK2) relations, or, more generally, with the (CK2) relations imposed at a given subset $Y \subseteq \mathrm{Reg}(F)$. This is the motivating idea behind the main concepts of this section. We start by reminding the reader of a basic idea in graph theory, one which we will need to modify and expand upon in order to make it useful in our context.

Definition 1.6.1 A *graph homomorphism* $\varphi: F = (F^0, F^1, r_F, s_F) \to E = (E^0, E^1, r_E, s_E)$ is a pair of maps $\varphi^0: F^0 \to E^0$ and $\varphi^1: F^1 \to E^1$ such that $r_E(\varphi^1(e)) = \varphi^0(r_F(e))$ and $s_E(\varphi^1(e)) = \varphi^0(s_F(e))$ for every $e \in F^1$.

As the observation made above about the embedding of R_2 into R_3 demonstrates, a graph homomorphism from F to E need not induce a homomorphism of algebras $L_K(F) \to L_K(E)$. However, the following additional conditions on a graph homomorphism will allow such an extension to the algebra level.

Definition 1.6.2 We consider the category $\mathscr{G}$, defined as follows. The objects of $\mathscr{G}$ are pairs (E, X), where E is a graph and $X \subseteq \mathrm{Reg}(E)$. If $(F, Y), (E, X) \in \mathrm{Ob}(\mathscr{G})$, then $\psi = (\psi^0, \psi^1) : (F, Y) \to (E, X)$ is a morphism in $\mathscr{G}$ if

(1) $\psi : F \to E$ is a graph homomorphism for which $\psi^0 : F^0 \to E^0$ and $\psi^1 : F^1 \to E^1$ are injective,
(2) $\psi^0(Y) \subseteq X$, and
(3) for all $v \in Y$, ψ^1 restricts to a bijection $\psi^1 : s_F^{-1}(v) \to s_E^{-1}(\psi^0(v))$.

We note that a morphism $\psi : (F, Y) \to (E, X)$ in $\mathscr{G}$ depends not only on the underlying graphs F and E, but also on the distinguished sets of vertices Y and X.

Lemma 1.6.3 *Suppose* $\psi = (\psi^0, \psi^1) : (F, Y) \to (E, X)$ *is a morphism in* $\mathscr{G}$. *Then there exists a homomorphism of K-algebras* $\overline{\psi} : C_K^Y(F) \to C_K^X(E)$.

Proof We define $\overline{\psi} : C_K^Y(F) \to C_K^X(E)$ as the extension of ψ on F^0 and F^1. We define $\overline{\psi}(f^*) = \psi(f)^*$ for all $f \in F^1$. As F^0, F^1, and $(F^1)^*$ generate $C_K^Y(F)$ as an algebra, this will yield a K-algebra homomorphism with domain $C_K^Y(F)$, once we show that the defining relations on $C_K^Y(F)$ are preserved.

The idempotent and orthogonality properties of relation (V) are preserved by $\overline{\psi}$ because ψ^0 is injective. (Note that if $v \neq w$ in F^0 then $\overline{\psi}(vw) = \overline{\psi}(0)$, while $\overline{\psi}(v)\overline{\psi}(w) = 0$ using injectivity.) That relations (E1) and (E2) are preserved by $\overline{\psi}$ follows from the hypothesis that ψ is a graph homomorphism. That (CK1) is preserved by $\overline{\psi}$ follows because ψ^1 is injective (using an argument similar to the one given for relation (V)). Finally, the condition that ψ^1 restricts to a bijection from $s_F^{-1}(v)$ onto $s_E^{-1}(\psi^0(v))$ for every $v \in Y$ yields the preservation of (CK2) under $\overline{\psi}$ at the elements of Y. Thus, we get the desired extension of ψ to an algebra homomorphism $\overline{\psi}: C_K^Y(F) \to C_K^X(E)$. □

Proposition 1.6.4 *The category $\mathcal{G}$ has arbitrary direct limits. Moreover, for any field K, the assignment $(E, X) \mapsto C_K^X(E)$ extends to a continuous functor from the category $\mathcal{G}$ to the category K-alg of not-necessarily-unital K-algebras.*

Proof We first show that $\mathcal{G}$ admits direct limits. Let I be an upward directed partially ordered set, and let $\{(E_i, X_i)_{i\in I}, (\varphi_{ji})_{i,j\in I, j\geq i}\}$ be a directed system in $\mathcal{G}$. (So for each $j \geq i$ in I, $\varphi_{ji} : (E_i, X_i) \to (E_j, X_j)$ is a morphism in $\mathcal{G}$.) For $s = 0, 1$, set $E^s = \bigsqcup_{i\in I} E_i^s / \sim$, where $\sim$ is the equivalence relation on $\bigsqcup_{i\in I} E_i^s$ given by the following: For $\alpha \in E_i^s$ and $\beta \in E_j^s$, set $\alpha \sim \beta$ if and only if there is an index $k \in I$ such that $i \leq k$ and $j \leq k$ and $\varphi_{ki}^s(\alpha) = \varphi_{kj}^s(\beta)$. Observe that $E = (E^0, E^1)$ is a graph in a natural way, and there are injective graph homomorphisms $\psi_i = (\psi_i^0, \psi_i^1): E_i \to E$ such that $E^s = \bigcup_{i\in I} \psi_i^s(E_i^s)$, $s = 0, 1$. Note that E^s is the direct limit of (E_i^s, φ_{ji}^s) in the category of sets. Now define $X = \bigcup_{i\in I} \psi_i^0(X_i)$. We see that ψ_i defines a graph homomorphism from E_i to E for all $i \in I$, such that $\psi_i = \psi_j \circ \varphi_{ji}$ for all $j \geq i$. Clearly ψ_i satisfies conditions (1) and (2) in Definition 1.6.2. To check condition (3), take any vertex v in X_i, for $i \in I$. Then $s_E^{-1}(\psi_i^0(v)) = \bigcup_{j\geq i} \psi_j^1(s_{E_j}^{-1}(\varphi_{ji}^0(v)))$. But since for $j \geq i$ the map φ_{ji}^1 induces a bijection between $s_{E_i}^{-1}(v)$ and $s_{E_j}^{-1}(\varphi_{ji}^0(v))$, and $\psi_i^1 = \psi_j^1 \circ \varphi_{ji}^1$, it follows that

$$\psi_j^1(s_{E_j}^{-1}(\varphi_{ji}^0(v))) = \psi_j^1(\varphi_{ji}^1(s_{E_i}^{-1}(v))) = \psi_i^1(s_{E_i}^{-1}(v)),$$

so that ψ_i^1 induces a bijection from $s_{E_i}^{-1}(v)$ onto $s_E^{-1}(\psi_i^0(v))$. This gives (3) of Definition 1.6.2, and shows that each ψ_i is a morphism in the category $\mathcal{G}$.

We now check that $((E, X), \psi_i)$ is the direct limit of the directed system $((E_i, X_i), \varphi_{ji})$. Let $\{\gamma_i: (E_i, X_i) \to (G, Z) \mid i \in I\}$ be a compatible family of morphisms in $\mathcal{G}$. Define $\gamma: E \to G$ by the rule

$$\gamma^s(\psi_i(\alpha)) = \gamma_i^s(\alpha),$$

for $\alpha \in E_i^s$, $s = 0, 1$. It is obvious that γ is the unique graph homomorphism from E to G such that $\gamma_i = \gamma \circ \psi_i$ for all $i \in I$. Since, for $v \in E_i^0$, ψ_i^1 induces a bijection from $s_{E_i}^{-1}(v)$ onto $s_E^{-1}(\psi_i^0(v))$, and γ_i^1 induces a bijection from

$s_{E_i}^{-1}(v)$ onto $s_G^{-1}(\gamma_i^0(v))$, it follows that γ^1 induces a bijection from $s_E^{-1}(\psi_i^0(v))$ onto $s_G^{-1}(\gamma_i^0(v)) = s_G^{-1}(\gamma^0(\psi_i^0(v)))$. This shows that γ defines a morphism in the category $\mathscr{G}$, and clearly γ is the unique object in the category $\mathscr{G}$ such that $\gamma_i = \gamma \circ \psi_i$ for all $i \in I$, showing that (E, X) is the direct limit of $((E_i, X_i), \varphi_{ji})$.

If $\psi: (F, Y) \to (E, X)$ is a morphism in $\mathscr{G}$, then there is an induced K-algebra homomorphism $\overline{\psi}: C_K^Y(F) \to C_K^X(E)$ by Lemma 1.6.3, and clearly the assignment $\psi \mapsto \overline{\psi}$ is functorial. Let

$$((E_i, X_i)_{i\in I}, (\varphi_{ji})_{i,j\in I, j\geq i})$$

be a directed system in $\mathscr{G}$. Let $((E, X), \psi_i)$ be the direct limit in $\mathscr{G}$ of the directed system $((E_i, X_i), \varphi_{ji})$. We have to check that $(C_K^X(E), \overline{\psi_i})$ is the direct limit of the directed system $(C_K^{X_i}(E_i), \overline{\varphi_{ji}})$. Let $\gamma_i: C_K^{X_i}(E_i) \to A$ be a compatible family of K-algebra homomorphisms, where A is a K-algebra. Define $\gamma: C_K^X(E) \to A$ by the rule

$$\gamma(\psi_i^s(\alpha)) = \gamma_i(\alpha)\,, \quad \gamma(\psi_i^s(\alpha)^*) = \gamma_i(\alpha^*)\,,$$

for $\alpha \in E_i^s$, $i \in I$, $s = 0, 1$. We have to check that relations (V), (E1), (E2), and (CK1) are preserved by γ, and that relation (CK2) at all the vertices in X is also preserved by γ. It is straightforward to check (using appropriate injectivity hypotheses) that relations (V), (E1), (E2) and (CK1) are satisfied. Let $w \in X$. Then there is a $v \in X_i$, for some $i \in I$, such that $w = \psi_i^0(v)$. Since ψ_i^1 induces a bijection from $s_{E_i}^{-1}(v)$ onto $s_E^{-1}(\psi_i^0(v)) = s_E^{-1}(w)$, we get

$$\gamma(w) = \gamma(\psi_i^0(v)) = \gamma_i(v) = \sum_{e\in s_{E_i}^{-1}(v)} \gamma_i(e)\gamma_i(e^*) = \sum_{e\in s_{E_i}^{-1}(v)} \gamma(\psi_i^1(e))\gamma(\psi_i^1(e)^*) = \sum_{f\in s_E^{-1}(w)} \gamma(f)\gamma(f^*).$$

This shows that relation (CK2) at $w \in X$ is preserved by γ. It follows that γ is a well-defined K-algebra homomorphism. For $i \in I$, the maps γ_i and $\gamma \circ \overline{\psi_i}$ agree on the generators $E_i^0 \cup E_i^1 \cup (E_i^1)^*$ of $C_K^{X_i}(E_i)$, so we get $\gamma_i = \gamma \circ \overline{\psi_i}$. This shows that $(C_K^X(E), \overline{\psi_i})$ is the direct limit of the directed system $(C_K^{X_i}(E_i), \overline{\varphi_{ji}})$, as desired. □

Although morphisms in $\mathscr{G}$ give rise to algebra homomorphisms between the associated relative Cohn path algebras as per the previous result, and although the morphisms in $\mathscr{G}$ are injective maps by definition, the induced algebra homomorphisms need not be injective. For instance, the identity map gives rise to a morphism $\iota : (R_n, \emptyset) \to (R_n, \{v\})$ in $\mathscr{G}$, where v is the unique vertex of the rose with n petals graph R_n. However, the corresponding induced map is the canonical surjection $C_K(1, n) \to L_K(1, n)$, which is not injective (as the nonzero element $v - \sum_{i=1}^n e_i e_i^*$ of $C_K(1, n)$ is mapped to zero in $L_K(1, n)$).

However, by adding an additional condition to morphisms in $\mathscr{G}$, we can ensure that the induced algebra homomorphisms are injective.

Definition 1.6.5 Suppose $\psi = (\psi^0, \psi^1) : (F, Y) \to (E, X)$ is a morphism in $\mathscr{G}$. We say that ψ is *complete* if, for every $v \in F^0$,

$$\text{if } \psi^0(v) \in X \text{ and } s_F^{-1}(v) \neq \emptyset, \text{ then } v \in Y.$$

That is, ψ is complete if each of the vertices in X which are in $\mathrm{Im}(\psi^0)$, and which come from a non-sink in F, in fact come from Y. Note that a morphism ψ is complete if and only if $Y = (\psi^0)^{-1}(X) \cap \mathrm{Reg}(F)$.

We note that a complete morphism $\varphi : (F, \mathrm{Reg}(F)) \to (E, \mathrm{Reg}(E))$ is not in general the same as a *CK-morphism* as defined in [87], but the two ideas coincide when E is row-finite.

Lemma 1.6.6 *Suppose $\psi = (\psi^0, \psi^1) : (F, Y) \to (E, X)$ is a complete morphism in $\mathscr{G}$. Then the induced homomorphism $\overline{\psi} : C_K^Y(F) \to C_K^X(E)$ described in Lemma 1.6.3 is a monomorphism of K-algebras.*

Proof Using Corollary 1.5.12 and the notation there, for every regular vertex $v \in F^0$, if $\{e_1^v, \ldots, e_{n_v}^v\}$ is an enumeration of the elements of $s^{-1}(v)$, then a basis for $C_K^Y(F)$ is

$$\mathscr{B}''(F, Y) = \mathscr{B} \setminus \{\lambda e_{n_v}^v (e_{n_v}^v)^* \nu^* | r(\lambda) = r(\nu) = v \in Y\}.$$

If $v \in Y$, then the map ψ^1 induces a bijection from $s_F^{-1}(v) = \{e_1^v, \ldots, e_{n_v}^v\}$ onto $s_E^{-1}(\psi^0(v))$, so that $s_E^{-1}(\psi^0(v)) = \{\psi^1(e_1^v), \ldots, \psi^1(e_{n_v}^v)\}$. We take a corresponding basis $\mathscr{B}''(E, X)$ of $C_K^X(E)$ such that, for $v \in Y$, the enumeration $\{e_1^{\psi^0(v)}, \ldots, e_{n_v}^{\psi^0(v)}\}$ of the edges in $s_E^{-1}(\psi^0(v))$ is given by $e_i^{\psi^0(v)} = \psi^1(e_i^v)$, for $i = 1, \ldots, n_v$.

The injectivity conditions on ψ^0 and ψ^1 give that ψ extends to an injective map from $\mathrm{Path}(\widehat{F})$ to $\mathrm{Path}(\widehat{E})$. It is now clear that $\overline{\psi}$ restricts to an injective map from the basis $\mathscr{B}''(F, Y)$ of $C_K^Y(F)$ into a subset of the basis $\mathscr{B}''(E, X)$ of $C_K^X(E)$. Indeed, the role here of the completeness condition is in assuring that the images of the basis elements $\lambda e_i^v (e_i^v)^* \nu^*$, $i = 1, \ldots, n_v$, for v a regular vertex in F such that $v \notin Y$, belong to the basis $\mathscr{B}''(E, X)$ of $C_K^X(E)$ associated to (E, X). This is so because if $v \in \mathrm{Reg}(F) \setminus Y$, then $\psi^0(v) \notin X$ by the completeness of ψ, and so the elements $\overline{\psi}(\lambda e_i^v (e_i^v)^* \nu^*)$ belong to the basis $\mathscr{B}''(E, X)$.

Therefore $\overline{\psi}$ is injective, as desired. □

Definition 1.6.7 We say that a subgraph F of a graph E is *complete* if the inclusion map

$$(F, \mathrm{Reg}(F) \cap \mathrm{Reg}(E)) \to (E, \mathrm{Reg}(E))$$

is a (complete) morphism in the category $\mathscr{G}$. Less formally, F is a *complete* subgraph of E if for each $v \in F^0$, whenever $s_F^{-1}(v) \neq \emptyset$ and $0 < |s_E^{-1}(v)| < \infty$, then

$s_F^{-1}(v) = s_E^{-1}(v)$. In words, a subgraph F of a graph E is complete if, whenever v is a vertex in F which emits at least one edge in F and finitely many in E (and so also finitely many in F, because F is a subgraph of E), then the edges emitted at v in the subgraph F are precisely *all* of the edges emitted at v in the full graph E.

By Lemma 1.6.6, if F is a complete subgraph of E then we get an embedding

$$C_K^{\text{Reg}(F)\cap\text{Reg}(E)}(F) \hookrightarrow L_K(E) = C_K^{\text{Reg}(E)}(E).$$

If $\text{Reg}(F) \cap \text{Reg}(E) = \text{Reg}(F)$ (for instance, if E is row-finite), then a complete subgraph F of E yields that the canonical inclusion map $F \hookrightarrow E$ gives rise to an embedding of $L_K(F) \hookrightarrow L_K(E)$.

In the example given above, R_2 is not a complete subgraph of R_3. This is because $\text{Reg}(R_3) = \{v\} = \text{Reg}(R_2)$, so that $\text{Reg}(R_2) \cap \text{Reg}(R_3) = \{v\}$; and the inclusion map from $s_{R_2}^{-1}(v) \to s_{R_3}^{-1}(v)$ is not a bijection. In contrast, the inclusion morphism $(R_2, \emptyset) \hookrightarrow (R_3, \emptyset)$ *is* a complete morphism in $\mathcal{G}$. On the other hand, consider the infinite rose graph R_∞, and let R_n be any finite subgraph of R_∞. Then R_n is a complete subgraph of R_∞, since $\text{Reg}(R_n) \cap \text{Reg}(R_\infty) = \{v\} \cap \emptyset = \emptyset$, and the morphism $(R_n, \emptyset) \hookrightarrow (R_\infty, \emptyset)$ is complete.

The following definition generalizes Definition 1.6.7, and it will be useful later on.

Definition 1.6.8 Let E be a graph and let S be a subset of $\text{Reg}(E)$. We say that a subgraph F of a graph E is *S-complete* if the inclusion map

$$(F, \text{Reg}(F) \cap S) \to (E, S)$$

is a complete morphism in the category $\mathcal{G}$. Thus, F is an *S-complete* subgraph of E if for each $v \in S$, we have $s_F^{-1}(v) = s_E^{-1}(v)$ whenever $s_F^{-1}(v) \neq \emptyset$.

We note that the literature contains alternate definitions of the notion of a *complete* subgraph of a graph, see e.g., [12]. However, the notion of completeness is identical across all definitions whenever the given graph is row-finite.

The notion of a complete morphism in $\mathcal{G}$, and the attendant notion of a complete subgraph, will allow us to produce homomorphisms from various relative Cohn path algebras over appropriately chosen subgraphs F of E to the Leavitt path algebra $L_K(E)$. This will in turn, by an application of Theorem 1.5.18, allow us to realize any Leavitt path algebra $L_K(E)$ as a direct limit of algebras, each of which is itself the Leavitt path algebra of a finite graph built from E.

Lemma 1.6.9 *Every object (E, X) of $\mathcal{G}$ is a direct limit in the category $\mathcal{G}$ of a directed system of the form $\{(F_i, X_i) \mid i \in I\}$, for which each F_i is a finite graph and all the maps $(F_i, X_i) \to (E, X)$ are complete morphisms in $\mathcal{G}$.*

Proof Clearly, E is the set theoretic union of its finite subgraphs. Let G be a finite subgraph of E. Define a finite subgraph F of E as follows:

$$F^0 = G^0 \cup \{r_E(e) \mid e \in E^1 \text{ and } s_E(e) \in G^0 \cap X\}$$

and

$$F^1 = \{e \in E^1 \mid s_E(e) \in G^0 \cap X\}.$$

Now notice that the set of vertices in $F^0 \cap X$ that emit edges in F is precisely the set $G^0 \cap X$, and if v is one of these vertices, then $s_E^{-1}(v) = s_F^{-1}(v)$. This shows that the inclusion map $(F, \mathrm{Reg}(F) \cap X) \hookrightarrow (E, X)$ is a complete morphism in $\mathscr{G}$. In particular, any finite subgraph G of E gives rise to a finite complete subobject $(F, \mathrm{Reg}(F) \cap X)$ of (E, X).

Since the union of a finite number of finite complete subobjects of (E, X) is again a finite complete subobject of (E, X), it follows that (E, X) is the direct limit in the category $\mathscr{G}$ of the directed family of its finite complete subobjects $(F, \mathrm{Reg}(F) \cap X)$. □

Now applying Lemma 1.6.9, Proposition 1.6.4 and Lemma 1.6.6, we have established the following useful result.

Theorem 1.6.10 *Let E be an arbitrary graph and K any field. Let X be any subset of* $\mathrm{Reg}(E)$. *Then as objects in the category K-alg, we have*

$$C_K^X(E) = \varinjlim_F \{C_K^{\mathrm{Reg}(F)\cap X}(F)\},$$

where $(F, \mathrm{Reg}(F) \cap X)$ ranges over all finite complete subobjects of (E, X) (i.e., F ranges over all X-complete subgraphs of E). Moreover, each of the homomorphisms $C_K^{\mathrm{Reg}(F)\cap X}(F) \to C_K^X(E)$ is injective. In particular,

$$L_K(E) = \varinjlim_F \{C_K^{\mathrm{Reg}(F)\cap \mathrm{Reg}(E)}(F)\},$$

where F ranges over all finite complete subgraphs of E, with all homomorphisms $C_K^{\mathrm{Reg}(F)\cap\mathrm{Reg}(E)}(F) \to L_K(E)$ being injective.

We are now in position to establish the aforementioned result regarding direct limits.

Corollary 1.6.11 *Let E be an arbitrary graph and K any field. Let X be any subset of* $\mathrm{Reg}(E)$. *Then $C_K^X(E)$ is the direct limit in the category K-alg of subalgebras, each of which is isomorphic to the Leavitt path algebra of a finite graph. In particular, $L_K(E)$ is the direct limit of unital subalgebras (with not-necessarily-unital transition homomorphisms), each of which is isomorphic to the Leavitt path algebra of a finite graph.*

Proof This follows directly from Theorems 1.6.10 and 1.5.18. □

To clarify the ideas of the previous two results, we present the following examples.

Example 1.6.12 Let $C_{\mathbb{N}}$ be the *infinite clock graph* pictured here.

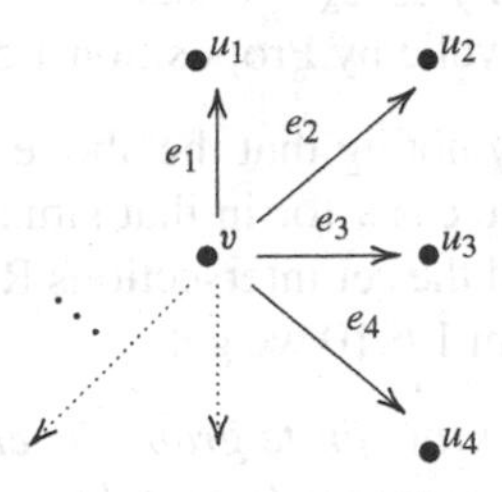

In this example, we have $L_K(C_{\mathbb{N}}) \cong \varinjlim_{n \in \mathbb{N}} C_K(C_n)$, where $C_K(C_n) = C_K^{\emptyset}(C_n) \cong L_K(C_n(\emptyset))$ is the Cohn path algebra of the n-edges clock C_n.

Example 1.6.13 We let $R_{\mathbb{N}}$ denote the *rose with* $\mathbb{N}$ *petals* graph having one vertex and $\mathbb{N}$ loops:

$$R_{\mathbb{N}} \; = \; \bullet^{v} \quad (e_1, e_2, e_3, \ldots) \; .$$

Here we have $L_K(R_{\mathbb{N}}) \cong \varinjlim_{n \in \mathbb{N}} C_K(R_n)$, where $C_K(R_n) = C_K^{\emptyset}(R_n) \cong L_K(R_n(\emptyset))$ is the Cohn path algebra of the n-edges rose. (See Example 1.5.20 for a description of the graph $R_n(\emptyset)$.)

Example 1.6.14 Let $A_{\mathbb{N}}$ be the *infinite line graph*

$$A_{\mathbb{N}} \; = \; \bullet^{v_1} \xrightarrow{e_1} \bullet^{v_2} \xrightarrow{e_2} \bullet^{v_3} \cdots\cdots$$

Here we have $L_K(A_{\mathbb{N}}) \cong \varinjlim_{n \in \mathbb{N}} L_K(A_n)$, because the graph $A_{\mathbb{N}}$ is row-finite (see Corollary 1.6.16 below). In this situation the transition homomorphisms $L_K(A_n) \to L_K(A_{n+1})$ can be identified with the maps $\mathrm{M}_n(K) \to \mathrm{M}_{n+1}(K)$ (cf. Proposition 1.3.5) that send an $n \times n$ matrix B to the $(n+1) \times (n+1)$ matrix B' consisting of B in the upper left $n \times n$ corner, and 0 elsewhere. This yields that $L_K(A_{\mathbb{N}}) \cong \mathrm{M}_{\mathbb{N}}(K)$, the (non-unital) K-algebra of $\mathbb{N} \times \mathbb{N}$ matrices consisting of those matrices having at most finitely many nonzero entries. (This isomorphism will also follow from Theorem 2.6.14 below.)

As a consequence of the results in this section which will prove to be quite useful later, we offer the following.

Proposition 1.6.15 *Let E be any acyclic graph. Then $L_K(E)$ is the direct limit, with injective transition homomorphisms, of algebras $\{L_K(F_i) \mid i \in I\}$, where each F_i is a finite acyclic graph.*

Proof As subgraphs of E, the graphs F which arise in Theorem 1.6.10 are necessarily acyclic. But $C_K^{\mathrm{Reg}(F)\cap\mathrm{Reg}(E)}(F) \cong L_K(F(\mathrm{Reg}(F) \cap \mathrm{Reg}(E)))$ by Theorem 1.5.18, and $F(\mathrm{Reg}(F) \cap \mathrm{Reg}(E))$ is acyclic by Proposition 1.5.21(i). □

We conclude this section by noting that the above direct limit construction may be streamlined in the row-finite case, for in that situation the regular vertices of E are precisely the non-sinks, and the set intersections $\mathrm{Reg}(F) \cap \mathrm{Reg}(E)$ are precisely the sets $\mathrm{Reg}(F)$. So by Theorem 1.6.10 we get

Corollary 1.6.16 *Let E be any row-finite graph. Then $L_K(E)$ is the directed union of unital subalgebras (with not-necessarily-unital transition homomorphisms), each of which is isomorphic to the Leavitt path algebra of a finite complete subgraph of E.*

1.7 A Brief Retrospective on the History of Leavitt Path Algebras

A brief retrospective on the subject's genesis is in order here. (A much fuller account may be found in [1].) The accomplishments achieved during the initial investigation by Leavitt in the late 1950s and early 1960s into the structure of non-IBN rings were followed up by P.M. Cohn's work (see e.g., [64]) in the mid-1960s on the algebras $U_{1,n}$ (herein denoted $C_K(1,n)$), and by Bergman's work in the mid-1970s on the $\mathcal{V}$-monoid question. The algebras $L_K(1,n)$ and $C_K(1,n)$ did not become the subject of intense interest again until more than a quarter century later, when they were dusted off and studied anew in [20, 25, 29]. (Perhaps this hiatus of interest was due to Cohn's remark in [64] that these algebras "... may be regarded as pathological rings"?) As noted previously, the algebras $C_K(1,n) \cong C_K(R_n)$ stand in relation to the more general Cohn path algebras in precisely the same way that the Leavitt algebras $L_K(1,n) \cong L_K(R_n)$ stand in relation to the more general Leavitt path algebras.

Working in a different corner of the mathematical universe, Cuntz in the late 1970s investigated a class of C^*-algebras arising from a natural question in physics, the now-so-called *Cuntz algebras* $\mathcal{O}_n$ (see [68]). Subsequently, Cuntz and Krieger in [71] realized that the Cuntz algebras are specific cases of a more general C^*-algebra structure which could be associated with any finite 0/1 matrix, the now-so-called *Cuntz–Krieger C^*-algebras*. (The names Cuntz and Krieger give rise to the letters which comprise the notation (CK1) and (CK2); this notation is now standardly used in both the algebraic and analytic literature to describe the appropriate conditions on the algebras.) Subsequently, it was realized that the Cuntz–Krieger algebras were themselves specific cases of an even more general C^*-algebra structure, the *graph C^*-algebras* defined in [155] and then initially investigated in depth in [106].

Using the 20/20 vision provided by the passage of a few years' time, it is fair to say that there were two seminal papers which served as the launching pad for the study of Leavitt path algebras: [5, 31]. The work for both of these articles was initiated in 2004, but the two groups of authors did not become aware of the others' efforts until Spring 2005, at which time it was immediately clear that the algebras under study in these two articles were identical. It is interesting to note that although the topic discussed in both [5, 31] is the then-newly-described notion of Leavitt path algebras, the results in the two articles are completely disjoint. Indeed, the former contains results for Leavitt path algebras which mimic some of the corresponding graph C^*-algebra results (e.g., regarding simplicity of the algebras). In fact, the construction given in [5] was motivated directly by interpreting the C^*-algebra equations displayed in Definitions 1.4.4 from a purely algebraic point of view. (The analogous interpretation relating $L_{\mathbb{C}}(1,n)$ and $\mathcal{O}_n$ had already been noted in [29].) On the other hand, [31] contains results describing Bergman's construction in the specific setting of graph monoids, as well as theretofore unknown information about the $\mathcal{V}$-monoid of the graph C^*-algebras. The common, historically appropriate name "Leavitt path algebras" which now describes these structures was then agreed upon by the two groups of authors while [5, 31] were in press.

The results presented in this opening chapter are meant to give the reader both an historical overview of the subject and a foundation for results which will be presented in subsequent chapters. The results described in Sects. 1.1 through 1.4 have by now resided in the literature for a number of years, and are for the most part well known. On the other hand, the main ideas of Sects. 1.5 and 1.6 are contributions to the theory which either make their first appearance in the literature here, or made their appearance in literature motivated in part by pre-publication versions of this book.

Again donning our historical 20/20 lenses, it seems clear now that Cohn's aforementioned "pathological rings" observation missed the mark rather significantly. As we hope will become apparent to the reader throughout this book, in fact these rings are quite natural, structurally quite interesting, and really quite beautiful.

Chapter 2
Two-Sided Ideals

In this chapter we investigate the ideal structure of Leavitt path algebras. In the introductory paragraphs we present many of the graph-theoretic ideas that will be useful throughout the subject. There is a natural $\mathbb{Z}$-grading on $L_K(E)$, which we discuss in Sect. 2.1. With this grading so noted, we will see in subsequent sections that the graded ideals with respect to this grading play a fundamental structural role. In Sect. 2.2 we consider the Reduction Theorem. Important consequences of this result include the two Uniqueness Theorems (also presented in Sect. 2.2), as well as various structural results about Leavitt path algebras (which comprise Sect. 2.3). In Sect. 2.4 we show that the quotient of a Leavitt path algebra by a graded ideal is itself isomorphic to a Leavitt path algebra. In Sect. 2.5 we show that the graded ideals of a Leavitt path algebra arise as ideals generated from data given by prescribed subsets of the graph E. Specifically, in the Structure Theorem for Graded Ideals (Theorem 2.5.8), we establish a precise relationship between graded ideals and explicit sets of idempotents. In the row-finite case, these sets of idempotents consist of hereditary saturated sets of vertices, while in the more general case additional sets of idempotents (arising from breaking vertices) are necessary. In addition, we show that a graded ideal viewed as an algebra in its own right is isomorphic to a Leavitt path algebra.

With a description of the graded ideals having been obtained, we focus in the remainder of the chapter on the structure of all ideals. We start in Sect. 2.6 by considering the socle of a Leavitt path algebra. Along the way, we obtain a description of the finite-dimensional Leavitt path algebras. In Sect. 2.7 we identify the ideal generated by the set of those vertices which connect to a cycle having no exits. The denouement of Chap. 2 occurs in Sect. 2.8, in which we present the Structure Theorem for Ideals (Theorem 2.8.10), an explicit description of the entire ideal lattice of $L_K(E)$ (including both the graded and non-graded ideals) for an arbitrary graph E and field K. This key result weaves the Structure Theorem for Graded Ideals together with the analysis of the ideal investigated in the previous section. A number of ring-theoretic results follow almost immediately from the

G. Abrams et al., *Leavitt Path Algebras*, Lecture Notes in Mathematics 2191,
DOI 10.1007/978-1-4471-7344-1_2

Structure Theorem for Ideals, including the Simplicity Theorem; we present those in Sect. 2.9.

Notation 2.0.1 For a ring or algebra R and subset $X \subseteq R$, we denote by $I(X)$ the ideal of R generated by X.

While only very basic graph-theoretic ideas and terminology were needed to define the Leavitt path algebras, additional graph-theoretic concepts will play a huge role in analyzing the structure of these algebras. We collect many of those in the following.

Definitions 2.0.2 Let $E = (E^0, E^1, r, s)$ be an arbitrary graph.

(i) Let $\mu = e_1 e_2 \cdots e_n \in \text{Path}(E)$. If $n = \ell(\mu) \geq 1$, and if $v = s(\mu) = r(\mu)$, then μ is called a *closed path based at* v.
(ii) A *closed simple path based at* v is a closed path $\mu = e_1 e_2 \cdots e_n$ based at v, such that $s(e_j) \neq v$ for every $j > 1$. We denote by $CSP(v)$ the set of all such paths.
(iii) If $\mu = e_1 e_2 \cdots e_n$ is a closed path based at v and $s(e_i) \neq s(e_j)$ for every $i \neq j$, then μ is called a *cycle based at* v. Note that a cycle is a closed simple path based at any of its vertices, but not every closed simple path based at v is a cycle, because a closed simple path may visit some of its vertices (other than v) more than once.
(iv) Suppose $\mu = e_1 e_2 \cdots e_n$ is a cycle based at the vertex v. Then for each $1 \leq i \leq n$, the path $\mu_i = e_i e_{i+1} \cdots e_n e_1 \cdots e_{i-1}$ is a cycle based at the vertex $s(e_i)$. (In particular, $\mu_1 = \mu$.) The *cycle of* μ is the collection of cycles $\{\mu_i\}$ based at $s(e_i)$.
(v) A *cycle* c is a set of paths consisting of the cycle of μ for μ some cycle based at a vertex v.
(vi) The *length of a cycle* c is the length of any of the paths in c. In particular, a cycle of length 1 is called a *loop*. (We note that the definition of the word *cycle* is somewhat non-standard, but will serve our purposes well here.)
(vii) A (directed) graph E is said to be *acyclic* if it does not have any closed paths based at any vertex of E, equivalently if it does not have any cycles based at any vertex of E.

Definition 2.0.3 A graph E satisfies *Condition* (K) if for each $v \in E^0$ which lies on a closed simple path, there exist at least two distinct closed simple paths α, β based at v.

Definition 2.0.4 Let $E = (E^0, E^1, r, s)$ be a graph. We define a preorder $\leq$ on E^0 given by:

$$w \leq v \text{ if there is a path } \mu \in \text{Path}(E) \text{ such that } s(\mu) = v \text{ and } r(\mu) = w.$$

(We will sometimes equivalently write $v \geq w$ in this situation.) If $v \in E^0$ then the *tree of* v, denoted $T(v)$, is the set

$$T(v) = \{w \mid w \in E^0, v \geq w\}.$$

(This notation is standard in the context of Leavitt path algebras; note, however, that $T(v)$ need not be a "tree" in the sense of undirected graphs, as $T(v)$ may indeed contain closed paths.) If $X \subseteq E^0$, we define $T(X) := \bigcup_{v \in X} T(v)$.

Note that $T(X)$ is the smallest hereditary subset of E^0 containing X.

Definitions 2.0.5 Let E be a graph, and $H \subseteq E^0$.

(i) We say H is *hereditary* if whenever $v \in H$ and $w \in E^0$ for which $v \geq w$, then $w \in H$.
(ii) We say H is *saturated* if whenever $v \in \text{Reg}(E)$ has the property that $\{r(e) \mid e \in E^1, s(e) = v\} \subseteq H$, then $v \in H$. (In other words, H is saturated if, for any non-sink vertex v which emits a finite number of edges in E, if *all* of the range vertices $r(e)$ for those edges e having $s(e) = v$ are in H, then v must be in H as well.)

We denote by $\mathscr{H}_E$ (or simply by $\mathscr{H}$ when the graph E is clear) the set of those subsets of E^0 which are both hereditary and saturated.

We refer back to the graph E given in Example 1.2.8. We see that the set $S_1 = \{v_3\}$ is hereditary (trivially), but not saturated, since the vertex v_2 emits all of its edges (there is only one) into S_1, but v_2 itself is not in S_1. However, the set $S_2 = \{v_2, v_3\}$ is both hereditary and saturated: while v_1 emits edges into S_2, not *all* of the edges emitted from v_1 have ranges in S_2.

Definition 2.0.6 If X is a subset of E^0, then the *hereditary saturated closure of* X, denoted $\overline{X}$, is the smallest hereditary and saturated subset of E^0 containing X. Since the intersection of hereditary (resp., saturated) subsets of E^0 is again hereditary (resp., saturated), $\overline{X}$ is well defined.

We denote by $S(X)$ the set of all vertices obtained by applying the saturated condition among the elements of X, that is,

$$S(X) := \{v \in \text{Reg}(E) \mid \{r(e) \mid s(e) = v\} \subseteq X\} \cup X.$$

For $X \subseteq E^0$, the hereditary saturated closure of X may be inductively constructed as follows.

Lemma 2.0.7 *Let X be a nonempty subset of vertices of a graph E. We define $X_0 := T(X)$, and for $n \geq 0$ we define inductively $X_{n+1} := S(X_n)$. Then $\overline{X} = \bigcup_{n\geq 0} X_n$.*

Proof It is immediate to see that every hereditary and saturated subset of E^0 containing X must contain $\bigcup_{n\geq 0} X_n$. Note that every X_n is hereditary (it is easy to show that if $Y \subseteq E^0$ is hereditary, then so is $S(Y)$), which implies that $\bigcup_{n\geq 0} X_n$ is hereditary as well. We now show that $\bigcup_{n\geq 0} X_n$ is saturated. Take $v \in \text{Reg}(E)$ such that $r(s^{-1}(v)) \subseteq \bigcup_{n\geq 0} X_n$; since $X_n \subseteq X_{n+1}$ and $r(s^{-1}(v))$ is a finite subset, there exists an $N \in \mathbb{N}$ such that $r(s^{-1}(v)) \subseteq X_N$, hence $v \in X_{N+1}$ as required. □

We finish the introduction to this chapter by describing how the path algebra $K\widehat{E}$ of K over the extended graph $\widehat{E}$ can be endowed with an involution.

Lemma 2.0.8 *Let E be an arbitrary graph and K any field. Let* $^{-} : K \to K$ *be an involution on K. Then the following map can be extended to a unique involution* $* : K\widehat{E} \to K\widehat{E}$:

(i) $(kv)^* = \overline{k}v$ *for every* $k \in K$ *and* $v \in E^0$.
(ii) $(k\gamma)^* = \overline{k}\gamma^*$ *for every* $k \in K$ *and* $\gamma \in \text{Path}(E)$.
(iii) $(k\gamma^*)^* = \overline{k}\gamma$ *for every* $k \in K$ *and* $\gamma \in \text{Path}(E)$.

In particular, $(KE)^* = KE^*$.

Proof Define the map $\rho : E^0 \cup E^1 \cup (E^1)^* \to (K\widehat{E})^{op}$ by setting $\rho(v) = v$, $\rho(e) = e^*$, and $\rho(e^*) = e$ for $v \in E^0$ and $e \in E^1$. It is easy to see that ρ is compatible with the relations (V), (E1) and (E2) in $K\widehat{E}$, and hence ρ can be extended in a unique way to a homomorphism of K-algebras $\rho : K\widehat{E} \to (K\widehat{E})^{op}$. This homomorphism ρ is precisely the involution in the statement. □

Corollary 2.0.9 *Let E be an arbitrary graph, let* $X \subseteq \text{Reg}(E)$, *and let K be any field. Let* $^{-} : K \to K$ *be an involution on K. Then there is a unique involution* $* : C_K^X(E) \to C_K^X(E)$ *satisfying the three properties of Lemma 2.0.8.*

Consequently, taking the involution on K to be the identity map, we have that $C_K^X(E)$ *is isomorphic to its opposite ring* $C_K^X(E)^{op}$ *as K-algebras. In particular,* $L_K(E) \cong L_K(E)^{op}$ *as K-algebras.*

2.1 The $\mathbb{Z}$-Grading

One of the most important properties of the class of Leavitt path algebras is that each $L_K(E)$ is a $\mathbb{Z}$-graded K-algebra. As we shall see, this grading provides the key ingredient which allows us to obtain many structural results about Leavitt path algebras, as well as to streamline proofs of additional results.

In this section we will explore the natural $\mathbb{Z}$-grading on $L_K(E)$ (the one induced by the length of paths). Of particular importance will be the structure of the zero component of any Leavitt path algebra relative to this grading.

Definitions 2.1.1 Let G be a group and A an algebra over a field K. We say that A is *G-graded* if there exists a family $\{A_\sigma\}_{\sigma \in G}$ of K-subspaces of A such that

$$A = \bigoplus_{\sigma \in G} A_\sigma \text{ as } K\text{-spaces, and } A_\sigma \cdot A_\tau \subseteq A_{\sigma\tau} \text{ for each } \sigma, \tau \in G.$$

An element x of A_σ is called a *homogeneous element of degree* σ. An ideal I of a G-graded K-algebra A is said to be a *graded ideal* if $I \subseteq \sum_{\sigma \in G}(I \cap A_\sigma)$, or, equivalently, if

$$y = \sum_{\sigma \in G} y_\sigma \in I \text{ implies } y_\sigma \in I \text{ for every } \sigma \in G.$$

Remark 2.1.2 Let e denote the identity element of the group G. It is straightforward to show that if A is a G-graded ring, and X is a subset of A_e, then the ideal $I(X)$ of A generated by X is a graded ideal.

It is easy to prove that the quotient of a G-graded algebra $A = \bigoplus_{\sigma \in G} A_\sigma$ by a graded ideal I is a G-graded algebra, with the natural grading induced by that of A. Specifically, consider the projection map $A \to A/I$ via $a \mapsto \overline{a}$, and denote A/I by $\overline{A}$. Then, using the graded property of I, for any $\sigma \in G$ the homogeneous component $\overline{A}_\sigma$ of $\overline{A}$ of degree σ is $\overline{A}_\sigma := \overline{A_\sigma}$. Hence

$$\overline{A} = \bigoplus_{\sigma \in G} \overline{A_\sigma} .$$

In general, not every ideal in a Leavitt path algebra is graded (see, e.g., Examples 2.1.7). It will be shown in Sect. 2.4 that graded ideals can be obtained from specified subsets of vertices. Concretely, Leavitt path algebras all of whose ideals are graded will be shown to coincide with the exchange Leavitt path algebras; equivalently, to coincide with those Leavitt path algebras whose associated graph satisfies Condition (K).

We recall here that for an arbitrary graph E and field K the Leavitt path algebra $L_K(E)$ can be obtained as a quotient of the Cohn path algebra $C_K(E)$ by the ideal I generated by $\{v - \sum_{e \in s^{-1}(v)} ee^* \mid v \in \operatorname{Reg}(E)\}$ (Proposition 1.5.5). We establish that the Cohn path algebra has a natural $\mathbb{Z}$-grading given by the length of the monomials, which thereby will induce a $\mathbb{Z}$-grading on $L_K(E)$. (Although we derive the grading on $L_K(E)$ from the grading on $C_K(E)$, a more direct proof may also be produced.)

Definition 2.1.3 Let E be an arbitrary graph and K any field. For any $v \in E^0$ and $e \in E^1$, define $\deg(v) = 0$, $\deg(e) = 1$ and $\deg(e^*) = -1$. For any monomial $kx_1 \cdots x_m$, with $k \in K$ and $x_i \in E^0 \cup E^1 \cup (E^1)^*$, define $\deg(kx_1 \cdots x_m) = \sum_{i=1}^m \deg(x_i)$. Finally, for any $n \in \mathbb{Z}$ define

$$A_n := \operatorname{span}_K(\{x_1 \cdots x_m \mid x_i \in E^0 \cup E^1 \cup (E^1)^* \text{ with } \deg(x_1 \cdots x_m) = n\}).$$

Proposition 2.1.4 *With the notation of Definition 2.1.3, $K\widehat{E} = \bigoplus_{n \in \mathbb{Z}} A_n$ as K-subspaces, and this decomposition defines a $\mathbb{Z}$-grading on the path algebra $K\widehat{E}$.*

Proof By Remark 2.1.2, the ideal I generated by the relations (V), (E1) and (E2) is graded, hence $K\widehat{E}$, which is isomorphic to $K\langle E^0 \cup E^1 \cup (E^1)^* \rangle / I$, is graded as in the indicated decomposition. □

Corollary 2.1.5 *Let E be an arbitrary graph and K any field.*

(i) *For any subset X of* Reg(E), *the Cohn path algebra $C_K^X(E)$ of E relative to X is a $\mathbb{Z}$-graded K-algebra with the grading induced by the length of paths.*
(ii) *$C_K(E) = \bigoplus_{n \in \mathbb{Z}} C_n$, where*

$$C_n := \operatorname{span}_K(\{\gamma\lambda^* \mid \gamma, \lambda \in \operatorname{Path}(E) \text{ and } \ell(\gamma) - \ell(\lambda) = n\}),$$

defines a $\mathbb{Z}$-grading on the Cohn path algebra $C_K(E)$.

(iii) $L_K(E) = \bigoplus_{n\in\mathbb{Z}} L_n$, *where*

$$L_n := \mathrm{span}_K(\{\gamma\lambda^* \mid \gamma, \lambda \in \mathrm{Path}(E) \text{ and } \ell(\gamma) - \ell(\lambda) = n\}),$$

defines a $\mathbb{Z}$*-grading on the Leavitt path algebra* $L_K(E)$.

Proof Items (ii) and (iii) are particular cases of (i), hence we will prove only this case. By definition (see Definition 1.5.9), the relative Cohn path algebra $C_K^X(E) = K\widehat{E}/I$, where I is the K-algebra ideal of $K\widehat{E}$ generated by relations of the forms (V), (E1), (E2), and (CK1), and by the idempotents $\{q_v \mid v \in X\}$, where $q_v = v - \sum_{e\in s^{-1}(v)} ee^*$. Proposition 2.1.4 establishes that the path algebra $K\widehat{E}$ is $\mathbb{Z}$-graded. But I is generated by homogeneous elements of degree 0, hence it is a graded ideal by Remark 2.1.2; consequently, the quotient $K\widehat{E}/I$ is a $\mathbb{Z}$-graded algebra. □

Remark 2.1.6 This remark will turn out to be quite useful in understanding the ideal structure of general Leavitt path algebras. There is a natural $\mathbb{Z}$-grading on the Laurent polynomial algebra $A = K[x, x^{-1}]$, given by setting $A_i = Kx^i$ for all $i \in \mathbb{Z}$. Furthermore, it is well known (and easy to prove) that the set of units in $K[x, x^{-1}]$ consists of the set $\{kx^i \mid k \in K^\times, i \in \mathbb{Z}\}$. Consequently, the only graded ideals of $K[x, x^{-1}]$ are the two ideals $\{0\}$ and $K[x, x^{-1}]$ itself.

Moreover, there are infinitely many non-graded ideals in $K[x, x^{-1}]$, since every nontrivial ideal of $K[x, x^{-1}]$ is generated by a unique element of the form $1 + k_1x + \cdots + k_nx^n$ with $k_n \neq 0$.

Consider a field K and a group G. Given two G-graded K-algebras $A = \oplus_{\sigma\in G}A_\sigma$ and $B = \oplus_{\sigma\in G}B_\sigma$, a K-algebra homomorphism f from A into B is said to be a *graded homomorphism* if $f(A_\sigma) \subseteq B_\sigma$ for every $\sigma \in G$. It is easy to show that $\mathrm{Ker}(f)$ is a graded ideal of A in this case. If there exists a K-algebra isomorphism $f : A \to B$ for which both f and f^{-1} are graded homomorphisms, then we say that A and B are *graded isomorphic*.

Examples 2.1.7 We demonstrate how the $\mathbb{Z}$-grading on $L_K(E)$ manifests in two fundamental cases.

First, let A_n be the oriented n-line graph $\bullet \longrightarrow \bullet \cdots\cdots \bullet \longrightarrow \bullet$ of Notation 1.3.1. In Proposition 1.3.5 we established that $L_K(A_n) \cong \mathrm{M}_n(K)$, by writing down an explicit isomorphism φ between these two algebras. For each integer t with $-(n-1) \leq t \leq n-1$ we consider the K-subspace A_t of $A = \mathrm{M}_n(K)$ consisting of those elements $(a_{i,j})$ for which $a_{i,j} = 0$ for each pair i, j having $i - j \neq t$. (Less formally, A_t consists of the elements of the tth-superdiagonal of A.) For $|t| \geq n$ we set $A_t = \{0\}$. Then it is easy to see (and well known) that $\oplus_{t\in\mathbb{Z}}A_t$ is a $\mathbb{Z}$-grading of $\mathrm{M}_n(K)$. Furthermore, $\varphi : L_K(A_n) \to \mathrm{M}_n(K)$ is a graded isomorphism with respect to this grading.

Now let R_1 be the graph ${}^v\bullet\!\circlearrowleft^e$, also of Notation 1.3.1. In Proposition 1.3.4 we showed that $L_K(R_1) \cong K[x, x^{-1}]$, via an isomorphism which takes v to 1 and e to x. With the usual grading on $K[x, x^{-1}]$ (described in Remark 2.1.6), this isomorphism is clearly graded. This immediately implies that there are infinitely many non-graded ideals in $L_K(R_1)$, to wit, any ideal generated by a non-monomial expression in e and/or e^*. For instance, $I(v+e)$ is such an ideal. The only graded ideals of $L_K(R_1)$ are $L_K(R_1)$ itself, and $\{0\}$.

We showed in Chap. 1 that the path K-algebra KE over a graph E, as well as the path K-algebra KE^* over the graph E^*, can be seen as subalgebras of the Cohn path algebra $C_K(E)$ (Corollary 1.5.7) and of the Leavitt path algebra $L_K(E)$ (Corollary 1.5.13). In fact, both KE and KE^* are graded subalgebras of both $C_K(E)$ and $L_K(E)$.

Lemma 2.1.8 *Let E be an arbitrary graph and K any field.*

(i) *The canonical map $K\widehat{E} \to C_K(E)$ is a $\mathbb{Z}$-graded K-algebra homomorphism. The restrictions $KE \to C_K(E)$ and $KE^* \to C_K(E)$ are $\mathbb{Z}$-graded K-algebra monomorphisms.*
(ii) *The canonical map $K\widehat{E} \to L_K(E)$ is a $\mathbb{Z}$-graded K-algebra homomorphism. The restrictions $KE \to L_K(E)$ and $KE^* \to L_K(E)$ are $\mathbb{Z}$-graded K-algebra monomorphisms.*

Proof The canonical projections given in Corollary 1.5.7 and in Corollary 1.5.13 are K-algebra monomorphisms sending homogeneous elements of degree n into elements of the same degree. □

The proof of the following result is easy, so we omit it.

Lemma 2.1.9 *Let E be an arbitrary graph and K any field. Let I be the ideal of the Cohn path algebra generated by the set $\{v - \sum_{e\in s^{-1}(v)} ee^* \mid v \in \mathrm{Reg}(E)\}$. Then $L_K(E)$ and $C_K(E)/I$ are $\mathbb{Z}$-graded isomorphic K-algebras.*

Lemma 2.1.9 is a particular case of

Proposition 2.1.10 *Let E be an arbitrary graph and K any field. Let X be any subset of* $\mathrm{Reg}(E)$. *Then $C_K^X(E)$ and $L_K(E(X))$ are $\mathbb{Z}$-graded isomorphic K-algebras.*

Proof By reconsidering the proof of Theorem 1.5.18, it is clear that the given isomorphism indeed respects the grading. □

For the remainder of this section we will focus on the structure of the *zero components* $(C_K(E))_0$ of $C_K(E)$ and $(L_K(E))_0$ of $L_K(E)$ with respect to the grading described above. As we shall see, these subrings will play important roles in the sequel. Let S be a subset of $\mathrm{Reg}(E)$. Given $k \in \mathbb{Z}^+$, let X be a finite set of paths of E of length $\leq k$. For $0 \leq i \leq k$, let X_i be the set of initial paths of elements of X of length i, and let Y_i be the set of edges which appear in position i in a path of an element of X. That is,

$$X_i = \{\lambda \in \mathrm{Path}(E) \mid |\lambda| = i, \text{ and there exists } \lambda' \in \mathrm{Path}(E) \text{ such that } \lambda\lambda' \in X\}, \text{ and}$$

$$Y_i = \{e \in E^1 \mid \text{there exists } \lambda, \gamma \in \mathrm{Path}(E) \text{ such that } |\lambda| = i-1, \text{ and } \lambda e\gamma \in X\}.$$

Note that X_0 is the set of source vertices of paths in X. For a path λ of length $\geq i$, denote by λ_i the initial segment of λ of length i, so that $\lambda = \lambda_i\lambda'$, with $|\lambda_i| = i$.

Definitions 2.1.11 Let S, X, X_i, Y_i, and k be as above. We say that X is an *S-complete subset* of Path(E) if the following conditions are satisfied:

(i) All the paths in X of length $< k$ end in a sink.
(ii) For every $\lambda \in X$, every $i < |\lambda|$ such that $r(\lambda_i) \in S$ and every $e \in s^{-1}(r(\lambda_i))$, we have that $\lambda_i e = \gamma_{i+1}$ for some $\gamma \in X$.
(iii) For any $\lambda \in X_i$ $(1 \leq i < k)$ and any $e \in Y_{i+1}$ such that $r(\lambda) = s(e)$, we have $\lambda e \in X_{i+1}$.

Recall that we defined the notion of an S-complete subgraph in Chap. 1 (see Definition 1.6.8). This notion should not be confused with the just defined concept of an S-complete subset of paths of a graph.

There is a natural way to build S-complete finite subsets of Path(E) from S-complete finite subgraphs of E, as follows. The goal is to extend the paths in the S-complete finite subgraph to either paths of length k, or to paths of length less than k which end in a sink, in a specifically described way.

Proposition 2.1.12 *Let F be a finite S-complete subgraph of E and $k \geq 1$. Then there exists an S-complete subset of* Path(E) *of paths of length $\leq k$ which contains all the paths of length k of F, as well as all the paths of length $< k$ of F which end in a sink of E. More precisely, there is a finite S-complete subgraph F' of E containing F such that X is the set of all paths of F' of length k starting at a vertex of F together with the set of all paths of F' of length $< k$ starting at a vertex of F and ending in a sink of E.*

Proof For a vertex v of E with $v \in (E^0 \setminus (\mathrm{Sink}(E) \cup S)) \cap (\mathrm{Sink}(F) \cup (E^0 \setminus F^0))$, we choose and fix some $e_v \in s_E^{-1}(v)$.

For each $v \in E^0$ and each $t \geq 1$, we denote by $\Gamma(v,t)$ the set of all paths of length $\leq t$ which satisfy the following conditions:

1. All paths in $\Gamma(v,t)$ start at v.
2. The paths in $\Gamma(v,t)$ either have length t, or have length $< t$ and end in a sink of E.
3. If $\alpha_1\alpha_2\cdots\alpha_s \in \mathrm{Path}(E)$ (where each $\alpha_i \in E^1$) belongs to $\Gamma(v,t)$, then for each i such that $s(\alpha_i) \in (E^0 \setminus S) \cap (\mathrm{Sink}(F) \cup (E^0 \setminus F^0))$ we have $\alpha_i = e_{s(\alpha_i)}$. Moreover, for each i such that $s(\alpha_i) \in F^0 \setminus \mathrm{Sink}(F)$, we have $\alpha_i \in F^1$.

The idea here is that we extend paths of length less than k arbitrarily in vertices of S, by using edges in F whenever we can; while we extend such paths by a predetermined edge if the vertex does not belong to S, is not a sink in E, and we cannot extend it by using edges in F. Observe that $\Gamma(v,t)$ is finite. Now note the following:

(a) Every path λ in $\Gamma(v,s)$, with $s < t$, can be extended to a path τ in $\Gamma(v,t)$, i.e., there is a path λ' such that $\lambda\lambda' \in \Gamma(v,t)$.
(b) If $\gamma \in \Gamma(v,t)$ and γ' is an initial segment of γ of positive length s, then $\gamma' \in \Gamma(v,s)$.

(c) If $\gamma \in \Gamma(v,t)$ and γ' is a final segment of γ of positive length s, then $\gamma' \in \Gamma(s(\gamma'), s)$.

Let $\Gamma^{(1)}$ denote the set of paths of F of length k together with the paths of F of length $< k$ which end in a sink of E.

Let $\Gamma^{(2)}$ denote the set of paths of length $\le k$ consisting of all paths of the form $\lambda\mu$, where λ is a path of F of length $< k$ which ends in a sink of F which is not a sink in E, and $\mu \in \Gamma(r(\lambda), k - |\lambda|)$.

Let X be the (disjoint) union of $\Gamma^{(1)}$ and $\Gamma^{(2)}$. To complete the proof, we need to check that X is an S-complete subset of Path(E). Observe that

$$X = \bigcup_{v \in F^0} \Gamma(v,k).$$

Condition (i) in the definition of an S-complete subset is obviously satisfied. For condition (ii), let $\lambda \in X$, $i < |\lambda|$ such that $r(\lambda_i) \in S$, and $e \in s^{-1}(r(\lambda_i))$. Note that $\lambda_i e \in \Gamma(s(\lambda_i e), i+1)$, so by observation (a) $\lambda_i e$ can be extended to a path $\gamma \in \Gamma(s(\lambda), k)$. If γ is a path of F then $\gamma \in \Gamma^{(1)}$. Otherwise we have $\gamma \in \Gamma^{(2)}$.

To finish, we check condition (iii). Let $\lambda \in X$, $1 \le i < k$, and $e \in Y_{i+1}$ such that $r(\lambda_i) = s(e)$. Then $\lambda_i \in \Gamma(s(\lambda), i)$, and $e\mu \in \Gamma(s(e), k-i)$ for a certain path μ (because $e \in Y_{i+1}$). Therefore $\lambda_i e\mu \in X$, so that $\lambda_i e \in X_{i+1}$, as desired.

The final statement is shown as follows. Let v be a vertex of E which appears as a non-final vertex of a path from X. If $v \in F^0 \setminus \mathrm{Sink}(F)$, then we set $s_{F'}^{-1}(v) = s_F^{-1}(v)$. If $v \in S$, then we set $s_{F'}^{-1}(v) = s_E^{-1}(v)$. If $v \in (E^0 \setminus (\mathrm{Sink}(E) \cup S)) \cap (\mathrm{Sink}(F) \cup (E^0 \setminus F^0))$, then we set $s_{F'}^{-1}(v) = \{e_v\}$. The graph F' is the smallest subgraph of E containing F and all these edges. □

Definition 2.1.13 A *matricial K-algebra* is a finite direct product of full matrix algebras (of finite size) over a field K.

Let S be a subset of Reg(E), and let X be an S-complete finite subset of Path(E) consisting of paths of length $\le k$. We define

$$\mathscr{G}(X) = \mathrm{span}_K(\lambda\mu^* \mid \lambda, \mu \in X, |\lambda| = |\mu|).$$

Proposition 2.1.14 *Let E be an arbitrary graph and K any field. Let S be a subset of* Reg(E). *Let X be an S-complete finite subset of* Path(E) *consisting of paths of length $\le k$. For $1 \le i \le k$, we consider the following K-subspaces $\mathscr{F}_i(X)$ of $C_K^S(E)$:*

$\mathscr{F}_i(X)$ is the K-linear span in $C_K^S(E)$ of the elements $\lambda(v - \sum_{e \in Y_i, s(e)=v} ee^*)\mu^*$,

where $\lambda, \mu \in X_{i-1}$, $r(\lambda) = r(\mu) = v \notin S$, and $Y_i \cap s^{-1}(v) \neq \emptyset$. We set

$$\mathscr{F}(X) = \mathscr{G}(X) + \sum_{i=1}^{k} \mathscr{F}_i(X).$$

Then $\mathscr{F}(X)$ is a matricial K-algebra. Moreover, $(C_K^S(E))_0$ is the direct limit of the subalgebras $\mathscr{F}(X)$, where X ranges over all the S-complete finite subsets of $\mathrm{Path}(E)$.

Proof We will show:

(1) for every $1 \le i \le k$, $\mathscr{F}_i(X)$ is a matricial K-algebra, and
(2) for $i \neq j$ we have $\mathscr{F}_i(X) \cdot \mathscr{F}_j(X) = 0$. In particular, the sum $\mathscr{F}(X) = \sum_{i=1}^k \mathscr{F}_i(X)$ is a direct sum.

To establish these two statements, write an element $\lambda(v - \sum_{e\in Y_i, s(e)=v} ee^*)\mu^*$ in $\mathscr{F}(X)$ as $\lambda\tau_i(v)\mu^*$, where $\tau_i(v) := v - \sum_{e\in Y_i, s(e)=v} ee^*$. To show (1) for $1 \le i \le k$, observe that if $\lambda\tau_i(v)\mu^*$ and $\gamma\tau_i(w)\eta^*$ belong to $\mathscr{F}_i(X)$, and $v \neq w$ then we have

$$\lambda\tau_i(v)\mu^* \cdot \gamma\tau_i(w)\eta^* = 0.$$

If $v = w$ then

$$\lambda\tau_i(v)\mu^* \cdot \gamma\tau_i(v)\eta^* = \delta_{\mu,\gamma}\lambda\tau_i(v)\eta^*.$$

It follows that $\mathscr{F}_i(X) = \bigoplus_v \mathscr{F}_{i,v}(X)$, where $\mathscr{F}_{i,v}(X)$ is the linear span of the set of elements of the form $\lambda\tau_i(v)\mu^*$. Moreover, $\mathscr{F}_{i,v}(X)$ is a matrix algebra over K of size $|X_{i-1}|$. This shows (1).

Now assume that $i \neq j$ and that $\alpha = \lambda\tau_i(v)\mu^*$ and $\beta = \gamma\tau_j(w)\eta^*$ belong to $\mathscr{F}_i(X)$ and $\mathscr{F}_j(X)$ respectively. Assume for convenience that $j > i$. Then $\alpha\beta = 0$ unless $\gamma = \mu\gamma'$, with $|\gamma'| = j - i > 0$, in which case

$$\alpha \cdot \beta = \lambda\tau_i(v)\gamma'\tau_j(w)\eta^*.$$

Write $\gamma' = f\gamma''$. Then $f \in Y_i$ and $s(f) = r(\mu) = v$ and thus

$$\tau_i(v)\gamma' = (v - \sum_{e\in Y_i, s(e)=v} ee^*)f\gamma'' = (f - f)\gamma'' = 0.$$

It follows that $\alpha\beta = 0$. This shows that $\sum_{i=1}^k \mathscr{F}_i(X)$ is a direct sum.

The K-vector space $\mathscr{G}(X)$ is also a matricial K-algebra, indeed

$$\mathscr{G}(X) = \Big[\bigoplus_{i=0}^{k-1} \bigoplus_{v\in \mathrm{Sink}(E)} \mathscr{G}_{i,v}(X)\Big] \bigoplus \Big[\bigoplus_{v\in E^0} \mathscr{G}_{k,v}(X)\Big],$$

where $\mathscr{G}_{i,v}(X)$ is the K-linear span of the set of elements of the form $\lambda\mu^*$, where $\lambda, \mu \in X$, $|\lambda| = |\mu| = i$ and $r(\lambda) = r(\mu) = v$. (This property relies on condition (i) in the definition of an S-complete subset of $\mathrm{Path}(E)$.) It is easy to show that the above sum is direct and also that each $\mathscr{G}_{i,v}(X)$ is a finite matrix K-algebra of size the number of elements of X with the prescribed conditions on length and range.

The proof that $\mathscr{G}(X) \cdot \mathscr{F}_i(X) = 0$ for all i is similar to the above. Hence we get the direct sum

$$\mathscr{F}(X) = \mathscr{G}(X) \bigoplus (\bigoplus_{i=1}^{k} \mathscr{F}_i(X)).$$

We now describe the transition homomorphisms $\mathscr{F}(X) \to \mathscr{F}(X')$, for appropriate pairs of S-complete finite subsets X, X' of Path(E). Suppose that X is an S-complete finite subset of paths of length $\leq k$ and that X' is an S-complete finite subset of paths of length $\leq \ell$. Then we write $X \leq X'$ if $k \leq \ell$ and every path in X can be extended to a path in X', that is, for each λ in X there is a path λ' such that $\lambda\lambda'$ belongs to X'. Observe that only paths of length k can be properly extended. The condition $X \leq X'$ implies that $X_i \subseteq X'_i$ for $1 \leq i \leq k$. Also $X < X'$ implies $k < \ell$.

To describe the transition homomorphism $\mathscr{F}(X) \to \mathscr{F}(X')$ for $X < X'$, we need to specify a rule that allows us eventually to write any of the generators of $\mathscr{F}(X)$ as a linear combination of the generators in $\mathscr{F}(X')$. Let us write $\tau_i(v)$ and $\tau'_i(v)$ for the corresponding elements $v - \sum_{e \in Y_i, s(e)=v} ee^*$ and $v - \sum_{e \in Y'_i, s(e)=v} ee^*$, respectively.

We first describe the map on $\mathscr{G}(X)$. Let v be a vertex in E, and suppose that $\lambda, \mu \in X_i$ and $r(\lambda) = r(\mu) = v$. If v is a sink then $\lambda\mu^*$ belongs to $\mathscr{F}_\ell(X')$, so the map is the identity in this case. If $v \in S$ then $i = k$ and

$$\lambda\mu^* = \lambda(\sum_{e \in s^{-1}(v)} ee^*)\mu^* = \sum_{e \in s^{-1}(v)} (\lambda e)(\mu e)^*.$$

Note that, for $e \in s^{-1}(v)$, λe and μe can be enlarged to a path in X' by the S-completeness of X' (condition (ii)). If $v \notin S$ then

$$\lambda\mu^* = \lambda(\sum_{e \in Y'_{k+1}} ee^*)\mu^* + \lambda\tau'_{k+1}(v)\mu^*.$$

Note that $\lambda\tau'_{k+1}(v)\mu^* \in \mathscr{F}_{k+1}(X')$ and that the paths $\lambda e, \mu e$, with $e \in Y'_{k+1}$, can be enlarged to paths in X', again by the S-completeness of X' (condition (iii)). In this way, an inductive procedure gives the description of the transition mapping $\mathscr{G}(X) \to \mathscr{F}(X')$.

Now let $\lambda\tau_i(v)\mu^*$ be a generating element of $\mathscr{F}_i(X)$, for $1 \leq i \leq k$. Then

$$\lambda\tau_i(v)\mu^* = \lambda\tau'_i(v)\mu^* + \sum_{f \in Y'_i \setminus Y_i} (\lambda f)(\mu f)^*,$$

and $\lambda\tau'_i(v)\mu^* \in \mathscr{F}_i(X')$, whilst $\lambda f, \mu f$ can be enlarged to paths in X' for all $f \in Y'_i \setminus Y_i$. Thus we can proceed as above in order to obtain the image of $\lambda ff^*\mu^*$ in $\mathscr{F}(X')$. This allows us to describe the transition homomorphism $\mathscr{F}_i(X) \to \mathscr{F}(X')$.

Finally, let $a = \sum_{\lambda,\mu\in T,|\lambda|=|\mu|} k_{\lambda,\mu}\lambda\mu^*$ be an arbitrary element in $(C_K^S(E))_0$, where T is a finite set of paths in E. By Proposition 2.1.12, there is a finite S-complete subgraph F of E such that all the paths in T have all their edges in F. Let k be an upper bound for the length of the paths in T. By using Proposition 2.1.12, we can find an S-complete finite subset of Path(E) consisting of paths of length $\leq k$ such that all paths in T can be enlarged to paths in X. Now the above procedure enables us to write a as an element of $\mathscr{F}(X)$. This shows that $(C_K^S(E))_0$ is the direct limit of the subalgebras $\mathscr{F}(X)$, where X ranges over all the S-complete finite subsets of Path(E), and completes the proof. □

A foundational reference for the material in the remainder of this section is [89, Sect. 2.3]. Every injective K-algebra homomorphism

$$\phi\colon A = \mathrm{M}_{n_1}(K) \times \cdots \times \mathrm{M}_{n_r}(K) \longrightarrow B = \mathrm{M}_{m_1}(K) \times \cdots \times \mathrm{M}_{m_s}(K)$$

is conjugate to a block diagonal one, and so it is completely determined by its multiplicity matrix $M = (m_{ji}) \in \mathrm{M}_{s\times r}(\mathbb{Z}^+)$, which has the property that $\sum_{i=1}^r m_{ji}n_i \leq m_j$ for $j = 1, \ldots, s$. If ϕ is unital, then this inequality is an equality. Note that the injectivity hypothesis is equivalent to the statement that there is no zero column in the matrix M. For $i \in \{1, \ldots, r\}$, the integers m_{ji} can be computed as follows. Take a minimal idempotent e_i in the component $\mathrm{M}_{n_i}(K)$ of A. Then $\phi(e_i)$ can be written as $\phi(e_i) = \sum_{j=1}^s \sum_{m=1}^{m_{ji}} g_{j,m}^{(i)}$, where, for each j, $\{g_{j,1}^{(i)}, \ldots, g_{j,m_{ji}}^{(i)}\}$ are pairwise orthogonal minimal idempotents in the factor $\mathrm{M}_{m_j}(K)$ of B.

Definition 2.1.15 Let E be a finite graph. We denote by $A_E = (a_{v,w}) \in \mathrm{M}_{E^0\times E^0}(\mathbb{Z}^+)$ the *incidence* or *adjacency* matrix of E, where $a_{v,w} = |\{e \in E^1 \mid s(e) = v, r(e) = w\}|$. We let A_{ns} denote the matrix A_E with the zero-rows removed; that is, A_{ns} is the (not necessarily square) matrix obtained from A_E by removing the rows corresponding to the sinks of E.

We are now in position to give an explicit description of the zero component of the Leavitt path algebra of a finite graph.

Corollary 2.1.16 *Let E be a finite graph and K any field. For each $n \in \mathbb{Z}^+$ let $L_{0,n} \subseteq L_K(E)$ denote the K-linear span of elements of the form $\gamma\eta^*$, where $\gamma, \eta \in$ Path(E) for which $|\gamma| = |\eta| = n$ and $r(\gamma) = r(\eta)$, together with elements of the form $\gamma\eta^*$ where $\gamma, \eta \in$ Path(E) for which $|\gamma| = |\eta| < n$ and $r(\gamma) = r(\eta)$ is a sink in E. Then we have*

$$(L_K(E))_0 = \bigcup_{n\in\mathbb{Z}^+} L_{0,n}.$$

For each v in E^0, and each $n \in \mathbb{Z}^+$, we denote by $P(n, v)$ the set of paths γ in E such that $|\gamma| = n$ and $r(\gamma) = v$. Then

$$L_{0,n} \cong \Big[\prod_{m=0}^{n-1}\Big(\prod_{v\in\mathrm{Sink}(E)} \mathrm{M}_{|P(m,v)|}(K)\Big)\Big] \times \Big[\prod_{v\in E^0} \mathrm{M}_{|P(n,v)|}(K)\Big].$$

The transition homomorphism $L_{0,n} \to L_{0,n+1}$ *is the identity on the factors* $\prod_{v\in \mathrm{Sink}(E)} \mathrm{M}_{|P(m,v)|}(K)$, *for* $0 \le m \le n-1$, *and also on the factor* $\prod_{v\in \mathrm{Sink}(E)} \mathrm{M}_{|P(n,v)|}(K)$ *of the right-hand term of the displayed formula. The transition homomorphism*

$$\prod_{v\in E^0\setminus \mathrm{Sink}(E)} \mathrm{M}_{|P(n,v)|}(K) \to \prod_{v\in E^0} \mathrm{M}_{|P(n+1,v)|}(K)$$

has multiplicity matrix equal to A_{ns}^t.

Proof All these facts follow directly from the proof of Proposition 2.1.14. For instance, observe that for $v \in E^0 \setminus \mathrm{Sink}(E)$ and $\lambda \in P(n,v)$, we have that $\lambda\lambda^*$ is a minimal idempotent in the factor $\mathrm{M}_{|P(n,v)|}(K)$ of $L_{0,n}$ and that by the (CK2) relation

$$\lambda\lambda^* = \sum_{e\in s^{-1}(v)} (\lambda e)(\lambda e)^*,$$

so that, for $w \in E^0$, the multiplicity $m_{w,v}$ of the inclusion map

$$\prod_{v\in E^0\setminus \mathrm{Sink}(E)} \mathrm{M}_{|P(n,v)|}(K) \to \prod_{v\in E^0} \mathrm{M}_{|P(n+1,v)|}(K)$$

is precisely $a_{v,w}$, which shows that $M = A_{ns}^t$. □

We note that the K-subspaces $L_{0,n}$ described in the previous result form a filtration of $(L_K(E))_0$, given by the K-linear span of the paths $\gamma\nu^*$ such that $|\gamma| = |\nu| \le n$ and $r(\gamma) = r(\nu)$.

Example 2.1.17 Let $E = R_2$, with vertex v and edges e, f. Then for each $n \in \mathbb{Z}^+$ we have $|P(n,v)| = 2^n$. There are no sinks in E, so that $A_{ns}^t = A = (2)$. Thus $L_{0,n} \cong \mathrm{M}_{2^n}(K)$ for each $n \in \mathbb{Z}^+$, and the transition homomorphism from $L_{0,n}$ to $L_{0,n+1}$ takes an element $(m_{i,j})$ of $\mathrm{M}_{2^n}(K)$ to the element $(m_{i,j}I_2)$ of $\mathrm{M}_{2^{n+1}}(K)$, where I_2 is the 2×2 identity matrix. Thus $(L_K(R_2))_0 \cong \varinjlim_{n\in\mathbb{Z}^+} \mathrm{M}_{2^n}(K)$. (See also [2, Sect. 2] for further analysis of this direct limit.)

Example 2.1.18 Let E_T be the Toeplitz graph as presented in Example 1.3.6, and let $\mathscr{T} = \mathscr{T}_K$ denote the algebraic Toeplitz K-algebra $L_K(E_T)$. Then it is easy to see that $|P(n,u)| = |P(n,v)| = 1$ for all $n \in \mathbb{Z}^+$. In particular, $\mathscr{T}_{0,0} \cong K \times K$. By Corollary 2.1.16 we have that

$$\mathscr{T}_{0,n} \cong [\prod_{m=0}^{n-1} K] \times [K \times K] \cong K^{n+2}$$

for each $n \in \mathbb{N}$. The transition homomorphism from $\mathscr{T}_{0,n}$ to $\mathscr{T}_{0,n+1}$ takes $(r_0, \ldots, r_{n-1}, r_n, r_{n+1}) \in K^{n+2}$ to $(r_0, \ldots, r_{n-1}, r_n, r_{n+1}, r_{n+1}) \in K^{n+3}$. Thus $\mathscr{T}_0$

is isomorphic to the subring of the direct product $\prod_{m \in \mathbb{Z}^+} K$ consisting of those elements which are eventually constant.

2.2 The Reduction Theorem and the Uniqueness Theorems

The name of this section derives in part from the name given to Theorem 2.2.11, a result which will prove to be an extremely useful tool in a variety of contexts. For instance, we will see how it yields both Theorems 2.2.15 and 2.2.16 with only a modicum of additional effort. The Reduction Theorem 2.2.11 will also be key in establishing various ring-theoretic properties of an arbitrary Leavitt path algebra, among other uses.

Notation 2.2.1 For a cycle c based at the vertex v, we will use the following notation:

$$c^0 := v, \text{ and } c^{-n} := (c^*)^n \text{ for all } n \in \mathbb{N}.$$

Definitions 2.2.2 Let E be a graph, let $\mu = e_1 e_2 \cdots e_n$ be a path in E, and let $e \in E^1$.

(i) We say that e is an *exit* for μ if there exists an i $(1 \leq i \leq n)$ such that $s(e) = s(e_i)$ and $e \neq e_i$.
(ii) We say that E satisfies *Condition* (L) if every cycle in E has an exit.

Examples 2.2.3 Here is how Condition (L) manifests in the fundamental graphs of the subject.

(i) Let E be the graph R_n $(n \geq 2)$, with edges $\{e_1, e_2, \ldots, e_n\}$. Each e_i is a cycle (of length 1) in E, and these are the only cycles in E. Moreover, each e_j $(j \neq i)$ is an exit for e_i (since $s(e_i) = s(e_j)$ for all i, j). In particular, E satisfies Condition (L).
(ii) On the other hand, the cycle e consisting of the unique loop in the graph $E = R_1$ has no exit (and thus R_1 does not satisfy Condition (L)).
(iii) In the oriented n-line graph A_n, no element of Path(A_n) has an exit. However, A_n does satisfy Condition (L) vacuously, as A_n is acyclic.
(iv) In the Toeplitz graph E_T of Example 1.3.6, the edge f is an exit for the loop e (which is the unique cycle in E_T). So E_T satisfies Condition (L).

Notation 2.2.4 Let E be an arbitrary graph. We denote by $P_c(E)$ the set of all vertices v of E which are in cycles without exits; i.e., $v \in c^0$ for some cycle c having no exits.

Remark 2.2.5 If e is an edge of a path without exits, then $s^{-1}(s(e))$ is a singleton (necessarily e itself). As a result, the (CK2) relation at $s(e)$ reduces to the equation $s(e) = ee^*$.

We start by exploring the structure of a corner of a Leavitt path algebra at a vertex which lies in a cycle without exits.

Definition 2.2.6 Let E be an arbitrary graph and K any field. For every cycle c based at a vertex v in E, and every polynomial $p(x) = \sum_{i=m}^{n} k_i x^i \in K[x, x^{-1}]$ ($m \leq n$; $m, n \in \mathbb{Z}$), we denote by $p(c)$ the element

$$p(c) := \sum_{i=m}^{n} k_i c^i \in L_K(E)$$

(using Notation 2.2.1).

Lemma 2.2.7 *Let E be an arbitrary graph and K any field. If c is a cycle without exits based at a vertex v, then*

$$vL_K(E)v = \left\{ \sum_{i=m}^{n} k_i c^i \mid k_i \in K,\ m \leq n, m, n \in \mathbb{Z} \right\} \cong K[x, x^{-1}],$$

via an isomorphism that sends v to 1, c to x and c^* to x^{-1}.

Proof Write $c = e_1 \cdots e_n$, where $e_i \in E^1$. We establish first that any $\gamma \in \text{Path}(E)$ such that $s(\gamma) = v$ is of the form $c^m \tau_q$, where $m \in \mathbb{Z}^+$, $\tau_q = e_1 \cdots e_q$ for $1 \leq q < n$, $\tau_0 = v$, and $\deg(\gamma) = mn + q$. We proceed by induction on $\deg(\gamma)$. If $\deg(\gamma) = 1$ and $s(\gamma) = s(e_1)$ then $\gamma = e_1$ by Remark 2.2.5. Suppose now that the result holds for any $\lambda \in \text{Path}(E)$ with $s(\lambda) = v$, $\deg(\lambda) \leq sn + t$, and consider any $\gamma \in \text{Path}(E)$ with $s(\gamma) = v$ and $\deg(\gamma) = sn + t + 1$. We can write $\gamma = \gamma' f$ with $\gamma' \in \text{Path}(E)$, $s(\gamma') = v$, $f \in E^1$ and $\deg(\gamma') = sn + t$, so by the induction hypothesis $\gamma' = c^s e_1 \cdots e_t$. Since c has no exits, $s(f) = r(e_t) = s(e_{t+1})$ implies $f = e_{t+1}$. Thus $\gamma = \gamma' e_{t+1} = c^s e_1 \cdots e_{t+1}$.

Now let $\gamma, \lambda \in \text{Path}(E)$ with $s(\gamma) = s(\lambda) = v$. If $\deg(\gamma) = \deg(\lambda)$ and $\gamma\lambda^* \neq 0$, we have $\gamma\lambda^* = c^p e_1 \cdots e_k e_k^* \cdots e_1^* c^{-p} = v$ (using the result of the previous paragraph together with Remark 2.2.5). On the other hand, $\deg(\gamma) > \deg(\lambda)$ and $\gamma\lambda^* \neq 0$ imply $\gamma\lambda^* = c^{d+q} e_1 \cdots e_k e_k^* \cdots e_1^* c^{-q} = c^d$, $d \in \mathbb{N}$. In a similar way, from $\deg(\gamma) < \deg(\lambda)$ and $\gamma\lambda^* \neq 0$ follows $\gamma\lambda^* = c^q e_1 \cdots e_k e_k^* \cdots e_1^* c^{-q-d} = c^{-d}$, $d \in \mathbb{N}$.

For any $\alpha \in vL_K(E)v$, write $\alpha = \sum_{i=1}^{p} k_i \gamma_i \beta_i^* + kv$, with $k_i, k \in K$ and $\gamma_i, \beta_i \in \text{Path}(E)$ such that $s(\gamma_i) = s(\beta_i) = v$ for all $1 \leq i \leq p$. Then, using what has been established in the previous paragraphs, we get $\alpha = \sum_{i=0}^{p} k_i c^{m_i}$, where $\deg(\gamma_i \beta_i^*) = m_i n$ for some $m_i \in \mathbb{Z}$.

Define $\varphi \colon K[x, x^{-1}] \to L_K(E)$ by setting $\varphi(1) = v$, $\varphi(x) = c$ and $\varphi(x^{-1}) = c^*$. It is a straightforward routine to check that φ is a monomorphism of K-algebras with image $vL_K(E)v$, so that $vL_K(E)v$ is isomorphic to the K-algebra $K[x, x^{-1}]$. □

We note that the isomorphism φ of the previous result is a graded isomorphism precisely when the cycle c is a loop. Also, we note that Lemma 2.2.7 allows us to easily reestablish Proposition 1.3.4, namely, that $L_K(R_1)$ is isomorphic to $K[x, x^{-1}]$.

The following result provides a significant portion of the Reduction Theorem; effectively, it will allow us to "reduce" various elements of $L_K(E)$ to a nonzero scalar multiple of a vertex.

Lemma 2.2.8 *Let E be an arbitrary graph and K any field. Suppose that v is a vertex of E for which $T(v) \cap P_c(E) = \emptyset$; in other words, for every $w \in E^0$ for which $v \geq w$, w does not lie on a cycle without exits. Let $\alpha := kv + \sum_{i=1}^n k_i\tau_i \in KE$, where $n \in \mathbb{N}$, $k, k_i \in K^\times$ and $\tau_i \in \mathrm{Path(E)} \setminus \{v\}$ with $s(\tau_i) = r(\tau_i) = v$, for which $\tau_i \neq \tau_j$. Then there exists a $\gamma \in \mathrm{Path}(E)$, with $s(\gamma) = v$, such that $\gamma^*\alpha\gamma = kr(\gamma)$.*

Proof We may suppose that $0 < \deg(\tau_1) \leq \ldots \leq \deg(\tau_n)$. Since the τ_i's are paths starting and ending at v, and $T(v) \cap P_c(E) = \emptyset$, there exists a $\gamma \in \mathrm{Path}(E)$ such that $\tau_1 = \gamma\tau'$ (with $\tau' \in \mathrm{Path}(E)$), $s(\gamma) = v$ and $|s^{-1}(r(\gamma))| > 1$. For those values of i for which there exists a τ_i' such that $\tau_i = \gamma\tau_i'$ we have $\gamma^*\tau_i\gamma = \tau_i'\gamma$; otherwise $\gamma^*\tau_i\gamma = 0$. After reordering the subindices we get $\gamma^*\alpha\gamma = kr(\gamma) + \sum_{i=1}^m k_i\tau_i'\gamma$, with $m \leq n$. Let e be the initial edge of $\tau_1'\gamma$. Observe that $s(\tau_1') = r(\gamma)$, and $|s^{-1}(r(\gamma))| > 1$. So there exists an $f \in s^{-1}(r(\gamma))$ such that $f \neq e$. We have

$$f^*\gamma^*\alpha\gamma f = kr(f) + \sum_{i=2}^m k_i f^*\tau_i'\gamma f,$$

and, as an element of $L_K(E)$, $f^*\tau_i'\gamma f$ is either a path in real edges, or is zero. Moreover, $T(r(f)) \cap P_c(E) = \emptyset$ as $r(f) \leq v$ and $T(v) \cap P_c(E) = \emptyset$. Hence we have reached the same initial conditions, but using fewer summands. So continuing in this way we eventually produce a nonzero multiple of a vertex. □

Definitions 2.2.9 A monomial $e_1 \cdots e_m f_1^* \cdots f_n^*$ in a path algebra $K\widehat{E}$ over an extended graph $\widehat{E}$, where $e_i, f_j \in E^1$ and $m, n \in \mathbb{Z}^+$, is said to have *degree in ghost edges* (or simply *ghost degree*) equal to n. Monomials in KE are said to have degree in ghost edges equal to 0. The *degree in ghost edges* of an element of $K\widehat{E}$ of the form $\sum_{i=1}^n k_i\gamma_i\lambda_i^*$, with $k_i \in K^\times$, denoted $\mathrm{gdeg}(\sum_{i=1}^n k_i\gamma_i\lambda_i^*)$, is defined to be the maximum of the degree in ghost edges of the monomials $\gamma_i\lambda_i^*$.

Because the representation of an element $\alpha \in L_K(E)$ as an element of the form $\sum_{i=1}^n k_i\gamma_i\lambda_i^*$ is not uniquely determined, the direct extension of the notion of "degree in ghost edges" to elements of $L_K(E)$ is not well-defined. However, we define the degree in ghost edges of an element $\alpha \in L_K(E)$, also denoted $\mathrm{gdeg}(\alpha)$, to be the minimum of the degrees in ghost edges among all the representations of α as an expression $\sum_{i=1}^n k_i\gamma_i\lambda_i^*$ in $K\widehat{E}$ as above.

Lemma 2.2.10 *Let E be an arbitrary graph and K any field. Let α be an element of $L_K(E)$ with positive degree in ghost edges and let $e \in E^1$. Then $\mathrm{gdeg}(\alpha e) < \mathrm{gdeg}(\alpha)$.*

Proof Let $\alpha = \sum_{i=1}^n k_i\gamma_i\lambda_i^*$, $k_i \in K^\times$, be an expression of α in $L_K(E)$ with smallest degree in ghost edges. Note that if the degree in ghost edges of a monomial $\gamma_j\lambda_j^*$ is positive, then $\mathrm{gdeg}(\gamma_j\lambda_j^* e) < \mathrm{gdeg}(\gamma_j\lambda_j^*)$. The result follows. □

We now come to the key result of this section. Roughly speaking, this theorem says that any nonzero element of a Leavitt path algebra may be "reduced", via multiplication on the left and right by appropriate paths, to either a nonzero K-multiple of a vertex, or to a monic polynomial in a cycle without exits, or to both.

Theorem 2.2.11 (The Reduction Theorem) *Let E be an arbitrary graph and K any field. For any nonzero element $\alpha \in L_K(E)$ there exist $\mu, \eta \in \mathrm{Path}(E)$ such that either:*

(i) $0 \neq \mu^*\alpha\eta = kv$, *for some $k \in K^\times$ and $v \in E^0$, or*
(ii) $0 \neq \mu^*\alpha\eta = p(c)$, *where c is a cycle without exits and $p(x)$ is a nonzero polynomial in $K[x, x^{-1}]$.*

Proof The first step will be to show that for $0 \neq \alpha \in L_K(E)$ there exists an $\eta \in \mathrm{Path}(E)$ such that $0 \neq \alpha\eta \in KE$. Let $v \in E^0$ be such that $\alpha v \neq 0$ (such a vertex v exists by Lemma 1.2.12(v)). Write $\alpha v = \sum_{i=1}^{r} \alpha_i e_i^* + \alpha'$, where $\alpha_i \in L_K(E)r(e_i)$, $\alpha' \in (KE)v$, $e_i \in E^1$, $e_i \neq e_j$ for every $i \neq j$, and $s(e_i) = v$ for all $1 \leq i \leq r$.

Note that if $\mathrm{gdeg}(\alpha v) = 0$, then we are done.

Suppose otherwise that $\mathrm{gdeg}(\alpha v) > 0$. If $\alpha v e_j = 0$ for every $j \in \{1, \ldots, r\}$, then multiplying the equation $\alpha v = \sum_{i=1}^{r} \alpha_i e_i^* + \alpha'$ on the right by e_j gives $0 = \alpha v e_j = \alpha_j + \alpha' e_j$, so $\alpha_j = -\alpha' e_j$, and $\alpha v = \sum_{i=1}^{r} (-\alpha' e_i e_i^*) + \alpha' = \alpha'((\sum_{i=1}^{r} -e_i e_i^*) + v) \neq 0$. In particular, $0 \neq (\sum_{i=1}^{r} -e_i e_i^*) + v$ and $\alpha' \neq 0$. So by (CK2) there exists an $f \in s^{-1}(v) \setminus \{e_1, \ldots, e_r\}$. Now, by the structure of KE, $\alpha v f = \alpha' f \in KE \setminus \{0\}$, and we have finished the proof of the first step.

On the other hand, suppose that there exists a $j \in \{1, \ldots, r\}$ such that $\alpha v e_j \neq 0$. There is no loss of generality if we consider $j = 1$. Then $0 \neq \alpha v e_1 = \alpha_1 + \alpha' e_1 = (\alpha_1 + \alpha' e_1)r(e_1)$, where $\mathrm{gdeg}\,(\alpha_1 + \alpha' e_1) < \mathrm{gdeg}\,(\alpha v)$ by Lemma 2.2.10. Repeating this argument a finite number of times, we reach $\eta \in \mathrm{Path}(E)$ with $\alpha\eta \in KE \setminus \{0\}$.

Now pick $0 \neq \alpha \in L_K(E)$. By the previous paragraph, we know that there exists an $\eta \in \mathrm{Path}(E)$ such that $\beta := \alpha\eta \in KE \setminus \{0\}$. Write $\beta = \sum_{i=1}^{s} k_i \gamma_i$, with $k_i \in K^\times$, $\gamma_i \in \mathrm{Path}(E)$, and with $r(\gamma_i) = r(\eta) =: v$ for every i. We will prove the result by induction on s.

Suppose $s = 1$. If $\deg(\gamma_1) = 0$, then there is nothing to prove. If $\deg(\gamma_1) > 0$, then $\gamma_1^*\alpha\eta = \gamma_1^*\beta = k_1\gamma_1^*\gamma_1 = k_1 r(\gamma_1) \neq 0$.

Now suppose the result is true for any element having at most $s - 1$ summands. Write again $\beta = \sum_{i=1}^{s} k_i \gamma_i$, where $k_i \in K^\times$, $\gamma_i \in \mathrm{Path}(E)$, $\gamma_i \neq \gamma_j$ if $i \neq j$ and $\deg(\gamma_i) \leq \deg(\gamma_{i+1})$ for every $i \in \{1, \ldots, s-1\}$. Then $0 \neq \gamma_1^*\beta = k_1 v + \sum_{i=2}^{s} k_i \gamma_1^* \gamma_i$.

If $\gamma_1^*\gamma_i = 0$ for some $i \in \{2, \ldots, s\}$, then apply the induction hypothesis to get the result. Otherwise, $0 \neq \mu := \gamma_1^*\beta = k_1 v + \sum_{i=2}^{s} k_i \mu_i$, where the μ_i are paths starting and ending at v and satisfying $0 < \deg(\mu_2) \leq \ldots \leq \deg(\mu_s)$. If $T(v) \cap P_c(E) = \emptyset$, then by Lemma 2.2.8 there exists a path τ such that $\tau^*\gamma_1^*\alpha\eta\tau = \tau^*\mu\tau = k_1 r(\tau)$, and we are done. If $T(v) \cap P_c(E) \neq \emptyset$, then there is a path ρ starting at v such that $w := r(\rho)$ is a vertex in a cycle c without exits. In this case, $0 \neq \rho^*\gamma_1^*\alpha\eta\rho = \rho^*\mu\rho \in wL_K(E)w$, and by Lemma 2.2.7 the proof is complete.

□

We note that both cases in the Reduction Theorem 2.2.11 can occur simultaneously: for instance, in $L_K(R_1)$ we have $e^*e = v$, which is simultaneously a vertex as well as the base of a cycle without exits.

The conclusion we obtained in the first step of the proof of the Reduction Theorem, and a consequence of it, will be of great use later on, so we record them in the following two results.

Corollary 2.2.12 *Let E be an arbitrary graph and K any field. Let α be a nonzero element in $L_K(E)$.*

(i) *There exists an $\eta \in \text{Path}(E)$ such that $0 \neq \alpha\eta \in KE$.*
(ii) *If α is a homogeneous element of $L_K(E)$, then there exists an $\eta \in \text{Path}(E)$ such that $0 \neq \alpha\eta$ is a homogeneous element of KE.*

Corollary 2.2.13 *Let E be an arbitrary graph and K any field. Let α be a nonzero homogeneous element of $L_K(E)$. Then there exist $\mu, \eta \in \text{Path}(E)$, $k \in K^\times$, and $v \in E^0$ such that $0 \neq \mu^*\alpha\eta = kv$.*

In particular, every nonzero graded ideal of $L_K(E)$ contains a vertex.

Proof By Corollary 2.2.12(ii) there exists an $\eta \in \text{Path}(E)$ for which $0 \neq \alpha\eta$ is a homogeneous element in KE. So we may write $\alpha\eta = \sum_{i=1}^{n} k_i\beta_i$ where $k_i \in K^\times$, the β_i are distinct paths in E, and the lengths of the β_i are equal. But then $\beta_1^*\beta_1 = r(\beta_1)$, while $\beta_1^*\beta_i = 0$ for all $2 \leq i \leq n$ by Lemma 1.2.12(i). Thus $\beta_1^*\alpha\eta = k_1 r(\beta_1)$, as desired.

The particular statement follows immediately. □

We noted in Examples 2.1.7 that the Leavitt path algebra $L_K(R_1)$ contains infinitely many nontrivial non-graded ideals. Since the single vertex of R_1 acts as the identity element of $L_K(R_1)$, none of these ideals contains a vertex. The following result shows that the existence of ideals in $L_K(R_1)$ which do not contain any vertices is a consequence of the fact that the graph R_1 contains a cycle without exits.

Proposition 2.2.14 *Let E be a graph satisfying Condition* (L) *and K any field. Then every nonzero ideal of $L_K(E)$ contains a vertex.*

Proof Let I be a nonzero ideal of $L_K(E)$, and let α be a nonzero element in I. Since E satisfies Condition (L) then by the Reduction Theorem there exist $\mu, \eta \in \text{Path}(E)$ such that $0 \neq \mu^*\alpha\eta = kv$ with $v \in E^0$ and $k \in K^\times$. This implies $0 \neq v = k^{-1}\mu^*\alpha\eta \in L_K(E)IL_K(E) \subseteq I$. □

The converse of Proposition 2.2.14 is also true, as will be proved in Proposition 2.9.13.

Two results of fundamental importance which are direct consequences of the Reduction Theorem 2.2.11 are the following Uniqueness Theorems. These results can be considered as the analogs of the Gauge-Invariant Uniqueness Theorem [129, Theorem 2.2] and the Cuntz–Krieger Uniqueness Theorem [129, Theorem 2.4] for graph C^*-algebras; see Sect. 5.2 below.

Theorem 2.2.15 (The Graded Uniqueness Theorem) *Let E be an arbitrary graph and K any field. If A is a $\mathbb{Z}$-graded ring, and $\pi : L_K(E) \to A$ is a graded ring homomorphism with $\pi(v) \neq 0$ for every vertex $v \in E^0$, then π is injective.*

Theorem 2.2.16 (The Cuntz–Krieger Uniqueness Theorem) *Let E be an arbitrary graph which satisfies Condition (L), let K be any field, and let A be any K-algebra. If $\pi : L_K(E) \to A$ is a ring homomorphism with $\pi(v) \neq 0$ for every vertex $v \in E^0$, then π is injective.*

Proof of Theorems 2.2.15 and 2.2.16 We use the basic fact that the kernel of any ring homomorphism is an ideal of the domain. For the Graded Uniqueness Theorem, as π is a graded homomorphism we have that $\mathrm{Ker}(\pi)$ is a graded ideal of $L_K(E)$. Thus $\mathrm{Ker}(\pi)$ is either $\{0\}$ or contains a vertex, by Corollary 2.2.13. For the Cuntz–Krieger Uniqueness Theorem, we use Proposition 2.2.14 to conclude that $\mathrm{Ker}(\pi)$ is either $\{0\}$ or contains a vertex in this situation as well. Since the hypotheses of both statements presume that π sends vertices to nonzero elements, the only option is $\mathrm{Ker}(\pi) = \{0\}$ in both cases. □

We present now the first of many applications of the Uniqueness Theorems. Specifically, we use the Graded Uniqueness Theorem 2.2.15 to show that any finite matrix ring over a Leavitt path algebra is itself a Leavitt path algebra.

Definition 2.2.17 Given any graph E and positive integer n, we let M_nE denote the graph formed from E by taking each $v \in E^0$ and attaching a "head" of length $n-1$ at v of the form

$$\bullet^{v_{n-1}} \xrightarrow{e^v_{n-1}} \cdots \xrightarrow{e^v_3} \bullet^{v_2} \xrightarrow{e^v_2} \bullet^{v_1} \xrightarrow{e^v_1} \bullet^{v}$$

Example 2.2.18 If E is the graph

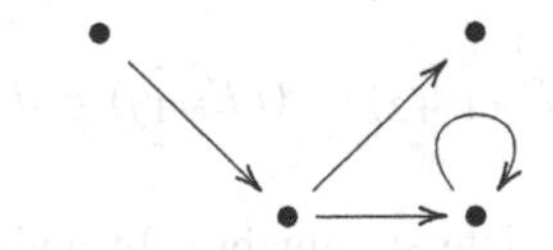

then M_3E is the graph

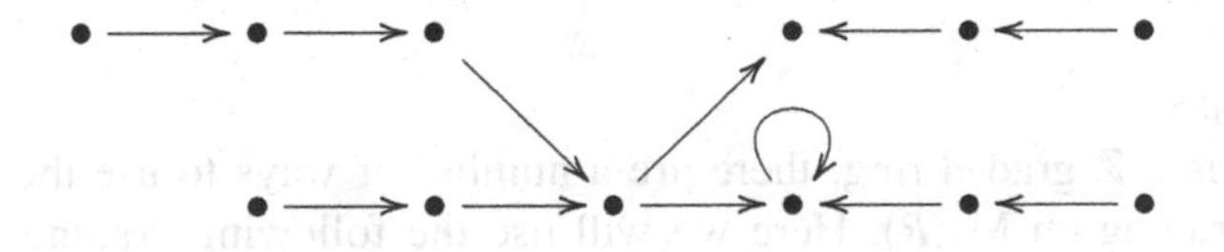

Proposition 2.2.19 *Let E be an arbitrary graph and K any field. Then there is an isomorphism of K-algebras*

$$L_K(M_nE) \cong \mathrm{M}_n(L_K(E)).$$

Proof For $1 \leq i,j \leq n$, we let $E_{i,j}$ denote the element of $\mathrm{M}_n(K)$ having 1 in the $(i,j)^{\text{th}}$ position and 0's elsewhere. For $a \in L_K(E)$ we let $aE_{i,j}$ denote the element of $\mathrm{M}_n(L_K(E))$ having a in the $(i,j)^{\text{th}}$ position and 0's elsewhere. Note that $(aE_{i,j})(bE_{k,l}) = abE_{i,j}E_{k,l}$ in $\mathrm{M}_n(L_K(E))$.

For each $v \in E^0, e \in E^1$, and $k \in \{1,\ldots,n-1\}$ define $Q_v, Q_{v_k}, T_e, T_e^*, T_{e_k^v}$, and $T_{e_k^v}^*$ by setting

$$Q_v = vE_{1,1},\quad Q_{v_k} = vE_{k+1,k+1},\quad T_e = eE_{1,1},\quad T_e^* = e^*E_{1,1},\quad T_{e_k^v} = vE_{k+1,k},\quad \text{and } T_{e_k^v}^* = vE_{k,k+1}.$$

It is straightforward to verify that $\{T_e, T_e^*, T_{e_k^v}, T_{e_k^v}^* \mid v \in E^0, e \in E^1, 1 \leq k \leq n-1\} \cup \{Q_v, Q_{v_k} \mid v \in E^0, 1 \leq k \leq n-1\}$ is an M_nE-family in $\mathrm{M}_n(L_K(E))$. Thus by the Universal Property 1.2.5 there exists a K-algebra homomorphism $\phi : L_K(M_nE) \to \mathrm{M}_n(L_K(E))$ for which

$$\phi(v) = Q_v, \phi(v_k) = Q_{v_k}, \phi(e) = T_e,\ \phi(e^*) = T_e^*,\ \phi(e_{k^v}) = T_{e_k^v},\ \text{and}\ \phi(e_{k^v}^*) = T_{e_k^v}^*.$$

To see that ϕ is onto, it suffices to show that $vE_{i,j}$ and $eE_{i,j}$ are in the K-subalgebra of $\mathrm{M}_n(L_K(E))$ generated by $\{Q_w, T_f, T_f^* \mid f \in M_nE^1, w \in M_nE^0\}$ for all $v \in E^0$, $e \in E^1$, and $1 \leq i,j \leq n$. Straightforward computations yield that

$$vE_{i,i} = \begin{cases} Q_v & \text{if } i = 1 \\ Q_{v_{i-1}} & \text{if } i \geq 2, \end{cases}$$

that for $i > j$ we have

$$vE_{i,j} = (vE_{i,i-1})(vE_{i-1,i-2})\cdots(vE_{j+1,j}) = T_{e_{i-1}^v}T_{e_{i-2}^v}\cdots T_{e_j^v},$$

and that for $i < j$ we have

$$vE_{i,j} = (vE_{i,i+1})(vE_{i+1,i+2})\cdots(vE_{j-1,j}) = T_{e_i^v}^*T_{e_{i+1}^v}^*\cdots T_{e_{j-1}^v}^*.$$

So each $vE_{i,j}$ is in the appropriate subalgebra. In addition, for any $e \in E^1$ and $1 \leq i,j \leq n$ we have

$$eE_{i,j} = (s(e)E_{i,1})(eE_{1,1})(r(e)E_{1,j}) = T_{e_{i-1}^v}\cdots T_{e_1^v}T_eT_{e_1^v}^*\cdots T_{e_{j-1}^v}^*.$$

Thus ϕ is onto.

When R is a $\mathbb{Z}$-graded ring, there are a number of ways to use the grading to build a $\mathbb{Z}$-grading on $\mathrm{M}_n(R)$. Here we will use the following grading on $\mathrm{M}_n(R)$: for $x \in R_t$, the degree of $xE_{i,j}$ is defined to be $t + (i - j)$. It is straightforward to establish that, with respect to this grading on $\mathrm{M}_n(L_K(E))$, the homomorphism ϕ

described above is in fact $\mathbb{Z}$-graded. (Note, for example, that in this grading we have $\deg(\phi(e_k^v)) = \deg(T_{e_k^v}) = \deg(vE_{k+1,k}) = 0 + ((k+1) - k) = 1 = \deg(e_{k^v})$.) Since for each vertex v in M_nE we have $\phi(v) \neq 0$, we conclude by the Graded Uniqueness Theorem 2.2.15 that ϕ is injective, and thus an isomorphism. □

The next result is similar in flavor to the two Uniqueness Theorems.

Proposition 2.2.20 *Let E be an arbitrary graph and K any field. Let A be a $\mathbb{Z}$-graded K-algebra and let $\pi : L_K(E) \to A$ be a (not necessarily graded) K-algebra homomorphism for which $\pi(v) \neq 0$ for every vertex $v \in E^0$, and for which π maps each cycle without exits in E to a nonzero homogeneous element of nonzero degree in A. Then π is injective.*

Proof By hypothesis, $\mathrm{Ker}(\pi)$ is an algebra ideal of $L_K(E)$ which does not contain vertices. If $\mathrm{Ker}(\pi)$ is nonzero, then by the Reduction Theorem $\mathrm{Ker}(\pi)$ contains a nonzero element $p(c)$, where $p(x) = \sum_{i=m}^{n} k_i x^i \in K[x, x^{-1}]$ and c is a cycle without exits. Let $q(x) = x^{-m}p(x) \in K[x]$; then $q(c) = c^{-m}p(c) = \sum_{i=0}^{n-m} k_{i+m}c^i \in \mathrm{Ker}(\pi)$. So $0 = \pi(q(c)) = q(\pi(c)) = \sum_{i=0}^{n-m} k_{i+m}\pi(c)^i$. But this is impossible since $\pi(c)$ is a nonzero homogeneous element of nonzero degree in A. □

We conclude the section by giving a direct application of the Graded Uniqueness Theorem, in which we demonstrate an embedding of Leavitt path algebras corresponding to naturally arising subgraphs F of a given graph E.

Definition 2.2.21 (The Restriction Graph) Let E be an arbitrary graph, and let H be a hereditary subset of E^0. We denote by E_H the *restriction graph*:

$$E_H^0 := H, \quad E_H^1 := \{e \in E^1 \mid s(e) \in H\},$$

and the source and range functions in E_H are simply the source and range functions in E, restricted to H.

Proposition 2.2.22 *Let E be an arbitrary graph and K any field. Let H be a hereditary subset of E^0.*

(i) Consider the assignment

$$v \mapsto v, \quad e \mapsto e, \quad \text{and } e^* \mapsto e^*$$

(for $v \in E_H^0$ and $e \in E_H^1$), which maps elements of $L_K(E_H)$ to elements of $L_K(E)$. Then this assignment extends to a $\mathbb{Z}$-graded monomorphism of Leavitt path algebras $\varphi : L_K(E_H) \to L_K(E)$.

(ii) If H is finite, then $\varphi(L_K(E_H)) = p_H L_K(E) p_H$, where $p_H := \sum_{v \in H} v \in L_K(E)$.

Proof (i) Consider these elements of $L_K(E)$: $a_v = v$, $a_e = e$, and $b_e = e^*$ for $v \in E_H^0, e \in E_H^1$. Then by definition we have that the set $\{a_v, a_e, b_e\}$ is an E_H-family in $L_K(E)$, so the indicated assignment extends to a K-algebra homomorphism $\varphi : L_K(E_H) \to L_K(E)$ by the Universal Property 1.2.5. That φ is a graded homomorphism is clear from the definition of the grading on $L_K(E_H)$ and $L_K(E)$. That φ is a monomorphism then follows from an application of the Graded Uniqueness Theorem 2.2.15.

(ii) We show that (ii) follows from (i). Since every element in $L_K(E)$ is a K-linear combination of elements of the form $\gamma\lambda^*$ with $\gamma, \lambda \in \text{Path}(E)$, then every element in $p_H L_K(E) p_H$ is a K-linear combination of elements $\gamma\lambda^*$, with $\gamma, \lambda \in \text{Path}(E)$ having $s(\gamma), s(\lambda) \in H$. Thus $\gamma\lambda^* \in \text{Im}(\varphi)$. The containment $\text{Im}(\varphi) \subseteq p_H L_K(E) p_H$ is immediate using that p_H is the multiplicative identity of $L_K(E_H)$. □

2.3 Additional Consequences of the Reduction Theorem

As part of the power of the Reduction Theorem 2.2.11 we will see that every Leavitt path algebra is semiprime, semiprimitive, and nonsingular. Numerous additional applications of the Reduction Theorem will be presented throughout the sequel.

Recall that a ring R is said to be *semiprime* if, for every ideal I of R, $I^2 = 0$ implies $I = 0$. A ring R is said to be *semiprimitive* if the Jacobson radical $J(R)$ of R is zero.

Proposition 2.3.1 *Let E be an arbitrary graph and K any field. Then the Leavitt path algebra $L_K(E)$ is semiprime.*

Proof Let I be a nonzero ideal of $L_K(E)$, and consider a nonzero element $\alpha \in I$. By the Reduction Theorem 2.2.11, there exist $\gamma, \lambda \in \text{Path}(E)$ such that $\gamma^*\alpha\lambda = kv$ or $\gamma^*\alpha\lambda = p(c) \in wL_K(E)w$, where $k \in K^\times$, $v, w \in E^0$, $c \in P_c(E)$ and $w \in c^0$. Then $kv \in I$ or $p(c) \in I$. Observe that since $(kv)^2 = k^2 v \neq 0$ and $(p(c))^2 \neq 0$ (use that $wL_K(E)w$ has no nonzero zero divisors, by Lemma 2.2.7), $I^2 \neq 0$ and hence $L_K(E)$ is semiprime. □

Proposition 2.3.2 *Let E be an arbitrary graph and K any field. Then the Leavitt path algebra $L_K(E)$ is semiprimitive.*

Proof Denote by J the Jacobson radical of $L_K(E)$, and suppose there is a nonzero element $\alpha \in J$. By the Reduction Theorem 2.2.11, there exist $\mu, \eta \in \text{Path}(E)$ such that $0 \neq \mu^*\alpha\eta = kv$ or $\mu^*\alpha\eta = p(c) \in wL_K(E)w$, where $k \in K^\times$, $v, w \in E^0$, $c \in P_c(E)$ and $w \in c^0$. In the first case we would have $v \in J$, but this is not possible, as the Jacobson radical of any ring contains no nonzero idempotents. In the second case, let u denote $s(c)$. Then $\mu^*\alpha\eta$ is a nonzero element in $J \cap uL_K(E)u$, which coincides with the Jacobson radical of $uL_K(E)u$ by Jacobson [99, §III.7, Proposition 1]. But by Lemma 2.2.7 $uL_K(E)u \cong K[x, x^{-1}]$ which has zero Jacobson radical. In both cases we get a contradiction, hence $J = \{0\}$. □

We note that Proposition 2.3.2 indeed directly implies Proposition 2.3.1, as it is well known that any semiprimitive ring is semiprime. We have included

Proposition 2.3.1 simply to provide an additional example of the power of the Reduction Theorem.

We present here a second approach to establishing that every Leavitt path algebra is semiprimitive. This approach makes use of an extension of an unpublished result of Bergman [50] about the Jacobson radical of unital $\mathbb{Z}$-graded rings; this extension (the following result) may be of interest in its own right.

Lemma 2.3.3 *Let R be a $\mathbb{Z}$-graded ring that contains a set of local units consisting of homogeneous elements of degree* 0. *Then the Jacobson radical $J(R)$ of R is a graded ideal of R.*

Proof Given $x \in J(R)$, decompose x into its homogeneous components: $x = x_{-n} + \cdots + x_{-1} + x_0 + x_1 + \cdots + x_n$, where $n \in \mathbb{N}$ (and x_i can be zero). Let u be a homogeneous local unit (of degree 0) for each x_i, i.e., $ux_iu = x_i$. Then clearly $uxu = x$, and we get

$$x = uxu = ux_{-n}u + \cdots + ux_{-1}u + ux_0u + ux_1u + \cdots + ux_nu$$

is also a decomposition of x into its homogeneous components inside the unital ring uRu, so that $x_i = ux_iu$ for every $i \in \{-n, \ldots, -1, 0, 1, \ldots n\}$. As the corner uRu is also a $\mathbb{Z}$-graded ring, and as $J(uRu) = uJ(R)u$, the displayed equation yields a decomposition of the element x in the Jacobson radical of uRu, which is a graded ideal of the $\mathbb{Z}$-graded unital ring uRu (see [120, 2.9.3 Corollary], or the aforementioned unpublished result of Bergman). Therefore every x_i is in $J(uRu)$, and, consequently, in $J(R)$. □

A second proof of Proposition 2.3.2 By Lemma 2.3.3 and Corollary 2.2.13, if the Jacobson radical of $L_K(E)$ were nonzero, then it would contain a vertex, hence a nonzero idempotent, which is impossible. □

Definitions 2.3.4 Let R be a ring and $x \in R$. The *left annihilator of x in R*, denoted by $\mathrm{lan}_R(x)$ (or more simply by $\mathrm{lan}(x)$ if the ring R is understood), is the set $\{r \in R \mid rx = 0\}$. A left ideal I of R is said to be *essential* if $I \cap I' \neq 0$ for every nonzero left ideal I' of R. In this situation we write $I \triangleleft_e^l R$. The set

$$Z_l(R) = \{x \in R \mid \mathrm{lan}(x) \triangleleft_e^l R\},$$

which is an ideal of R (see [108, Corollary 7.4]), is called the *left singular ideal* of R. The ring R is called *left nonsingular* if $Z_l(R) = \{0\}$. *Right nonsingular* rings are defined similarly, while *nonsingular* means that R is both left and right nonsingular.

To overcome the lack of a unit element in a ring or algebra, and to translate problems from a non-unital context to a unital one, local rings at elements provide a very useful tool. This notion was first introduced in the context of associative algebras in [80]. We refer the reader to [84] for a fuller account of the transfer of various properties between rings and their local rings at elements.

Definition 2.3.5 Let R be a ring and let $a \in R$. The *local ring* of R *at* a is defined as $R_a = aRa$, with sum inherited from R, and product given by $axa \cdot aya = axaya$.

Notice that if e is an idempotent in the ring R, then the local ring of R at e is just the corner eRe. The following result can be found in [84].

Lemma 2.3.6 *Let R be a semiprime ring. Then:*

(i) *If $a \in Z_l(R)$, then $Z_l(R_a) = R_a$.*
(ii) $Z_l(R_a) \subseteq Z_l(R)$.
(iii) *R is left nonsingular if and only if R_a is left nonsingular for every $a \in R$.*

Proposition 2.3.7 *Let E be an arbitrary graph and K any field. Then $L_K(E)$ is nonsingular.*

Proof Suppose that the left singular ideal $Z_l(L_K(E))$ contains a nonzero element α. By the Reduction Theorem there exist $\gamma, \mu \in \mathrm{Path}(E)$ such that $0 \neq \gamma^*\alpha\mu \in Kv$ for some vertex $v \in E^0$, or $0 \neq \gamma^*\alpha\mu \in uL_K(E)u \cong K[x, x^{-1}]$ (by Lemma 2.2.7), where u is a vertex in a cycle without exits. Since, for any ring R, $Z_l(R)$ is an ideal of R and does not contain nonzero idempotents, the first case cannot happen.

In the second case, denote by β the nonzero element $\gamma^*\alpha\mu \in Z_l(L_K(E))$, and for notational convenience denote $L_K(E)$ by L. Then, by Lemma 2.3.6(i) (which can be applied due to Proposition 2.3.1), $Z_l(L_\beta) = L_\beta$. It is not difficult to see that $L_\beta = (L_u)_\beta$, and therefore, $Z_l((L_u)_\beta) = (L_u)_\beta$. Note that $L_u \cong K[x, x^{-1}]$, which is a nonsingular ring. This implies, by Lemma 2.3.6(iii), that every local algebra of L_u at an element is left nonsingular; in particular, $L_\beta = Z_l(L_\beta) = 0$. Now the semiprimeness of L yields $\beta = 0$, a contradiction.

The right nonsingularity of $L_K(E)$ follows from Corollary 2.0.9. □

2.4 Graded Ideals: Basic Properties and Quotient Graphs

In this section we present a description of the graded ideals of a Leavitt path algebra. The main goal here (Theorem 2.4.8) is to show that every graded ideal can be constructed from a hereditary saturated subset of E^0, possibly augmented by a set of *breaking vertices* (cf. Definition 2.4.4). With this information in hand, we then proceed to analyze the quotient algebra $L_K(E)/I$ for a graded ideal I. Specifically, we show in Theorem 2.4.15 that there exists a graph F for which $L_K(E)/I \cong L_K(F)$ as $\mathbb{Z}$-graded K-algebras.

This introductory analysis of the graded ideal structure will provide a foundation for the remaining results of Chap. 2. Looking forward, we will use the ideas of this section to explicitly describe the lattice of graded ideals of $L_K(E)$ in terms of graph-theoretic properties; to show how graded ideals of $L_K(E)$ are themselves Leavitt path algebras in their own right; and how the graded ideals, together with various sets of cycles in E and polynomials in $K[x]$, provide complete information about the lattice of *all* ideals of $L_K(E)$.

We start by presenting a description of the elements in the ideal generated by a hereditary subset of vertices.

Lemma 2.4.1 *Let E be an arbitrary graph and K any field. Let H be a hereditary subset of E^0. Then the ideal $I(H)$ of $L_K(E)$ consists of elements of $L_K(E)$ of the form*

$$I(H) = \left\{ \sum_{i=1}^{n} k_i \gamma_i \lambda_i^* \mid n \geq 1, k_i \in K, \gamma_i, \lambda_i \in \text{Path}(E) \textit{ such that } r(\gamma_i) = r(\lambda_i) \in H \right\}.$$

Moreover, if $\overline{H}$ denotes the saturated closure of H, then $I(H) = I(\overline{H})$.

Proof Let J denote the set presented in the display. To see that J is an ideal of $L_K(E)$ we need to show that for every element of the form $\alpha\beta^*$, where $r(\alpha) = r(\beta) = u \in H$, and for every $a, b \in L_K(E)$, we have $a\alpha u\beta^* b \in J$. Taking into account statements (i) and (iii) of Lemma 1.2.12, it is enough to prove that $\gamma\lambda^* u\mu\eta^* \in J$ for every $\gamma, \lambda, \mu, \eta \in \text{Path}(E)$ and $u \in H$.

If $\gamma\lambda^* u\mu\eta^* = 0$ we are done. Suppose otherwise that $\gamma\lambda^* u\mu\eta^* \neq 0$. By Lemma 1.2.12(i), $\gamma\lambda^* u\mu\eta^* = \gamma\mu'\eta^*$ if $\mu = \lambda\mu'$, or $\gamma\lambda^* u\mu\eta^* = \gamma(\lambda')^*\eta^*$ if $\lambda = \mu\lambda'$. Note that $u = s(\mu)$ and H hereditary imply $r(\mu) \in H$, therefore, $r(\mu') = r(\mu) \in H$ in the first case, and $r(\lambda') = r(\mu) \in H$ in the second case, which imply $\gamma\lambda^* u\mu\eta^* \in J$ in both cases. This shows that J is an ideal of $L_K(E)$; as it contains H and must be contained in every ideal containing H, it must coincide with $I(H)$.

Now we prove $I(H) = I(\overline{H})$. Clearly $I(H) \subseteq I(\overline{H})$. Conversely, we will show by induction that $H_n \subseteq I(H)$ for every $n \in \mathbb{Z}^+$ (where the notation H_n is as in Lemma 2.0.7). For $n = 0$ there is nothing to prove, as $H_0 = T(H) = H \subseteq I(H)$. Suppose $H_{n-1} \subseteq I(H)$ and take $u \in H_n$. Then $s^{-1}(u) = \{e_1, \ldots, e_m\}$, and so $\{r(e_i) | 1 \leq i \leq m\} = r(s^{-1}(u)) \subseteq H_{n-1}$, which is contained in $I(H)$ by the induction hypothesis. This means $u = \sum_{i=1}^m e_i e_i^* = \sum_{i=1}^m e_i r(e_i) e_i^* \in I(H)$ and the proof is complete. □

Corollary 2.4.2 *Let E be an arbitrary graph and K any field. Let H be a nonempty hereditary subset of E^0. Then for every nonzero homogeneous $x \in I(H)$ there exist $\alpha, \beta \in \text{Path}(E)$ such that $\alpha^* x\beta = ku$ for some $k \in K^\times$ and $u \in H$.*

Proof Given the nonzero homogeneous element $x \in I(H)$, apply Corollary 2.2.13 to choose $\lambda, \mu \in \text{Path}(E)$ such that $k^{-1}\lambda^* x\mu = v$ for some $k \in K^\times$ and $v \in E^0$. Since $x \in I(H)$ this equation gives that $v \in I(H)$. So by Lemma 2.4.1 we may write $v = \sum_{i=1}^m k_i' \lambda_i \mu_i^*$ with $k_i' \in K^\times$ and $\lambda_i, \mu_i \in \text{Path}(E)$ with $r(\lambda_i) = r(\mu_i) \in H$. Then $0 \neq r(\mu_1) = \mu_1^*\mu_1 = \mu_1^* v\mu_1 = k^{-1}\mu_1^*\lambda^* x\mu\mu_1 \in H$, so that $r(\mu_1) = u$, $\mu_1^*\lambda^* = \alpha^*$ and $\mu\mu_1 = \beta$ satisfy the assertion. □

The following result demonstrates the natural, fundamental connection between the (CK1) and (CK2) condition on the elements of $L_K(E)$ on the one hand, and the ideal structure of $L_K(E)$ on the other. Recall the definition of the set $\mathscr{H}_E$ of hereditary saturated subsets of E given in Definitions 2.0.5.

Lemma 2.4.3 *Let E be an arbitrary graph and K any field. Let I be an ideal of $L_K(E)$. Then $I \cap E^0 \in \mathscr{H}_E$.*

Proof Let $v, w \in E^0$ be such that $v \geq w$, and $v \in I$. So there exists a path $p \in \text{Path}(E)$ with $v = s(p)$ and $w = r(p)$. Then Lemma 1.2.12(i) implies that $w = p^*p = p^*vp \in I$. This shows that $I \cap E^0$ is hereditary.

Now let $u \in \text{Reg}(E)$, and suppose $r(e) \in I$ for every $e \in s^{-1}(u)$. By (CK2), $u = \sum_{e\in s^{-1}(u)} ee^* = \sum_{e\in s^{-1}(u)} er(e)e^* \in I$. Thus $I \cap E^0$ is saturated. □

One eventual goal in our study of the graded ideals in a Leavitt path algebra is the Structure Theorem for Graded Ideals 2.5.8. The idea is to associate with each graded ideal of $L_K(E)$ some data inherent in the underlying graph. The previous lemma establishes a first connection of this type. The following graph-theoretic idea will provide a key ingredient in this association.

Definitions 2.4.4 Let E be an arbitrary graph and K any field. Let H be a hereditary subset of E^0, and let $v \in E^0$. We say that v is a *breaking vertex of H* if v belongs to the set

$$B_H := \{v \in E^0 \setminus H \mid v \in \text{Inf}(E) \text{ and } 0 < |s^{-1}(v) \cap r^{-1}(E^0 \setminus H)| < \infty\}.$$

In words, B_H consists of those vertices of E which are infinite emitters, which do not belong to H, and for which the ranges of the edges they emit are all, except for a finite (but nonzero) number, inside H. For $v \in B_H$, we define the element v^H of $L_K(E)$ by setting

$$v^H := v - \sum_{e\in s^{-1}(v)\cap r^{-1}(E^0\setminus H)} ee^*.$$

We note that any such v^H is homogeneous of degree 0 in the standard $\mathbb{Z}$-grading on $L_K(E)$. For any subset $S \subseteq B_H$, we define $S^H \subseteq L_K(E)$ by setting $S^H = \{v^H \mid v \in S\}$.

Of course a row-finite graph contains no breaking vertices, so that this concept does not play a role in the study of the Leavitt path algebras arising from such graphs.

Remark 2.4.5 Let E be an arbitrary graph. It is easy to show both that $B_\emptyset = \emptyset$, and that $B_{E^0} = \emptyset$. The latter is trivial, while the former follows by noting that $|s^{-1}(v) \cap r^{-1}(E^0 \setminus \emptyset)| = \infty$ for any $v \in \text{Inf}(E)$.

To clarify the concept of a breaking vertex, we revisit the infinite clock graph $C_\mathbb{N}$ of Example 1.6.12.

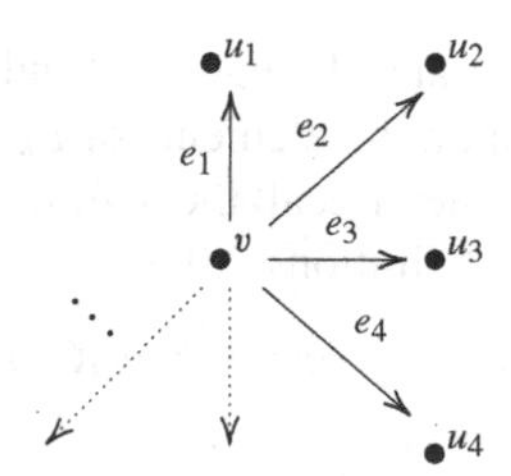

Let U denote the set $\{u_i \mid i \in \mathbb{N}\} = C^0_{\mathbb{N}} \setminus \{v\}$. Let H be a subset of U. Since the elements of H are sinks in E, H is clearly hereditary. If $U \setminus H$ is infinite, or if $H = U$, then $B_H = \emptyset$. On the other hand, if $U \setminus H$ is finite, then $B_H = \{v\}$, and in this situation, $v^H = v - \sum_{\{i \mid r(e_i) \in U \setminus H\}} e_i e_i^*$.

For any hereditary subset H of a graph E, and for any $S \subseteq B_H$, the ideal $I(H \cup S^H)$ of $L_K(E)$ is graded, as it is generated by elements of $L_K(E)$ of degree zero (see Remark 2.1.2). We describe this ideal more explicitly in the following result.

Lemma 2.4.6 *Let E be an arbitrary graph and K any field. Let H be a hereditary subset of vertices of E, and S a subset of B_H. Then*

$$I(H \cup S^H) = \mathrm{span}_K(\{\gamma\lambda^* \mid \gamma, \lambda \in \mathrm{Path}(E) \textit{ such that } r(\gamma) = r(\lambda) \in H\})$$
$$+ \mathrm{span}_K(\{\alpha v^H \beta^* \mid \alpha, \beta \in \mathrm{Path}(E) \textit{ and } v \in S\}).$$

Moreover, the first summand equals $I(H)$, while the second summand (call it J) is a subalgebra of $L_K(E)$ for which $I(S^H) \subseteq I(H) + J$.

Proof Clearly $I(H \cup S^H) = I(H) + I(S^H)$. Moreover, by virtue of Lemma 2.4.1, the first summand in the displayed formula of the statement coincides with $I(H)$.

Now we study $I(S^H)$. Take $v \in S$, and denote the set $s^{-1}(v) \cap r^{-1}(E^0 \setminus H)$ by $\{f_1, \dots, f_n\}$, where $n \in \mathbb{N}$. For each $e \in E^1$ we compute $e^* v^H$ and $v^H e$. If $s(e) \neq v$, then $e^* v^H = v^H e = 0$. Otherwise, if $s(e) = v$, we distinguish two cases. If $e = f_j$ for some j, then $e^* v^H = e^*(v - \sum_{i=1}^n f_i f_i^*) = f_j^* v - f_j^* f_j f_j^* = f_j^* - f_j^* = 0$, and, in addition, $v^H e = (v - \sum_{i=1}^n f_i f_i^*) e = v f_j - f_j f_j^* f_j = f_j - f_j = 0$. If on the other hand $e \notin \{f_1, \dots, f_m\}$, then $e \in s^{-1}(v) \cap r^{-1}(H)$, so that $r(e) \in H$, and $e^* v^H = e^* = r(e) e^* \in I(H)$. Similarly, $v^H e = e = e r(e) \in I(H)$. This means that for $\alpha, \beta \in \mathrm{Path}(E)$ we have either $\alpha^* v^H = 0$, or $\alpha^* v^H = \alpha^* \in I(H)$; similarly, either $v^H \beta = 0$ or $v^H \beta = \beta \in I(H)$. In either case the resulting product is in $I(H)$, and so $I(S^H) \subseteq I(H) + J$. To see that J is a subalgebra, apply the previous calculation and use that $H v^H = v^H H = 0$ for every $v \in S$. This finishes our proof because $I(H) + J \subseteq I(H) + I(S^H)$. □

Here is a useful application of Lemma 2.4.6.

Proposition 2.4.7 *Let E be an arbitrary graph and K any field. Let $\{H_i\}_{i \in \Lambda}$ be a family of hereditary pairwise disjoint subsets of a graph E. Then*

$$I\left(\overline{\bigsqcup_{i \in \Lambda} H_i}\right) = I\left(\bigsqcup_{i \in \Lambda} H_i\right) = \bigoplus_{i \in \Lambda} I(H_i) = \bigoplus_{i \in \Lambda} I\left(\overline{H_i}\right).$$

Proof The final equality follows from Lemma 2.4.1. It is easy to see that the union of any family of hereditary subsets is again hereditary, hence $H := \bigsqcup_{i \in \Lambda} H_i$ is a hereditary subset of E^0. Thus the first equality also follows from Lemma 2.4.1.

By Lemma 2.4.1 every element x in $I(H)$ can be written as $x = \sum_{l=1}^n k_l \alpha_l \beta_l^*$, where $k_l \in K^\times$, $\alpha_l, \beta_l \in \mathrm{Path}(E)$ and $r(\alpha_l) = r(\beta_l) \in H$. Separate the vertices appearing as ranges of the α_l's depending on the H_i's they belong to, and apply again

Lemma 2.4.1. This gives $x \in \sum_{i\in\Lambda} I(H_i)$, so that $I(H) \subseteq \sum_{i\in\Lambda} I(H_i)$. The containment $\sum_{i\in\Lambda} I(H_i) \subseteq I(H)$ is clear.

So all that remains is to show that the sum $\sum_{i\in\Lambda} I(H_i)$ is direct. If this is not the case, there exists a $j \in \Lambda$ such that $I(H_j) \cap \sum_{j\neq i\in\Lambda} I(H_i) \neq 0$. Since for every l, $I(H_l)$ is a graded ideal, we get that $I(H_j) \cap \sum_{j\neq i\in\Lambda} I(H_i)$ is a graded ideal as well, so there exists a nonzero homogeneous element $y \in I(H_j) \cap \sum_{j\neq i\in\Lambda} I(H_i)$. By Corollary 2.4.2 there exist $\alpha, \beta \in \text{Path}(E)$ and $k \in K^\times$ such that $0 \neq k^{-1}\alpha^* y\beta = w \in H_j$. Observe that w also belongs to $I(\bigcup_{j\neq i\in\Lambda} H_i)$. Write $w = \sum_{l=1}^n k_l\alpha_l\beta_l^*$, with $k_l \in K^\times, \alpha_l, \beta_l \in \text{Path}(E)$, and $r(\alpha_l) = r(\beta_l) \in \bigcup_{j\neq i\in\Lambda} H_i$. Then $0 \neq r(\beta_1) = \beta_1^*\beta_1 = \beta_1^* w\beta_1 \in \bigcup_{j\neq i\in\Lambda} H_i$. On the other hand, $s(\alpha_1) = w \in H_j$ implies (since H_j is a hereditary set) $r(\alpha_1) \in H_j$; therefore, $r(\alpha_1) = r(\beta_1) \in H_j \cap \left(\bigcup_{j\neq i\in\Lambda} H_i\right)$, a contradiction. □

We now deepen the connection between graded ideals of $L_K(E)$ and various subsets of E^0.

Theorem 2.4.8 *Let E be an arbitrary graph and K any field. Then every graded ideal I of $L_K(E)$ is generated by $H \cup S^H$, where $H = I \cap E^0 \in \mathscr{H}_E$, and $S = \{v \in B_H \mid v^H \in I\}$.*

In particular, every graded ideal of $L_K(E)$ is generated by a set of homogeneous idempotents.

Proof It is immediate to see that $I(H \cup S^H) \subseteq I$. Now we show $I \subseteq I(H \cup S^H)$. As I is a graded ideal, it is enough to consider nonzero homogeneous elements of the form $\alpha = \alpha v$ of I, where $v \in E^0$.

We will prove $\alpha v \in I(H \cup S^H)$ by induction on the degree in ghost edges of the elements in I (recall Definitions 2.2.9). Suppose first gdeg $(\alpha) = 0$. Then, $\alpha = \sum_{i=1}^m k_i\gamma_i$, with $k_i \in K^\times$, $m \in \mathbb{N}$, and $\gamma_i \in \text{Path}(E)$ with $r(\gamma_i) = v$. As α is a homogeneous element, we may consider those γ_i's having the same degree (i.e., length) as that of α. Moreover, we may suppose all the γ_i's are distinct, hence $\gamma_i^*\gamma_j = 0$ for $i \neq j$ by Lemma 1.2.12(i). Then for every j, $k_j^{-1}\gamma_j^*\alpha v = k_j^{-1}\gamma_j^*\left(\sum_{i=1}^m k_i\gamma_i\right) = k_j^{-1}k_j\gamma_j^*\gamma_j = r(\gamma_j) = v \in I \cap E^0 = H$. This means $\alpha v \in I(H) \subseteq I(H \cup S^H)$.

We now suppose the result is true for appropriate elements of $L_K(E)$ having degree in ghost edges strictly less than $n \in \mathbb{N}$, and prove the result for gdeg $(\alpha v) = n$. Write $\alpha v = \sum_{i=1}^m \mu_i e_i^* + \lambda$, with $\mu_i \in L_K(E)$, $e_i \in E^1$ and $\lambda \in KE$, in such a way that this is a representation of αv of minimal degree in ghost edges.

If $\lambda = 0$ then for every i we have $\alpha v e_i = \mu_i$, which is in $I(H \cup S^H)$ by the induction hypothesis, and we have finished. Hence, we may assume that $\lambda \neq 0$.

As α is homogeneous, we may choose μ_i and λ to be homogeneous as well. Write $\lambda = \sum_{l=1}^n k_l\lambda_l$ for some $k_l \in K^\times$ and λ_l distinct paths of the same length. We first observe that v cannot be a sink because $e_i^* = e_i^* v$ implies $v = s(e_i)$ for every i; in particular, $s^{-1}(v) \neq \emptyset$. Choose $f \in s^{-1}(v)$. If $e_i^*f = 0$ for every i, then $\alpha vf = \lambda f$, which is in $I(H \cup S^H)$ by the previous case. Otherwise, suppose $e_j^*f \neq 0$ for some j. By (CK1) this happens precisely when $f = e_j$, and hence $\alpha vf = \left(\sum_{i=1}^m \mu_i e_i^* + \lambda\right)f = \mu_j e_j^* f + \lambda f = \mu_j + \lambda f$, which lies in $I(H \cup S^H)$ by the induction hypothesis. (Note that the induction hypothesis can be applied because $\mathrm{gdeg}\,(\mu_j + \lambda f) < \mathrm{gdeg}\,(\alpha v)$.) In any case, $\alpha vf \in I(H \cup S^H)$. Now, if v is not an infinite emitter then $\alpha v = \alpha \sum_{f \in s^{-1}(v)} ff^* \in I(H \cup S^H)$. If v is an infinite emitter, then either $v \in H$, in which case $\alpha v \in I(H \cup S^H)$, or $v \notin H$, in which case $v \in B_H$, as follows. For any $f \in s^{-1}(v) \cap r^{-1}(E^0 \setminus H)$, observe that f must coincide with some e_i because otherwise $\alpha f = \sum_{i=1}^m \mu_i e_i^* f + \lambda f = \lambda f \in I$ would imply $r(f) = f^*f = f^* k_1^{-1} \lambda_1^* \lambda f \in I \cap E^0 = H$, a contradiction. Thus $s^{-1}(v) \cap r^{-1}(E^0 \setminus H) \subseteq \{e_i \mid 1 \leq i \leq m\}$, and so $v \in B_H$.

Now write $\alpha v = \alpha v^H + \alpha \sum_{\{f \in s^{-1}(v) \cap r^{-1}(E^0 \setminus H)\}} ff^*$. Since $\alpha f \in I(H \cup S^H)$ for all $f \in s^{-1}(v) \cap r^{-1}(E^0 \setminus H)$, to show that $\alpha v \in I(H \cup S^H)$, it is enough to show that $v \in S$. We compute

$$e_i^* v^H = e_i^*(v - \sum_{f \in s^{-1}(v) \cap r^{-1}(E^0 \setminus H)} ff^*) = \begin{cases} 0 & \text{if } \ e_i \in s^{-1}(v) \cap r^{-1}(E^0 \setminus H) \\ e_i^* v & \text{if } \ e_i \in s^{-1}(v) \cap r^{-1}(H). \end{cases}$$

In the second of these two cases, $s(e_i^*) = r(e_i) \in H$. In either case $e_i^* v^H \in I(H)$.

But $\alpha v^H \in I$ and $e_i^* v^H \in I$ imply $\lambda v^H \in I$, hence $k_1 v^H = \lambda_1^*(\lambda v^H) \in I$, therefore $v^H \in I$ and so $v \in S$ as desired. □

Proposition 2.4.9 is an immediate consequence of Theorem 2.4.8.

Proposition 2.4.9 *Let E be a row-finite graph and K any field. Then every graded ideal I of $L_K(E)$ is generated by a hereditary and saturated subset of E^0, specifically, $I = I(I \cap E^0)$.*

Let $(A, *)$ be an algebra with involution. An ideal I of A is said to be *self-adjoint* if $y^* \in I$ whenever $y \in I$. Not every ideal in a Leavitt path algebra is self-adjoint. For instance, consider an arbitrary field K and let E be the graph R_1. Then the ideal I of $L_K(E)$ generated by $v + e + e^3$ is not self-adjoint, as follows. Identify $L_K(R_1)$ and $K[x, x^{-1}]$ via the isomorphism given in Proposition 1.3.4. Our statement rephrased says that $I(1 + x + x^3)$ is not a self-adjoint ideal, which is clear as otherwise we would have $1 + x^{-1} + x^{-3} \in I(1 + x + x^3)$, which would give $x^3(1 + x^{-1} + x^{-3}) = 1 + x^2 + x^3 \in I$, which is impossible by an observation made in Remark 2.1.6.

By observing that any ideal in an arbitrary graded ring with involution which is generated by a set of self-adjoint elements is necessarily self-adjoint, we record this consequence of Theorem 2.4.8.

Corollary 2.4.10 *Let E be an arbitrary graph and K any field. If I is a graded ideal of $L_K(E)$, then $I = I(X)$ for some set X of homogeneous self-adjoint idempotents in $L_K(E)$. Specifically, every graded ideal of a Leavitt path algebra is self-adjoint.*

The converse to Corollary 2.4.10 does not hold. For instance, the ideal $I = I(v + e)$ of $L_K(R_1)$ is self-adjoint, as $v + e^* = e^*(v + e) \in I$. However, I is not graded, as noted in Examples 2.1.7. Indeed, this same behavior is exhibited by any ideal of $L_K(R_1)$ of the form $I(p(e))$, where $p(x) \in K[x, x^{-1}]$ is not homogeneous and has the property that $p(x)^* = x^n p(x)$ for some integer n.

In the next section we will strengthen Theorem 2.4.8 to show that in fact there is a bijection between the graded ideals of $L_K(E)$ and pairs of the form (H, S^H). In order to establish that distinct pairs of this form correspond to distinct graded ideals, we analyze the K-algebras which arise as quotients of a Leavitt path algebra by graded ideals. As we shall see, such quotients turn out to be Leavitt path algebras in their own right.

Definition 2.4.11 (The Quotient Graph by a Hereditary Subset) Let E be an arbitrary graph, and let H be a hereditary subset of E^0. We denote by E/H the *quotient graph of E by H*, defined as follows:

$$(E/H)^0 = E^0 \setminus H, \quad \text{and} \quad (E/H)^1 = \{e \in E^1 \mid r(e) \notin H\}.$$

The range and source functions for E/H are defined by restricting the range and source functions of E to $(E/H)^1$.

We anticipate the following result with a brief discussion. We will show that the quotient algebra $L_K(E)/I(H \cup S^H)$ is isomorphic to a relative Cohn path algebra for the quotient graph E/H (with respect to an appropriate subset of vertices), and then subsequently apply Proposition 2.1.10. The intuitive idea underlying Theorem 2.4.12 is as follows. Let H be a hereditary saturated subset of E^0. Then the breaking vertices B_H of H are precisely the infinite emitters in E which become regular vertices in E/H. If $S \subseteq B_H$, and we consider the ideal $I(H \cup S^H)$ of $L_K(E)$, then we are imposing relation (CK2) only on the vertices corresponding to S in the quotient ring $L_K(E)/I(H \cup S^H)$. So it is natural to expect that the quotient $L_K(E)/I(H \cup S^H)$ will be a relative Cohn path algebra with respect to the set $X = (\text{Reg}(E) \setminus H) \cup S$.

Theorem 2.4.12 *Let E be an arbitrary graph and K any field. Let $H \in \mathcal{H}_E$, $S \subseteq B_H$, and $X = (\text{Reg}(E) \setminus H) \cup S$. Then there exists a $\mathbb{Z}$-graded isomorphism of K-algebras*

$$\overline{\Psi} : L_K(E)/I(H \cup S^H) \to C_K^X(E/H).$$

Proof We consider the assignment (which we denote by Ψ) of elements of the set $E^0 \cup E^1 \cup (E^1)^*$ with specific elements of $C_K^X(E/H)$ given as follows: for each $v \in E^0$ and $e \in E^1$,

$$\Psi(v) = \begin{cases} v & \text{if } v \notin H \\ 0 & \text{otherwise,} \end{cases} \quad \Psi(e) = \begin{cases} e & \text{if } r(e) \notin H \\ 0 & \text{otherwise,} \end{cases} \quad \text{and} \quad \Psi(e^*) = \begin{cases} e^* & \text{if } s(e^*) \notin H \\ 0 & \text{otherwise.} \end{cases}$$

Using this assignment, a set of straightforward computations yields that the collection

$$\{\Psi(v), \Psi(e), \Psi(e^*) \mid v \in E^0, e \in E^1\}$$

is an E-family in $C_K^X(E/H)$. So by the Universal Property of $L_K(E)$ 1.2.5 there is a unique extension of Ψ to a K-algebra homomorphism

$$\Psi : L_K(E) \to C_K^X(E/H).$$

We note that Ψ is indeed a $\mathbb{Z}$-graded homomorphism, as clearly Ψ preserves the grading of each of the generators of $L_K(E)$. By the definition of E/H, it is immediate that Ψ is surjective. Furthermore, Ψ is clearly 0 on $I(H)$. But we also have that $\Psi(v^H) = 0$ for $v \in S$, because $S \subseteq X$. Consequently, there is an induced map

$$\overline{\Psi} : L_K(E)/I(H \cup S^H) \to C_K^X(E/H).$$

We now define an inverse map for $\overline{\Psi}$. The map Φ is defined as follows: for $v \in (E/H)^0$ and $e \in (E/H)^1$, set

$$\Phi(v) = v + I(H \cup S^H), \quad \Phi(e) = e + I(H \cup S^H), \quad \text{and} \quad \Phi(e^*) = e^* + I(H \cup S^H).$$

By the Universal Property of $C_K^X(E/H)$ 1.5.10, Φ extends to a K-algebra homomorphism $\Phi : C_K^X(E/H) \to L_K(E)/I(H \cup S^H)$. It is then straightforward to verify that the compositions $\Phi \circ \overline{\Psi}$ and $\overline{\Psi} \circ \Phi$ give the identity on the canonical generators, and therefore give the identity on the corresponding algebras. □

Here are two specific consequences of Theorem 2.4.12.

Corollary 2.4.13 *Let K be any field.*

(i) *Suppose E is a row-finite graph, and $H \in \mathcal{H}_E$. Then $L_K(E)/I(H) \cong L_K(E/H)$ as $\mathbb{Z}$-graded K-algebras.*
(ii) *If E is an arbitrary graph and $H \in \mathcal{H}_E$, then*

$$L_K(E)/I(H \cup B_H^H) \cong C_K^{\text{Reg}(E/H)}(E/H) = L_K(E/H).$$

Proof (i) In this case $S = \emptyset$, so that $X = \text{Reg}(E) \setminus H$, and thus $C_K^X(E/H) = L_K(E/H)$. Now apply Theorem 2.4.12.

(ii) We set $S = B_H$. Then $X = (\text{Reg}(E) \setminus H)) \cup B_H = \text{Reg}(E/H)$, so that Theorem 2.4.12 yields the isomorphism. □

Theorem 2.4.12 gives a description of the quotient of a Leavitt path algebra by a graded ideal as a relative Cohn path algebra. But by defining a new type of quotient graph, we can in fact describe the quotient of a Leavitt path algebra by a graded ideal as the Leavitt path algebra over this new graph.

Definition 2.4.14 (The Quotient Graph Incorporating Breaking Vertices) Let E be an arbitrary graph, $H \in \mathcal{H}_E$, and $S \subseteq B_H$. We denote by $E/(H, S)$ the *quotient* *graph of E by* (H, S), defined as follows:

$$(E/(H,S))^0 = (E^0 \setminus H) \cup \{v' \mid v \in B_H \setminus S\},$$
$$(E/(H,S))^1 = \{e \in E^1 \mid r(e) \notin H\} \cup \{e' \mid e \in E^1 \text{ and } r(e) \in B_H \setminus S\},$$

and range and source maps in $E/(H,S)$ are defined by extending the range and source maps in E when appropriate, and in addition setting $s(e') = s(e)$ and $r(e') = r(e)'$.

We note that the quotient graph E/H given in Definition 2.4.11 is precisely the graph $E/(H, B_H)$ in the context of this broader definition. (In particular, we point out that E/H is *not* the same as $E/(H, \emptyset)$.)

With this definition, and using Theorems 2.4.12 and 1.5.18, we get the following.

Theorem 2.4.15 *Let E be an arbitrary graph and K any field. Then the quotient of $L_K(E)$ by a graded ideal of $L_K(E)$ is $\mathbb{Z}$-graded isomorphic to a Leavitt path algebra. Specifically, there is a $\mathbb{Z}$-graded K-algebra isomorphism*

$$\overline{\Psi} : L_K(E)/I(H \cup S^H) \to L_K(E/(H,S)),$$

where $\overline{\Psi}$ is defined as in Theorem 2.4.12.

Proof By Theorem 2.4.12, we have $L_K(E)/I(H \cup S^H) \cong C_K^X(E/H)$, where $X = (\text{Reg}(E) \setminus H) \cup S$. But then $\text{Reg}(E/H) \setminus X = B_H \setminus S$. Therefore, the graph $(E/H)(X)$ from Definition 1.5.16 coincides with the quotient graph $E/(H,S)$, and Theorem 1.5.18 gives that $C_K^X(E/H) \cong L_K(E/(H,S))$ naturally, thus yielding the result. □

We close this section with another consequence of Theorem 2.4.12.

Corollary 2.4.16 *Let E be an arbitrary graph and K any field. Suppose $H \in \mathcal{H}_E$ and let $S \subseteq B_H$.*

(i) *$I(H \cup S^H) \cap E^0 = H$. In particular, $I(H) \cap E^0 = H$.*
(ii) *$S = \{v \in B_H \mid v^H \in I(H \cup S^H)\}$.*

Proof (i) The containment $H \subseteq I(H \cup S^H)$ is clear. Conversely, for $v \in E^0 \setminus H$, we observe that $\Psi(v)$ is a nonzero element in $C_K^X(E/H)$, where Ψ is the isomorphism given in Theorem 2.4.12. Thus $v \notin I(H \cup S^H)$.

(ii) The containment $S \subseteq \{v \in B_H \mid v^H \in I(H \cup S^H)\}$ is clear. For the reverse containment, observe that in a manner analogous to that used in the proof of (i) we have $\Psi(v^H) \neq 0$ for any $v \in B_H \setminus S$. This shows that $v^H \notin I(H \cup S^H)$, as required. □

2.5 The Structure Theorem for Graded Ideals, and the Internal Structure of Graded Ideals

In the previous section we have developed much of the machinery which will allow us to achieve the main goal of the current section, the Structure Theorem for Graded Ideals (Theorem 2.5.8), which gives a complete description of the lattice of graded ideals of a Leavitt path algebra in terms of specified subsets of E^0.

Definition 2.5.1 Let E be an arbitrary graph and K any field. Denote by $\mathscr{L}_{gr}(L_K(E))$ the lattice of graded ideals of $L_K(E)$, with order given by inclusion, and supremum and infimum given by the usual operations of ideal sum and intersection.

Remark 2.5.2 Let E be an arbitrary graph. We define in $\mathscr{H}_E$ a partial order by setting $H \leq H'$ if $H \subseteq H'$. Using this ordering, $\mathscr{H}_E$ is a complete lattice, with supremum $\vee$ and infimum $\wedge$ in $\mathscr{H}_E$ given by setting $\vee_{i\in\Gamma} H_i := \overline{\cup_{i\in\Gamma} H_i}$ and $\wedge_{i\in\Gamma} H_i := \cap_{i\in\Gamma} H_i$, respectively.

Definition 2.5.3 Let E be an arbitrary graph. We set

$$\mathscr{S} = \bigcup_{H\in\mathscr{H}_E} \mathscr{P}(B_H),$$

where $\mathscr{P}(B_H)$ denotes the set of all subsets of B_H. We denote by $\mathscr{T}_E$ the subset of $\mathscr{H}_E \times \mathscr{S}$ consisting of pairs of the form (H, S), where $S \in \mathscr{P}(B_H)$. We define in $\mathscr{T}_E$ the following relation:

$$(H_1, S_1) \leq (H_2, S_2) \quad \text{if and only if} \quad H_1 \subseteq H_2 \text{ and } S_1 \subseteq H_2 \cup S_2.$$

The following comments, which explain why the relation $\leq$ in $\mathscr{T}_E$ has been defined as above, will help clarify the proof of the upcoming proposition. For a graph E, a hereditary saturated subset H of E^0, and a breaking vertex $v \in B_H$, define

$$A(v, H) := s^{-1}(v) \cap r^{-1}(E^0 \setminus H).$$

Note that $A(v, H)$ is a finite nonempty subset of E^1.

Now suppose that H_1 and H_2 are hereditary saturated subsets of vertices in E, with $H_1 \subseteq H_2$. Let $v \in B_{H_1}$. Since $H_1 \subseteq H_2$ then $v \in B_{H_2}$, unless it happens to be the case that $r(s^{-1}(v)) \subseteq H_2$ (since by definition a breaking vertex for a set must emit at least one edge whose range is outside the set). If $v \in B_{H_2}$, then write

$$A(v, H_1) = A(v, H_2) \sqcup B,$$

where $B = \{e \in A(v, H_1) \mid r(e) \in H_2\}$. In this case we have

$$v^{H_1} = v^{H_2} - \sum_{e\in B} ee^*. \tag{2.1}$$

Proposition 2.5.4 *Let E be an arbitrary graph. For $(H_1, S_1), (H_2, S_2) \in \mathscr{T}_E$, we have*

$$(H_1, S_1) \leq (H_2, S_2) \iff I(H_1 \cup S_1{}^{H_1}) \subseteq I(H_2 \cup S_2^{H_2}).$$

In particular, $\leq$ is a partial order on $\mathscr{T}_E$.

Proof For notational convenience, set $I(H_i, S_i) := I(H_i \cup S_i^{H_i})$ for $i = 1, 2$.

Suppose that $I(H_1, S_1) \subseteq I(H_2, S_2)$. Then $H_1 \subseteq H_2$ by Corollary 2.4.16(i). Now let $v \in S_1$. We will show that $v \in H_2 \cup S_2$. If on the one hand $r(s^{-1}(v)) \subseteq H_2$ then we have

$$v = v^{H_1} + \sum_{e \in A(v, H_1)} ee^* \in I(H_1, S_1) + I(H_2) \subseteq I(H_2, S_2),$$

so that $v \in H_2$ (by again invoking Corollary 2.4.16(i)). If on the other hand there is some $e \in s^{-1}(v)$ such that $r(e) \notin H_2$, then necessarily $v \notin H_2$ (since H_2 is hereditary). So, since we already know that $H_1 \subseteq H_2$, we see that $v \in B_{H_2}$. Moreover, we have, by (2.1),

$$v^{H_2} = v^{H_1} + \sum_{e \in B} ee^* \in I(H_1, S_1) + I(H_2) \subseteq I(H_2, S_2).$$

Hence $v \in S_2$ by Corollary 2.4.16(ii). So we have shown $S_1 \subseteq H_2 \cup S_2$, which yields $(H_1, S_1) \leq (H_2, S_2)$ by definition.

Conversely, suppose that $(H_1, S_1) \leq (H_2, S_2)$. This gives in particular that $I(H_1) \subseteq I(H_2)$, so we only need to check that $v^{H_1} \in I(H_2, S_2)$ for $v \in S_1$. So let $v \in S_1$. If on the one hand $r(s^{-1}(v)) \subseteq H_2$, then $v \in H_2$ because $S_1 \subseteq H_2 \cup S_2$ and $v \notin S_2$ (since $v \notin B_{H_2}$). If on the other hand there is some $e \in s^{-1}(v)$ such that $r(e) \notin H_2$, then $v \in B_{H_2}$ and, by (2.1) we have

$$v^{H_1} = v^{H_2} - \sum_{e \in B} ee^* \in I(H_2, S_2) + I(H_2) \subseteq I(H_2, S_2),$$

showing that $v^{H_1} \in I(H_2, S_2)$. Thus we obtain that $I(H_1, S_1) \subseteq I(H_2, S_2)$. □

For the proof of Proposition 2.5.6 we need to introduce a refinement of the definition of saturation which allows us to consider breaking vertices.

Definition 2.5.5 Let E be an arbitrary graph. Let H be a hereditary subset of E^0, and consider a subset $S \subseteq H \cup B_H$. The *S-saturation* of H is defined as the smallest hereditary subset H' of E^0 satisfying the following properties:

(i) $H \subseteq H'$.
(ii) H' is saturated.
(iii) If $v \in S$ and $r(s^{-1}(v)) \subseteq H'$, then $v \in H'$.

We denote by $\overline{H}^S$ the S-saturation of H.

To build the S-saturation of H we proceed as in Lemma 2.0.7. Concretely, for every $n \in \mathbb{Z}^+$ we define the hereditary subsets $\Lambda_n^S(H)$ inductively as follows. Let $\Lambda_0^S(H) := H$. For $n \geq 1$, we put

$$\Lambda_n^S(H) = \Lambda_{n-1}^S(H) \cup \{v \in E^0 \setminus \Lambda_{n-1}^S(H) \mid v \in \mathrm{Reg}(E) \cup S \text{ and } r(s^{-1}(v)) \subseteq \Lambda_{n-1}^S(H)\}.$$

It can be easily shown that $\overline{H}^S = \cup_{n \geq 0} \Lambda_n^S(H)$.

Proposition 2.5.6 *Let E be an arbitrary graph. Then with the partial order* $\leq$ *on* $\mathcal{T}_E$ *given in Definition 2.5.3,* $(\mathcal{T}_E, \leq)$ *is a complete lattice, with supremum* $\vee$ *and infimum* $\wedge$ *in* $\mathcal{T}_E$ *given by:*

$$(H_1, S_1) \vee (H_2, S_2) = (\overline{H_1 \cup H_2}^{S_1 \cup S_2},\ (S_1 \cup S_2) \setminus \overline{H_1 \cup H_2}^{S_1 \cup S_2}) \quad \textit{and}$$

$$(H_1, S_1) \wedge (H_2, S_2) = (H_1 \cap H_2,\ (S_1 \cap S_2) \cup ((S_1 \cup S_2) \cap (H_1 \cup H_2))).$$

Proof The fact that $\leq$ is a partial order is established in Proposition 2.5.4.

We first verify the displayed formula for the supremum. Observe that $(\overline{H_1 \cup H_2}^{S_1 \cup S_2},\ (S_1 \cup S_2) \setminus \overline{H_1 \cup H_2}^{S_1 \cup S_2}) \in \mathcal{T}_E$, and that it contains (H_i, S_i) for $i = 1, 2$.

To show minimality, let $(H, S) \in \mathcal{T}_E$ be such that $(H_i, S_i) \leq (H, S)$ for $i = 1, 2$. In order to show that $\overline{H_1 \cup H_2}^{S_1 \cup S_2} \subseteq H$, it suffices, by Definition 2.5.5, to prove that $\Lambda_n^{S_1 \cup S_2}(H_1 \cup H_2) \subseteq H$ for all $n \in \mathbb{Z}^+$. We do this inductively. For $n = 0$ this is clear by assumption. Now, assume $n \geq 1$ and that $\Lambda_{n-1}^{S_1 \cup S_2}(H_1 \cup H_2) \subseteq H$. Pick $v \in \Lambda_n^{S_1 \cup S_2}(H_1 \cup H_2)$. If $v \in \mathrm{Reg}(E)$, then v belongs to H because H is saturated. Now suppose $v \in S_1 \cup S_2$. By definition and the induction hypothesis, we have

$$r(s^{-1}(v)) \subseteq \Lambda_{n-1}^{S_1 \cup S_2}(H_1 \cup H_2) \subseteq H.$$

In particular, this implies $v \notin S$. Since $v \in S_1 \cup S_2 \subseteq H \cup S$, we conclude that $v \in H$, completing the induction step. The inclusion $(S_1 \cup S_2) \setminus \overline{H_1 \cup H_2}^{S_1 \cup S_2} \subseteq H \cup S$ is immediate.

Now we verify the indicated expression for the infimum, i.e., we will show that $(H_1 \cap H_2,\ (S_1 \cap S_2) \cup ((S_1 \cup S_2) \cap (H_1 \cup H_2)))$ is a lower bound for the pair $(H_1, S_1), (H_2, S_2)$, and is the maximal such. First, note that $(H_1 \cap H_2,\ (S_1 \cap S_2) \cup ((S_1 \cup S_2) \cap (H_1 \cup H_2))) \leq (H_i, S_i)$ for $i = 1, 2$. To see this, use $H_i \cap S_i = \emptyset$ for $i = 1, 2$, so that

$$(S_1 \cap S_2) \cup ((S_1 \cup S_2) \cap (H_1 \cup H_2)) = (S_1 \cap S_2) \cup (S_1 \cap H_2) \cup (S_2 \cap H_1).$$

Now, suppose $(H, S) \leq (H_i, S_i)$. Then $H \subseteq H_1 \cap H_2$ and $S \subseteq H_i \cup S_i$ and so

$$S \subseteq (H_1 \cup S_1) \cap (H_2 \cup S_2) = (H_1 \cap H_2) \cup (S_1 \cap S_2) \cup (S_1 \cap H_2) \cup (S_2 \cap H_1),$$

which by the formula above shows $(H, S) \leq (H_1 \cap H_2,\ (S_1 \cap S_2) \cup ((S_1 \cup S_2) \cap (H_1 \cup H_2)))$. □

The following examples clarify the notion of S-saturation.

Examples 2.5.7

(i) Let E be the following graph:

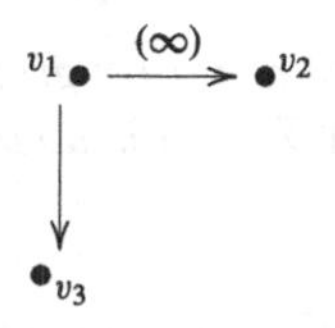

Let $H_1 = \{v_2\}$, $S_1 = \{v_1\}$; $H_2 = \{v_3\}$, $S_2 = \emptyset$. Note that $\overline{H_1 \cup H_2}$ does not contain the vertex v_1, which is not a breaking vertex for $H_1 \cup H_2$ as $r(s^{-1}(v_1)) \subseteq H_1 \cup H_2$. This is the reason why we have to consider the S-saturation, which is

$$\Lambda_1^{S_1}(H_1 \cup H_2) = \{v_1, v_2, v_3\},$$

and, consequently, the formula in Proposition 2.5.6 gives that $(H_1, S_1) \vee (H_2, S_2) = (E^0, \emptyset)$.

(ii) Let G be the following graph:

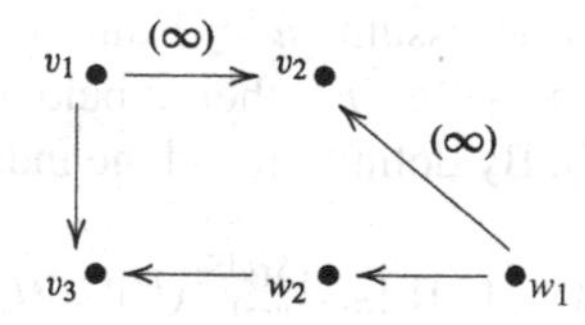

Let $H_1 = \{v_2\}$, $S_1 = \{v_1\}$; $H_2 = \{v_3\}$, $S_2 = \emptyset$. Then

$$\Lambda_1^{S_1}(H_1 \cup H_2) = \{v_1, v_2, v_3, w_2\} \quad \text{and} \quad \Lambda_2^{S_1}(H_1 \cup H_2) = \{v_1, v_2, v_3, w_2, w_1\}.$$

In this case, again the formula in Proposition 2.5.6 gives that $(H_1, S_1) \vee (H_2, S_2) = (G^0, \emptyset)$.

We now have all the pieces in place to achieve our previously stated goal, in which we give a precise description of the graded ideals of $L_K(E)$ in terms of specified subsets of E^0.

Theorem 2.5.8 (The Structure Theorem for Graded Ideals) *Let E be an arbitrary graph and K any field. Then the map φ given here provides a lattice isomorphism:*

$$\varphi : \mathscr{L}_{gr}(L_K(E)) \to \mathscr{T}_E \qquad \textit{via} \qquad I \mapsto (I \cap E^0, S)\,,$$

where $S = \{v \in B_H \mid v^H \in I\}$ *for* $H = I \cap E^0$. *The inverse* φ' *of* φ *is given by:*

$$\varphi' : \mathscr{T}_E \to \mathscr{L}_{gr}(L_K(E)) \quad \text{via} \quad (H, S) \mapsto I(H \cup S^H).$$

Proof By Lemma 2.4.3 and the definition of S, the map φ is well defined. The map φ' is clearly well defined. By Theorem 2.4.8 we get that $\varphi' \circ \varphi = \mathrm{Id}_{\mathscr{L}_{gr}(L_K(E))}$. On the other hand, Corollary 2.4.16 yields that $\varphi \circ \varphi' = \mathrm{Id}_{\mathscr{T}_E}$.

Now we prove that φ' preserves the order. Suppose that $(H_1, S_1), (H_2, S_2) \in \mathscr{T}_E$ are such that $(H_1, S_1) \leq (H_2, S_2)$. Then $H_1 \subseteq H_2$ and $S_1 \subseteq H_2 \cup S_2$. It is easy to see that $H_1 \subseteq I(H_2 \cup S^{H_2})$. Now we prove $S^{H_1} \subseteq I(H_2 \cup S^{H_2})$. Take $v^{H_1} \in S^{H_1}$. Then $v^{H_1} = v - \sum_{s(e)=v, r(e)\notin H_1} ee^*$ for some infinite emitter $v \in E^0$. We must distinguish two cases. First, if $v \in H_2$, then $v^{H_1} \in I(H_2) \subseteq I(H_2 \cup S^{H_2})$, while second, if $v \in S_2$, then

$$v^{H_1} = v^{H_2} - \sum_{\substack{s(e)=v \\ r(e)\in H_2\setminus H_1}} ee^* \in I(H_2 \cup S^{H_2}).$$

The final step is to show that φ preserves the order. To this end, consider two graded ideals I_1 and I_2 such that $I_1 \subseteq I_2$. Then $H_1 := I_1 \cap E^0 \subseteq H_2 := I_2 \cap E^0$. Now we show $S_1 \subseteq S_2$, where $S_i := \{v \in B_{H_i} \mid v^{H_i} \in I_i\}$, for $i = 1, 2$. Take $v \in S_1$. We again must distinguish two cases. Suppose first that for every $e \in E^1$ such that $s(e) = v$ we have $r(e) \in H_2$. Then

$$v = v^{H_1} + \sum_{\substack{s(e)=v \\ r(e)\in H_2\setminus H_1}} ee^* \in I_1 + I_2 = I_2,$$

and thus $v \in I_2 \cap E^0 = H_2$. On the other hand, suppose that there exists an $e \in E^1$ such that $s(e) = v$ and $r(e) \notin H_2$. Then $v \in B_{H_2}$ and

$$v^{H_2} = v^{H_1} + \sum_{\substack{s(e)=v \\ r(e)\in H_2\setminus H_1}} ee^* \in I_1 + I_2 = I_2.$$

This implies $v \in S_2$. We obtain that $S_1 \subseteq H_2 \cup S_2$ and hence that $(H_1, S_1) \leq (H_2, S_2)$. □

We record the Structure Theorem for Graded Ideals in the situation where the graph is row-finite.

Theorem 2.5.9 *Let E be a row-finite graph and K any field. The following map* φ *provides a lattice isomorphism:*

$$\varphi : \mathscr{L}_{gr}(L_K(E)) \to \mathscr{H}_E \quad \text{via} \quad \varphi(I) = I \cap E^0,$$

with inverse given by

$$\varphi' : \mathscr{H}_E \to \mathscr{L}_{gr}(L_K(E)) \qquad \textit{via} \qquad \varphi'(H) = I(H).$$

Example 2.5.10 The following is a description of all graded ideals of the Leavitt path algebra of the infinite clock graph $C_{\mathbb{N}}$ of Example 1.6.12. Recall that U denotes the set $\{u_i \mid i \in \mathbb{N}\}$ of all "non-center" vertices of $C_{\mathbb{N}}$. It is clear that the hereditary saturated subsets of $C_{\mathbb{N}}$ are $\emptyset$, $C_{\mathbb{N}}^0$, and subsets H of U. (Note that if v is in a hereditary subset H of $C_{\mathbb{N}}$, then necessarily $H = C_{\mathbb{N}}^0$.) For a subset H of U, there is a breaking vertex (namely, v) for H precisely when $U \setminus H$ is nonempty and finite. With this information in hand, we use Theorem 2.5.8 to conclude that a complete irredundant set of graded ideals of $L_K(C_{\mathbb{N}})$ is:

$$\{0\},\ L_K(C_{\mathbb{N}}),\ I(H) \text{ for } H \subseteq U, \quad \text{and} \quad I(H \cup \{v - \sum_{e \in r^{-1}(U \setminus H)} ee^*\}) \text{ for } H \subsetneqq U \text{ with } U \setminus H \text{ finite.}$$

Of interest are the following consequences of the Structure Theorem for Graded Ideals.

Corollary 2.5.11 *Let E be an arbitrary graph and K any field. Let J_1 and J_2 be graded ideals of $L_K(E)$. Then $J_1 \cdot J_2 = J_1 \cap J_2$.*

Proof The containment $J_1 \cdot J_2 \subseteq J_1 \cap J_2$ holds for any two-sided ideals in any ring. For the reverse containment, we use Theorem 2.5.8 to guarantee that we can write the graded ideal $J_1 \cap J_2$ as $I(H \cup S^H)$ for some $(H, S) \in \mathscr{T}_E$. So it suffices to show that each of the elements in the generating set $H \cup S^H$ of $J_1 \cap J_2$ is in $J_1 \cdot J_2$. But this follows immediately, as each of these elements is idempotent. □

Recall that a graded algebra A is said to be *graded noetherian* (resp., *graded artinian*) if A satisfies the ascending chain condition (resp., descending chain condition) on graded two-sided ideals. We need an observation which will be used more than once in the sequel.

Lemma 2.5.12 *Let E be an arbitrary graph. Then the following are equivalent.*

(1) *The lattice $\mathscr{T}_E$ satisfies the ascending (resp., descending) chain condition with respect to the partial order given in Definition 2.5.3.*
(2) *The lattice $\mathscr{H}_E$ satisfies the ascending (resp., descending) chain condition (under set inclusion), and, for each $H \in \mathscr{H}_E$, the corresponding set B_H of breaking vertices is finite.*

Proof We prove the ascending chain condition statement; the proof for the descending chain condition is essentially identical. So suppose the a.c.c. holds in $\mathscr{T}_E$. Let $H_1 \subseteq H_2 \subseteq \dots$ be an ascending chain of hereditary saturated subsets of vertices in E. Then we get an ascending chain $(H_1, \emptyset) \leq (H_2, \emptyset) \leq \dots$ in $\mathscr{T}_E$. By hypothesis, there is an integer n such that $(H_n, \emptyset) = (H_{n+1}, \emptyset) = \dots$. This implies that $H_n = H_{n+1} = \dots$, showing that the a.c.c. holds in $\mathscr{H}_E$. Let $H \in \mathscr{H}_E$. Then the corresponding set B_H of breaking vertices of H must be finite, since otherwise B_H would contain an infinite ascending chain of subsets $S_1 \subsetneqq S_2 \subsetneqq \dots$, and this would then give rise to a proper ascending chain $(H, S_1) \lneqq (H, S_2) \lneqq \dots$ in $\mathscr{T}_E$, contradicting the hypothesis that the a.c.c. holds in $\mathscr{T}_E$.

Conversely, suppose the a.c.c. holds in $\mathscr{H}_E$, and that B_H is a finite set for each $H \in \mathscr{H}_E$. Consider an ascending chain $(H_1, S_1) \leq (H_2, S_2) \leq \dots$ in $\mathscr{T}_E$. This gives rise to an ascending chain $H_1 \subseteq H_2 \subseteq \dots$ in $\mathscr{H}_E$, and so there is an integer n such that $H_i = H_n = H$ for all $i \geq n$. So from the nth term onwards, the given chain in $\mathscr{T}_E$ is of the form $(H, S_n) \leq (H, S_{n+1}) \leq \dots$, where $S_n, S_{n+1}, \dots$ are subsets of B_H. Observe that since $B_H \cap H = \emptyset$, it follows from the definition of $\leq$ on $\mathscr{T}_E$ that we have an ascending chain $S_n \subseteq S_{n+1} \subseteq \dots$. Since B_H is a finite set, there is a positive integer m such that $S_{n+m} = S_{n+m+i}$ for all $i \geq 0$. This establishes the a.c.c. in $\mathscr{T}_E$. □

Now combining the Structure Theorem for Graded Ideals with Lemma 2.5.12, we get

Proposition 2.5.13 *Let E be an arbitrary graph and K any field. Consider the standard $\mathbb{Z}$-grading on $L_K(E)$.*

(i) *$L_K(E)$ is graded artinian if and only if the set $\mathscr{H}_E$ satisfies the descending chain condition with respect to inclusion, and, for each $H \in \mathscr{H}_E$, the set B_H of breaking vertices is finite.*
(ii) *$L_K(E)$ is graded noetherian if and only if the set $\mathscr{H}_E$ satisfies the ascending chain condition with respect to inclusion, and, for each $H \in \mathscr{H}_E$, the set B_H of breaking vertices is finite.*

Corollary 2.5.14 *Let E be a finite graph and K any field. Then $L_K(E)$ is both graded artinian and graded noetherian.*

For another direct consequence of the Structure Theorem for Graded Ideals, recall that a graded algebra A is said to be *graded simple* if $A^2 \neq 0$, and A has no graded ideals other than 0 and A. Since $L_K(E)$ is a ring with local units for any graph E and field K, we have $L_K(E)^2 \neq 0$. Thus Theorem 2.5.8 immediately yields

Corollary 2.5.15 *Let E be an arbitrary graph and K any field. Then $L_K(E)$ is graded simple if and only if the only hereditary saturated subsets of E^0 are $\emptyset$ and E^0.*

We conclude our discussion of the graded ideals in Leavitt path algebras by establishing that every graded ideal in $L_K(E)$ is itself, up to isomorphism, the Leavitt path algebra of an explicitly-described graph. Because dealing with breaking vertices makes the proof of the result for arbitrary graded ideals less "visual", and because a number of our results in the sequel will rely only on this more specific setting, we start our analysis by considering graded ideals of the form $I(H)$ for $H \in \mathscr{H}_E$.

Definition 2.5.16 (The Hedgehog Graph for a Hereditary Subset) Let E be an arbitrary graph. Let H be a nonempty hereditary subset of E^0. We denote by $F_E(H)$ the set

$$F_E(H) = \{\alpha \in \text{Path}(E) \mid \alpha = e_1 \cdots e_n, \text{ with } s(e_1) \in E^0 \setminus H,\ r(e_i) \in E^0 \setminus H \text{ for all } 1 \leq i < n, \text{ and } r(e_n) \in H\}.$$

We denote by $\overline{F}_E(H)$ another copy of $F_E(H)$. If $\alpha \in F_E(H)$, we will write $\overline{\alpha}$ to refer to a copy of α in $\overline{F}_E(H)$. We define the graph ${}_HE = ({}_HE^0, {}_HE^1, s', r')$ as follows:

$$ {}_HE^0 = H \cup F_E(H), \qquad \text{and} \qquad {}_HE^1 = \{e \in E^1 \mid s(e) \in H\} \cup \overline{F}_E(H). $$

The source and range functions s' and r' are defined by setting $s'(e) = s(e)$ and $r'(e) = r(e)$ for every $e \in E^1$ such that $s(e) \in H$; and by setting $s'(\overline{\alpha}) = \alpha$ and $r'(\overline{\alpha}) = r(\alpha)$ for all $\overline{\alpha} \in \overline{F}_E(H)$.

Intuitively, $F_E(H)$ can be viewed as H, together with a new vertex corresponding to each path in E which ends at a vertex in H, but for which none of the previous edges in the path ends at a vertex in H. For every such new vertex, a new edge is added going into H. So the net effect is that in $F_E(H)$, the only paths entering the subgraph H have common length 1; pictorially, the situation evokes an image of the quills (edges into H) on the body of a hedgehog or porcupine (H itself), whence the name.

Remark 2.5.17 We note that, by construction, the cycles in the hedgehog graph ${}_HE$ are precisely the cycles in H. In particular, as H is hereditary, every cycle without exits in ${}_HE$ arises from a cycle without exits in H.

Example 2.5.18 Let E_T be the Toeplitz graph $e \circlearrowright \bullet^u \xrightarrow{f} \bullet^v$, and let H denote the hereditary subset $\{v\}$. Then $F_{E_T}(H) = \{e^n f \mid n \in \mathbb{Z}^+\}$, and ${}_HE_T$ is the following graph.

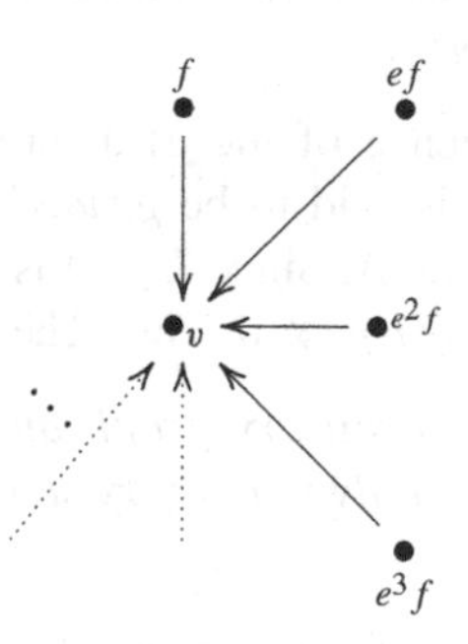

If I is an ideal of a ring R, then I itself may be viewed as a ring in its own right. (Of course I need not be unital, nor need it contain a set of local units, e.g., the ideal $2\mathbb{Z}$ of $\mathbb{Z}$.) Similarly, if I is an algebra ideal of a K-algebra A, then I may be viewed as a K-algebra in its own right. We note in this regard that the K-ideal $I(1+x)$ of $K[x, x^{-1}]$ does not contain any nonzero idempotents, hence $I(1+x)$ when viewed as a K-algebra cannot contain a set of local units. Using the identification established between $L_K(R_1)$ and $K[x, x^{-1}]$, this implies in particular that the ideal $I(v+e)$ of $L_K(R_1)$ cannot be isomorphic to the Leavitt path algebra of any graph, as any Leavitt path algebra is an algebra with local units. These comments provide context for the following result.

Theorem 2.5.19 *Let E be an arbitrary graph and K any field. Let H be a nonempty hereditary subset of E. Then $I(H)$, when viewed as a K-algebra, is K-algebra isomorphic to $L_K({}_HE)$.*

Proof We define a map $\varphi : \{u \mid u \in {}_HE^0\} \cup \{e \mid e \in {}_HE^1\} \cup \{e^* \mid e \in {}_HE^1\} \to I(H)$ by the following rule:

$$\varphi(v)=\begin{cases} v & \text{if } v\in H \\ \alpha\alpha^* & \text{if } v=\alpha\in F_E(H), \end{cases} \quad \varphi(e)=\begin{cases} e & \text{if } e\in E^1 \\ \alpha & \text{if } e=\overline{\alpha}\in \overline{F}_E(H), \end{cases} \quad \text{and } \varphi(e^*)=\begin{cases} e^* & \text{if } e\in E^1 \\ \alpha^* & \text{if } e=\overline{\alpha}\in \overline{F}_E(H). \end{cases}$$

Note that for distinct elements α, β in $F_E(H)$ we have $\alpha^*\beta = 0$, so $\{\varphi(u) \mid u \in {}_HE^0\}$ is a set of orthogonal idempotents in $I(H)$. Moreover, it is not difficult to establish that this set, jointly with $\{\varphi(e) \mid e \in {}_HE^1\}$ and $\{\varphi(e^*) \mid e \in {}_HE^1\}$, is an ${}_HE$-family in $I(H)$. So by the Universal Property 1.2.5, φ extends to a K-algebra homomorphism from $L_K({}_HE)$ into $I(H)$.

To see that φ is onto, by Lemma 2.4.1 it is enough to show that every vertex of H and every path α of E with $r(\alpha) \in H$ are in the image of φ. For any $v \in H$, $\varphi(v) = v$, so that this case is clear. Now, let $\alpha = \alpha_1 \cdots \alpha_n$ with $\alpha_i \in E^1$. If $s(\alpha_1) \in H$, then $\alpha = \varphi(\alpha_1)\cdots\varphi(\alpha_n)$. Suppose that $s(\alpha_1) \in E^0 \setminus H$ and $r(\alpha_n) \in H$. Then, there exists $1 \leq j \leq n-1$ such that $r(\alpha_j) \in E^0 \setminus H$ and $r(\alpha_{j+1}) \in H$. Thus, $\alpha = \alpha_1 \cdots \alpha_{j+1} \cdot \alpha_{j+2} \cdots \alpha_n$, where $\beta = \alpha_1 \cdots \alpha_{j+1} \in F_E(H)$. Hence, $\alpha = \varphi(\overline{\beta})\varphi(\alpha_{j+2})\cdots\varphi(\alpha_n)$.

To show injectivity, by Remark 2.5.17 we have that any cycle without exits in ${}_HE$ comes from a cycle without exits in E, where the vertices of the cycle are in H. So every cycle without exits in ${}_HE$ is mapped to a homogeneous nonzero element of nonzero degree in $I(H)$. The injectivity thereby follows by Proposition 2.2.20. □

In what follows, we will generalize Theorem 2.5.19 in Theorem 2.5.22 by showing that in fact *every* graded ideal in a Leavitt path algebra is isomorphic to a Leavitt path algebra.

Definition 2.5.20 (The Generalized Hedgehog Graph Construction, Incorporating Breaking Vertices) Let E be an arbitrary graph, H a nonempty hereditary subset of E, and $S \subseteq B_H$. We define

$$F_1(H,S) := \{\alpha \in \text{Path}(E) \mid \alpha = e_1 \cdots e_n,\ r(e_n) \in H \text{ and } s(e_n) \notin H \cup S\}, \text{ and}$$

$$F_2(H,S) := \{\alpha \in \text{Path}(E) \mid |\alpha| \geq 1 \text{ and } r(\alpha) \in S\}.$$

For $i = 1, 2$ we denote a copy of $F_i(H,S)$ by $\overline{F}_i(H,S)$. We define the graph ${}_{(H,S)}E$ as follows:

$${}_{(H,S)}E^0 := H \cup S \cup F_1(H,S) \cup F_2(H,S), \text{ and}$$

$${}_{(H,S)}E^1 := \{e \in E^1 \mid s(e) \in H\} \cup \{e \in E^1 \mid s(e) \in S \text{ and } r(e) \in H\} \cup \overline{F}_1(H,S) \cup \overline{F}_2(H,S).$$

The range and source maps for ${}_{(H,S)}E$ are described by extending r and s to ${}_{(H,S)}E^1$, and by defining $r(\overline{\alpha}) = \alpha$ and $s(\overline{\alpha}) = \alpha$ for all $\overline{\alpha} \in \overline{F}_1(H,S) \cup \overline{F}_2(H,S)$.

Remark 2.5.21 Here are some observations about the construction of the generalized hedgehog graph ${}_{(H,S)}E$.

(i) $F_1(H,S) \cap F_2(H,S) = \emptyset$.
(ii) Every cycle in E produces a cycle in ${}_{(H,S)}E$; moreover, cycles in ${}_{(H,S)}E$ come from cycles in E. Thus there is a bijection between the set of cycles in E and the set of cycles in ${}_{(H,S)}E$.
(iii) In the particular case $S = \emptyset$, we get:

$$F_1(H,\emptyset) = \{\alpha = e_1 \cdots e_n \in \text{Path}(E) \mid r(e_n) \in H \text{ and } s(e_n) \notin H\};$$

$$F_2(H,\emptyset) = \emptyset; \quad \text{and } {}_{(H,\emptyset)}E = {}_HE.$$

Thus Definition 2.5.20 indeed generalizes the construction of the graph ${}_HE$ given in Definition 2.5.16.

Theorem 2.5.22 *Let E be an arbitrary graph and K any field. Suppose H is a hereditary subset of E^0 and $S \subseteq B_H$. Then the graded ideal $I(H \cup S^H)$ of the Leavitt path algebra $L_K(E)$ is isomorphic as K-algebras to the Leavitt path algebra $L_K({}_{(H,S)}E)$.*

Proof Let $\varphi : \{v \mid v \in {}_{(H,S)}E^0\} \cup \{e \mid e \in {}_{(H,S)}E^1\} \cup \{e^* \mid e \in {}_{(H,S)}E^1\} \to I(H \cup S^H)$ be the map such that:

$$\varphi(v) = \begin{cases} v & \text{if } v \in H \\ v^H & \text{if } v \in S \\ \alpha\alpha^* & \text{if } v = \alpha \in F_1(H,S) \\ \alpha r(\alpha)^H \alpha^* & \text{if } v = \alpha \in F_2(H,S)\,, \end{cases}$$

$$\varphi(e) = \begin{cases} e & \text{if } e \in E^1 \\ \alpha & \text{if } e = \overline{\alpha} \in \overline{F}_1(H,S) \\ \alpha r(\alpha)^H & \text{if } e = \overline{\alpha} \in \overline{F}_2(H,S)\,, \end{cases} \quad \text{and} \quad \varphi(e^*) = \begin{cases} e^* & \text{if } e \in E^1 \\ \alpha^* & \text{if } e = \overline{\alpha} \in \overline{F}_1(H,S) \\ r(\alpha)^H \alpha^* & \text{if } e = \overline{\alpha} \in \overline{F}_2(H,S)\,. \end{cases}$$

It is not difficult to see that each of the elements $\varphi(v)$, $\varphi(e)$, and $\varphi(e^*)$ is an element of $I(H \cup S^H)$. In a manner similar to the proof of Theorem 2.5.19, one can show that the set

$$\{\varphi(v) \mid v \in {}_{(H,S)}E^0\} \cup \{\varphi(e) \mid e \in {}_{(H,S)}E^1\} \cup \{\varphi(e^*) \mid e \in {}_{(H,S)}E^1\}$$

is an ${}_{(H,S)}E$-family in $I(H \cup S^H)$. Consequently, by the Universal Property of $L_K({}_{(H,S)}E)$ 1.2.5, the map φ can be uniquely extended to a K-algebra homomorphism from $L_K({}_{(H,S)}E)$ to $I(H \cup S^H)$.

The injectivity of φ follows from Proposition 2.2.20. To show surjectivity, recall the description of the generators of $I(H \cup S^H)$ given in Lemma 2.4.6. Using this, the only two things we must show are that $\alpha \in \mathrm{Im}(\varphi)$ for every $\alpha \in \mathrm{Path}(E)$ such that $r(\alpha) \in H$, and that $\alpha v^H \in \mathrm{Im}(\varphi)$ for every $\alpha \in \mathrm{Path}(E)$ such that $r(\alpha) = v \in S$.

To show the first statement, take $\alpha = e_1 \cdots e_n$ as indicated. There are four cases to analyze. First, if $s(e_1) \in H$ then $s(e_i) \in H$ for all i and $e_i \in {}_{(H,S)}E^1$. Hence, $\varphi(\alpha) = \varphi(e_1) \cdots \varphi(e_n) = e_1 \cdots e_n = \alpha$, which proves $\alpha \in \mathrm{Im}(\varphi)$. Second, suppose $\alpha = fe_1 \cdots e_n$ with $r(f) = s(e_1) \in H$ and $s(f) \in S$. Then $f \in {}_{(H,S)}E^1$, $s(e_i) \in H$ and $e_i \in {}_{(H,S)}E^1$ for all i. Therefore, $\varphi(\alpha) = \varphi(f)\varphi(e_1) \cdots \varphi(e_n) = fe_1 \cdots e_n = \alpha$ and so $\alpha \in \mathrm{Im}(\varphi)$. In the third case, if $\alpha = f_1 \cdots f_m e_1 \cdots e_n$ with $r(f_m) = s(e_1) \in H$ and $s(f_m) \notin H \cup S$, then $\beta := f_1 \cdots f_m \in F_1(H,S)$ and $e_i \in {}_{(H,S)}E^1$ for all i, so $\varphi(\overline{\beta} e_1 \cdots e_n) = \varphi(\overline{\beta})\varphi(e_1) \cdots \varphi(e_n) = \beta e_1 \cdots e_n = \alpha$ and so $\alpha \in \mathrm{Im}(\varphi)$. Finally, if $\alpha = f_1 \cdots f_m g e_1 \cdots e_n$ with $r(g) = s(e) \in H$, $s(g) \in S$ and $m \geq 1$, then $\beta := f_1 \cdots f_m \in F_2(H,S)$, $g \in {}_{(H,S)}E^1$ and $e_i \in {}_{(H,S)}E^1$ for all i; therefore, $\varphi(\overline{\beta} g e_1 \cdots e_n) = \varphi(\overline{\beta})\varphi(g)\varphi(e_1) \cdots \varphi(e_n) = \beta g e_e \cdots e_n = \alpha$, which shows again $\alpha \in \mathrm{Im}(\varphi)$.

Now we verify that $\alpha v^H \in \mathrm{Im}(\varphi)$ for every $\alpha \in \mathrm{Path}(E)$ such that $r(\alpha) = v \in S$. If $|\alpha| = 0$ then $v := \alpha$ is a vertex in S and $\varphi(v) = v^H = \alpha v^H$ so that $\alpha v^H \in \mathrm{Im}(\varphi)$. If $|\alpha| \geq 1$ then $\alpha \in F_2(H,S)$ and $\varphi(\alpha) = \alpha r(\alpha)^H$. This shows $\alpha r(\alpha)^H \in \mathrm{Im}(\varphi)$, and the proof is complete. □

Corollary 2.5.23 *Let E be an arbitrary graph and K any field. Then every graded ideal of $L_K(E)$ is K-algebra isomorphic to a Leavitt path algebra.*

Proof Apply the Structure Theorem for Graded Ideals 2.5.8 with Theorem 2.5.22. □

Remark 2.5.24 We note that, except for the obvious trivial cases, the isomorphism established in Theorem 2.5.22 between the graded ideal $I(H \cup S^H)$ of $L_K(E)$ and the Leavitt path algebra $L_K(E/(H,S))$ is not a graded isomorphism with respect to the induced grading on $I(H \cup S^H)$ coming from $L_K(E)$. This is because if α is a path in $F_E(H)$ having $|\alpha| \geq 2$, then the equation $\varphi(\overline{\alpha}) = \alpha$ reveals that φ does not preserve the grading.

In summary, we have now shown that the graded ideals of $L_K(E)$ are "natural" in the context of Leavitt path algebras: by Theorem 2.4.15 every quotient of a Leavitt path algebra by a graded ideal is again a Leavitt path algebra, and by Theorem 2.5.22 every graded ideal of a Leavitt path algebra is itself a Leavitt path algebra. In contrast, the quotient of a graded algebra by a non-graded ideal is not a graded algebra with respect to an induced grading; see the comments subsequent to Remark 2.1.2. Moreover, once we develop a description of the structure of all ideals in a Leavitt path algebra, we will be able to prove that non-graded ideals are necessarily not isomorphic to Leavitt path algebras (see Corollary 2.9.11).

We close this section by establishing yet another consequence of the Structure Theorem for Graded Ideals 2.5.8.

Proposition 2.5.25 *Let $\{H_i\}_{i\in\Lambda}$ be a family of hereditary subsets of an arbitrary graph E and K any field. Then as ideals of $L_K(E)$ we have:*

(i) $I(\cap_{i\in\Lambda}\overline{H_i}) = \cap_{i\in\Lambda}I(H_i)$.
(ii) *If Λ is finite, then* $I(\cap_{i\in\Lambda}H_i) = \cap_{i\in\Lambda}I(H_i)$.

Proof (i) The containment $I(\cap_{i\in\Lambda}\overline{H_i}) \subseteq \cap_{i\in\Lambda}I(H_i)$ is clear because $I(H_i) = I\left(\overline{H_i}\right)$. Now we show the reverse containment. Observe first that since the intersection of graded ideals is a graded ideal, by the Structure Theorem for Graded Ideals 2.5.8 we get $\cap_{i\in\Lambda}I(\overline{H_i}) = I(H \cup S^H)$, where $H = \left(\cap_{i\in\Lambda}I(\overline{H_i})\right) \cap E^0 = \cap_{i\in\Lambda}\left(I(\overline{H_i}) \cap E^0\right) = \cap_{i\in\Lambda}\overline{H_i}$. Now, consider $v \in B_H$; we will see that $v^H \notin \cap_{i\in\Lambda}I(\overline{H_i})$. Since $v \notin H$, there is an $i \in \Lambda$ such that $v \notin \overline{H_i}$, hence $v \in B_{\overline{H_i}}$.

Write $\tilde{v}$ to denote either $v^{\overline{H_i}}$ (if $v \in B_{\overline{H_i}}$), or v (if $r(s^{-1}(v)) \subseteq \overline{H_i}$). Then we may write

$$\tilde{v} = v^H - \sum_{\substack{s(e)=v \\ r(e)\in\overline{H_i}\setminus H}} ee^*.$$

Since $\sum_{\{s(e)=v,r(e)\in\overline{H_i}\setminus H\}} ee^* \in I(\overline{H_i})$ and $v^H \in I(\overline{H_i})$, then $\tilde{v} \in I(\overline{H_i}) \cap E^0 = \overline{H_i}$, a contradiction. This implies $S = \emptyset$, giving the desired result.

(ii) When Λ is finite, then $\cap_{i\in\Lambda}\overline{H_i} = \overline{\cap_{i\in\Lambda}H_i}$, and consequently

$$I(\cap_{i\in\Lambda}H_i) = I(\overline{\cap_{i\in\Lambda}H_i}) = \cap_{i\in\Lambda}I(\overline{H_i}) = \cap_{i\in\Lambda}I(H_i).$$

□

2.6 The Socle

Because of its importance in the general theory, we present now a description of the socle of a Leavitt path algebra. Along the way, we will investigate various minimal left ideals of $L_K(E)$. This in turn will provide us with, among other things, an explicit description of the finite-dimensional Leavitt path algebras.

Definitions 2.6.1 Let E be an arbitrary graph. Recall that for $v \in E^0$, we say that *there exists a cycle at* v if v is a vertex lying on some cycle in E. Also, recall that for $v \in E^0$, $T(v)$ denotes the set $\{w \in E^0 \mid v \geq w\}$.

A vertex $v \in E^0$ is called a *bifurcation* vertex (or it is said that *there is a bifurcation at* v) if $|s_E^{-1}(v)| \geq 2$.

A vertex $u \in E^0$ is called a *line point* if there exist neither bifurcations nor cycles at any vertex of $T(u)$.

The set of line points of the graph E will be denoted by $P_l(E)$.

Remark 2.6.2 Vacuously, any sink in E is a line point. The set of line points $P_l(E)$ is always a hereditary subset of E^0, although it is not necessarily saturated.

If $u \in P_l(E)$, then $T(u)$ is a sequence $T(u) = \{u_1, u_2, u_3, \dots\}$, where $u = u_1$, and where, for all $i \in \mathbb{N}$, there exists a unique edge $e_i \in E^1$ with $s(e_i) = u_i, r(e_i) = u_{i+1}$. This sequence is finite precisely when there exists a sink w of E in $T(u)$, in which case w is the last element of the sequence. Intuitively, $T(u)$ is then essentially just a "directed line starting at u", from which the name "line point" derives.

Consequently, if u is a line point, then for each pair $u_i, u_j \in T(u)$ with $i < j$, there exists a unique path $p_{i,j}$ in E for which $s(p_{i,j}) = u_i$ and $r(p_{i,j}) = u_j$. In particular, the lack of bifurcations at any vertex in $T(u)$ together with the (CK2) relation yields that $p_{i,j}p_{i,j}^* = u_i$ for any pair $u_i, u_j \in T(u)$ for which $i \leq j$.

A key role in the subject is played by rings of the following form.

Notation 2.6.3 Let Γ be an infinite set, and let S be any unital ring. We denote by

$$\mathrm{M}_\Gamma(S)$$

the ring consisting of those square matrices M, with rows and columns indexed by Γ, with entries from S, for which there are at most finitely many nonzero entries in M.

Clearly any such ring $\mathrm{M}_\Gamma(S)$ contains a set of enough idempotents, consisting of the matrix units $\{e_{i,i} \mid i \in \Gamma\}$; this yields the set of local units in $\mathrm{M}_\Gamma(S)$ consisting of those matrices which equal 1_S in finitely many entries (i, i), and are 0 otherwise.

A subset $\{\epsilon_{\alpha,\beta} \mid \alpha, \beta \in \Gamma\}$ of an ideal T of a K-algebra R is called a *set of matrix units for* T if $T = \mathrm{span}_K(\{\epsilon_{\alpha,\beta}\})$, and $\epsilon_{\alpha,\beta}\epsilon_{\gamma,\kappa} = \delta_{\beta,\gamma}\epsilon_{\alpha,\kappa}$ for all $\alpha, \beta, \gamma, \kappa \in \Gamma$. In this case, $T \cong \mathrm{M}_\Gamma(K)$ as K-algebras, via an isomorphism sending $\epsilon_{\alpha,\beta}$ to the standard matrix element $e_{\alpha,\beta}$ (which is 1_K in row α, column β, and 0 elsewhere). The following result (which generalizes Proposition 1.3.5) allows us to explicitly describe the structure of the ideal $I(v)$ generated by a line point v. As a consequence of this description, we will be able to describe the structure of the socle of any Leavitt path algebra.

Lemma 2.6.4 *Let E be an arbitrary graph and K any field. Let v be a line point in E. Let Λ_v denote the set $F_E(T(v))$; that is, Λ_v is the set of paths $\alpha \in$ Path(E) for which $r(\alpha)$ meets $T(v)$ for the first time at $r(\alpha)$. Then*

$$I(v) \cong \mathrm{M}_{\Lambda_v}(K).$$

Proof We construct a set of matrix units in $I(v)$, indexed by Λ_v, as follows. Write $T(v) = \{v_1, v_2, \dots\}$ as in Remark 2.6.2. By Lemma 2.4.1 and the observations offered in Remark 2.6.2, each element in $I(v)$ is a K-linear combination of elements of the form $\alpha x_{i,j}\lambda^*$, where $\alpha, \lambda \in F_E(T(v))$, and $x_{i,j} = p_{i,j}$ if $i \leq j$, or $x_{i,j} = p^*_{j,i}$ if $j \leq i$. We denote such $\alpha x_{i,j}\lambda^*$ by $e_{\alpha,\lambda}$.

Again using Remark 2.6.2, we see that the set $\{x_{i,j} \mid i,j \in \mathbb{N}\}$ has the multiplicative property $x_{i,j}x_{k,\ell} = \delta_{j,k}x_{i,\ell}$ for all $i,j,k,\ell \in \mathbb{N}$. Using this, it is then straightforward to establish that the set $\{e_{\alpha,\lambda} \mid \alpha, \lambda \in \Lambda_v\}$ is a set of matrix units for $I(v)$. □

Corollary 2.6.5 *Let E be an arbitrary graph and K any field. Let v be a sink in E. Then* $I(v) \cong \mathrm{M}_{\Lambda_v}(K)$, *where* Λ_v *is the set of paths in E ending at v.*

Corollary 2.6.6 *Let K be any field. For any set* Λ *let* E_Λ *denote the graph with*

$$E^0_\Lambda = \{v\} \cup \{u_\lambda \mid \lambda \in \Lambda\} \quad and \quad E^1_\Lambda = \{f_\lambda \mid \lambda \in \Lambda\},$$

where $s(f_\lambda) = u_\lambda$ *and* $r(f_\lambda) = v$ *for all* $\lambda \in \Lambda$. *Then* $L_K(E_\Lambda) \cong \mathrm{M}_{\Lambda\cup\{v\}}(K)$. *In particular, by taking disjoint unions of graphs of this form, any direct sum of full matrix rings over K arises as the Leavitt path algebra of a graph. We note that* $E_{\mathbb{Z}^+}$ *is the graph arising in Example 2.5.18.*

With Example 1.6.12 in mind, we sometimes refer to $E_\mathbb{N}$ as the *infinite co-clock* graph.

Definitions 2.6.7 Let R be a ring. We say that a left ideal I of R is a *minimal left ideal* if $I \neq 0$ and I does not contain any left ideals of R other than 0 and I. (This is equivalent to saying that ${}_RI$ is a simple left R-module.) An idempotent $e \in R$ is called *left minimal* if Re is a minimal left ideal of R. The *left socle* of R is defined to be the sum of all the minimal left ideals of R (or is defined to be $\{0\}$ if R contains no minimal left ideals). The corresponding notions of *right minimal* and *right socle* are defined analogously.

Remark 2.6.8 It is well known that for any ring R, both the left socle and the right socle of R are two-sided ideals of R. For a semiprime ring R the left and right socles of R coincide; in this case, either of these is called the *socle* of R, and is denoted by $\mathrm{Soc}(R)$. In particular, for E an arbitrary graph and K any field, the two-sided ideal $\mathrm{Soc}(L_K(E))$ denotes the sum of the minimal left (or right) ideals of $L_K(E)$ (when such exist), or denotes $\{0\}$ (when $L_K(E)$ contains no minimal one-sided left ideals).

The following result is standard (see e.g., [109, Sect. 3.4]).

Lemma 2.6.9 *Let R be a semiprime ring, and* $e^2 = e \in R$. *Then Re is a minimal left ideal of R if and only if eRe is a division ring.*

The structure of left ideals generated by vertices lies at the heart of the description of the socle of a Leavitt path algebra. Here is a fundamental observation in that regard.

Lemma 2.6.10 *Let E be an arbitrary graph and K any field. Let $w \in E^0$. If there exists a bifurcation at w (i.e., if $|s^{-1}(w)| \geq 2$), then the left ideal $L_K(E)w$ is not minimal.*

Proof Suppose $e \neq f \in s^{-1}(w)$. Then ee^* and ff^* are nonzero elements of $L_K(E)w$. Since $ee^* \neq 0$, $L_K(E)ee^*$ is a nonzero submodule of $L_K(E)w$. But $ff^* \notin L_K(E)ee^*$, since otherwise we would have $ff^* = ree^*$ for some $r \in L_K(E)$, which upon multiplication on the right by ff^* and using (CK1) would give $ff^* = 0$, a contradiction. □

Proposition 2.6.11 *Let E be an arbitrary graph and K any field. A vertex v of E is a line point if and only if $L_K(E)v$ is a minimal left ideal of $L_K(E)$.*

Proof Suppose first that v is a line point. Since $L_K(E)$ is semiprime (Proposition 2.3.1), in order to show that $L_K(E)v$ is a minimal left ideal it suffices to show (by Lemma 2.6.9) that $vL_K(E)v$ is a division ring. To that end, consider an arbitrary nonzero element $a \in vL_K(E)v$. Then a will be of the form $a = v(\sum_{i=1}^n k_i\lambda_i\mu_i^*)v = \sum_{i=1}^n k_i(v\lambda_i\mu_i^*v)$, for $\lambda_i, \mu_i \in \text{Path}(E)$ such that $s(\lambda_i) = r(\mu_i^*) = v$ (so that $s(\mu_i) = v = s(\lambda_i)$), and $r(\lambda_i) = s(\mu_i^*) = r(\mu_i)$. But v is a line point, so by Remark 2.6.2 these two conditions give $\lambda_i = \mu_i$, and that $\lambda_i\mu_i^* = v$. So we get $a = \sum_{i=1}^n k_i \cdot v \in Kv$. This shows that $vL_K(E)v = Kv \cong K$.

Conversely, suppose $L_K(E)v$ is a minimal left ideal. We will see that no vertex in $T(v)$ has bifurcations, nor is any vertex in $T(v)$ the base of a cycle. We start by noting the following. For any $u \in T(v)$, let μ be a path such that $s(\mu) = v$ and $r(\mu) = u$. Then the map

$$\rho_\mu : L_K(E)v \to L_K(E)u \qquad av \mapsto av\mu = a\mu$$

is a nonzero epimorphism of left $L_K(E)$-modules, as for $\beta u \in L_K(E)u$ we have $\beta\mu^* \in L_K(E)v$, and $\rho_\mu(\beta\mu^*) = \beta\mu^*\mu = \beta u$. The minimality of $L_K(E)v$ implies that ρ_u is an isomorphism, so that $L_K(E)u$ must be minimal as well. In particular, by Lemma 2.6.9 $uL_K(E)u$ is a division ring.

With these observations, we conclude first (by Lemma 2.6.10) that there are no bifurcations at w for every $w \in T(v)$, and second (by Lemma 2.2.7) that w is not the base of a cycle without exits in E for every $w \in T(v)$. Thus v is a line point. □

Definition 2.6.12 For an arbitrary graph E and field K, we call a vertex $w \in E^0$ a *minimal* vertex if $L_K(E)w$ is a minimal left ideal of $L_K(E)$.

Lemma 2.6.13 *Let E be an arbitrary graph and K any field. Then there exists a family $\{H_i\}_{i\in\Gamma}$ of hereditary subsets of E^0 such that $P_l(E) = \bigsqcup_{i\in\Gamma} H_i$, and $I(H_i) = I(v_i)$ as ideals of $L_K(E)$ for every $v_i \in H_i$ and $i \in \Gamma$.*

Proof Define on $P_l(E)$ the following equivalence relation: for $u, v \in P_l(E)$, we say $u \equiv v$ if $I(u) = I(v)$. Let $\{H_i\}_{i\in\Gamma}$ be the set of all $\equiv$ equivalence classes.

We claim that each H_i is a hereditary subset of E^0. Indeed, suppose $u \in H_i$ and $v \in E^0$ such that $v = r(e)$ for some $e \in s^{-1}(u)$. Then $v \in P_l(E)$, as $P_l(E)$ is hereditary, and by hypothesis, $s^{-1}(u) = \{e\}$. This implies, by Remark 2.6.2 and (CK1), that $u = ee^* = eve^* \in I(v)$ and $v = e^*e = e^*ue \in I(u)$, hence $I(u) = I(v)$, and so $v \in H_i$. A similar argument holds for any $v \in T(u)$. The rest of the conditions in the statement are obviously fulfilled. □

We are now in position to describe the socle of a Leavitt path algebra.

Theorem 2.6.14 *Let E be an arbitrary graph and K any field. Decompose $P_l(E) = \bigsqcup_{i\in\Gamma} H_i$ as in Lemma 2.6.13. Then*

$$\mathrm{Soc}(L_K(E)) = I(P_l(E)) \cong \bigoplus_{i\in\Gamma} \mathrm{M}_{\Lambda_{v_i}}(K),$$

where for every $i \in \Gamma$, if v_i is an arbitrary element of H_i then $I(v_i) \cong \mathrm{M}_{\Lambda_{v_i}}(K)$ (with notation as in Lemma 2.6.4).

Proof We begin by showing $I(P_l(E)) = \mathrm{Soc}(L_K(E))$. Proposition 2.6.11 gives that $I(P_l(E)) \subseteq \mathrm{Soc}(L_K(E))$. To establish the reverse inclusion note that, since $\mathrm{Soc}(L_K(E))$ is generated by the minimal left ideals of $L_K(E)$, it suffices to show that $a \in I(P_l(E))$ for every a for which $L_K(E)a$ is a minimal left ideal of $L_K(E)$.

Use the Reduction Theorem 2.2.11 to find $\mu, \eta \in \mathrm{Path}(E)$ such that either $0 \neq \mu^* a\eta = kv$ for some $k \in K^\times$ and $v \in E^0$, or $0 \neq \mu^* a\eta \in wL_K(E)w$, where w is a vertex in a cycle without exits. The second option is not possible, since $wL_K(E)w$ is isomorphic as a K-algebra to $K[x, x^{-1}]$ by Lemma 2.2.7, and so if the second option were to hold we would have

$$\{0\} \neq w\mathrm{Soc}(L_K(E))w = \mathrm{Soc}(wL_K(E)w) \cong \mathrm{Soc}(K[x, x^{-1}]) = \{0\},$$

a contradiction.

Hence for some $v \in E^0$ and $k \in K^\times$ we have $\mu^* a\eta = kv$. By minimality of $L_K(E)a$, we get $L_K(E)\mu^* a = L_K(E)a$. Again by minimality, the nonzero surjection $\rho_\eta : L_K(E)\mu^* a \to L_K(E)v$ is an isomorphism. Thus $L_K(E)v \cong L_K(E)a$. In particular, $L_K(E)v$ is minimal, so that v is a line point by Proposition 2.6.11. But the isomorphism $L_K(E)v \cong L_K(E)a$ implies that $a = svt$ for some $s, t \in L_K(E)$, so that $a \in I(P_l(E))$.

In order to finish the proof of the theorem, we proceed as follows. We have $I(P_l(E)) = I(\bigsqcup_{i\in\Gamma} H_i) = \bigoplus_{i\in\Gamma} I(H_i)$ by Proposition 2.4.7. Now, use $I(H_i) = I(v_i)$ for any $v_i \in H_i$ (by construction), and apply Lemma 2.6.4. □

As a consequence of Theorem 2.6.14, we get

Corollary 2.6.15 *Let E be an arbitrary graph and K any field. The following are equivalent.*

(1) *E contains no line points.*
(2) *$L_K(E)$ has no minimal idempotents.*

Proof (1) implies (2) follows from the fact that if $L_K(E)$ has minimal idempotents, then $\mathrm{Soc}(L_K(E)) \neq \{0\}$, so that $P_l(E) \neq \emptyset$ by Theorem 2.6.14. That (2) implies (1) follows from Proposition 2.6.11. □

Examples 2.6.16 In general, the relative size of $\mathrm{Soc}(L_K(E))$ within $L_K(E)$ can run the gamut, even among the fundamental examples of Leavitt path algebras. For instance:

(i) Since for each $n \in \mathbb{N}$ there are no line points in the graph

$$R_n = \bullet \text{ (with } n \text{ loops)},$$

we conclude by Theorem 2.6.14 that $\mathrm{Soc}(L_K(R_n)) = \{0\}$. In particular, $\mathrm{Soc}(L_K(1,n)) = \{0\}$ for each of the Leavitt K-algebras $L_K(1,n)$. (We also recover the well-known, previously invoked fact that $\mathrm{Soc}(K[x,x^{-1}]) = \mathrm{Soc}(L_K(R_1)) = \{0\}$.)

(ii) Since in the graph

$$A_n = \bullet^{v_1} \xrightarrow{e_1} \bullet^{v_2} \xrightarrow{e_2} \bullet^{v_3} \cdots\cdots \bullet^{v_{n-1}} \xrightarrow{e_{n-1}} \bullet^{v_n}$$

we have that $I(v_n) = L_K(A_n)$ for the line point v_n, we conclude by Theorem 2.6.14 that $\mathrm{Soc}(L_K(A_n)) = L_K(A_n)$. (Of course this result is easy to see from first principles, since $L_K(A_n) \cong \mathrm{M}_n(K)$.)

(iii) Since in the Toeplitz graph

$$E_T = \circlearrowright \bullet \longrightarrow \bullet^v$$

the only line point is the vertex v, we conclude by Theorem 2.6.14 that $\mathrm{Soc}(L_K(E_T))$ is the ideal $I(v)$ of $L_K(E_T)$ generated by v. We see immediately that $\{0\} \subsetneqq \mathrm{Soc}(L_K(E_T)) \subsetneqq L_K(E_T)$.

Indeed, by Theorem 2.5.19, the ideal $I(v)$ is isomorphic to the Leavitt path algebra of the graph in Example 2.5.18, which in turn is isomorphic to $\mathrm{M}_{\mathbb{Z}^+}(K)$ by Corollary 2.6.6. Moreover, by Corollary 2.4.13(i) the quotient of $L_K(E_T)$ by the socle $I(v)$ is isomorphic to $L_K(E/\{v\}) \cong L_K(\circlearrowright \bullet) \cong K[x,x^{-1}]$.

We finish the section by giving the aforementioned key consequence of our newly developed tools, in which we describe the structure of all finite-dimensional Leavitt path algebras.

Theorem 2.6.17 (The Finite Dimension Theorem) *Let E be an arbitrary graph and K any field. The following conditions are equivalent.*

(1) *$L_K(E)$ is a finite-dimensional Leavitt path K-algebra.*
(2) *E is a finite and acyclic graph.*
(3) *$L_K(E)$ is K-algebra isomorphic to $\oplus_{i=1}^m \mathrm{M}_{n_i}(K)$, where $m = |\mathrm{Sink}(E)|$, and, for each $1 \le i \le m$, n_i is the number of different paths ending at the sink v_i.*

Proof (1) $\Rightarrow$ (2). Since $E^0 \cup E^1$ is a linearly independent set in $L_K(E)$ (apply Corollary 1.5.15), (1) implies that E must be finite. On the other hand, if c were a cycle in E, then applying Corollary 1.5.15 again would yield that $\{c^n\}_{n\in\mathbb{N}}$ is an independent set, contrary to the finite-dimensionality of $L_K(E)$.

(2) $\Rightarrow$ (3). We show that $L_K(E) = \oplus_{i=1}^m I(v_i)$, where $\{v_1, \ldots, v_m\} = \mathrm{Sink}(E)$. We note that in a finite acyclic graph E, there is a positive integer $b(E)$ for which every path in E has length at most $b(E)$. In addition, such a graph must contain at least one sink. Observe first that $\{\{v_i\}\}_{i=1}^m$ is a family of pairwise disjoint hereditary subsets of E. This implies, by Proposition 2.4.7, that $\sum_{i=1}^m I(v_i) = \oplus_{i=1}^m I(v_i)$.

Now consider an element $\alpha\beta^* \in L_K(E)$, with $\alpha, \beta \in \mathrm{Path}(E)$. If $r(\alpha) \in \mathrm{Sink}(E)$, then $\alpha\beta^* \in I(r(\alpha))$, which is one the $I(v_i)$'s. If this is not the case, then apply the (CK2) relation at $r(\alpha)$ to get

$$\alpha\beta^* = \alpha r(\alpha)\beta^* = \sum_{\{e\in s^{-1}(r(\alpha))\}} \alpha e e^* \beta^*.$$

If for every $e \in s^{-1}(r(\alpha))$ we have $r(e) \in \mathrm{Sink}(E)$, then we are done. Otherwise, rewrite every $r(e)$ which is not a sink as before, using (CK2). Since the graph is finite and acyclic, after at most $b(E)$ steps we have finished.

Finally, we note that m is exactly the cardinality of $\mathrm{Sink}(E)$, while by Corollary 2.6.5, n_i is the number of distinct paths ending in v_i.

(3) $\Rightarrow$ (1) is clear. □

We recall that a *matricial* K-algebra is a finite direct sum of full finite-dimensional matrix algebras over the field K.

Remark 2.6.18 The Finite Dimension Theorem 2.6.17 yields that the matricial Leavitt path K-algebras (Definition 2.1.13) coincide precisely with the finite-dimensional Leavitt path K-algebras. By Corollary 2.6.6, we see that every matricial K-algebra indeed arises as a Leavitt path K-algebra.

Definition 2.6.19 A *locally matricial K-algebra* is a direct limit of matricial K-algebras (with not-necessarily-unital transition homomorphisms).

Proposition 2.6.20 *Let E be an acyclic graph and K any field. Then $L_K(E)$ is locally matricial.*

Proof Write $L_K(E) = \varinjlim L_K(F_i)$, as in Proposition 1.6.15, where every F_i is a finite and acyclic graph. The result then follows, as each $L_K(F_i)$ is a matricial algebra by Theorem 2.6.17. □

Remark 2.6.21 The Finite Dimension Theorem 2.6.17 will play a central role in the theory of Leavitt path algebras. One immediate consequence is instructive. We see from Theorem 2.6.17 that the only information required to understand $L_K(E)$ up to

K-algebra isomorphism when E is a finite acyclic graph is the number of sinks in E, and the number of paths ending in each of those sinks. In particular, this allows us to construct isomorphic Leavitt path algebras from non-isomorphic graphs. For example, let

$$E = \bullet \longrightarrow \bullet \longrightarrow \bullet \qquad \text{and} \qquad F = \bullet \longrightarrow \bullet \longleftarrow \bullet \ .$$

Then E and F are clearly not isomorphic as directed graphs (for instance, F has a vertex of invalence 2, while E does not). However, by Theorem 2.6.17 we get

$$L_K(E) \cong L_K(F) \cong \mathrm{M}_3(K),$$

since both E and F contain exactly one sink, and in both E and F there are exactly three paths ending at that sink.

2.7 The Ideal Generated by the Vertices in Cycles Without Exits

For an arbitrary ring R, there are a number of ideals within R which merit special attention: the Jacobson radical of R, the socle of R, and the left singular ideal of R, to mention just a few. We have already identified these ideals (and others) in the context of Leavitt path algebras. However, there is an ideal that is specific to the context of Leavitt path algebras which plays a central role in the description of the lattice $\mathscr{L}_{id}(L_K(E))$ of all two-sided ideals of $L_K(E)$: the ideal $I(P_c(E))$ generated by those vertices which lie on a cycle without exits. We describe $I(P_c(E))$ in this section.

Just as the ideal generated by the line points has importance (as it coincides with the socle of the corresponding Leavitt path algebra), the ideal generated by the vertices which lie on cycles without exits will also have an important place in the theory. In this case, the cycles without exits will play a role similar to that of the line points. In addition, we will be able to view this ideal as the ideal generated by the primitive non-minimal idempotents in $L_K(E)$ (such idempotents are discussed further in Sect. 3.5). Recall from Notation 2.2.4 that

$$P_c(E) := \{v \in E^0 \mid v \text{ is the base of some cycle } c \text{ for which } c \text{ has no exits}\}.$$

Indeed, $P_c(E)$ may be viewed as the disjoint union $P_c(E) = \sqcup_{i \in \Upsilon}\{c_i^0\}$, where $\{c_i\}_{i\in\Upsilon}$ is the set of distinct cycles without exits in E (i.e., for which $c_i^0 \neq c_j^0$ for $i \neq j$). Note that although $P_c(E)$ is clearly hereditary, it is not necessarily saturated. For instance, in the graph

$$\bullet_u \longrightarrow \bullet_v \circlearrowleft$$

we have $P_c(E) = \{v\}$, which is a hereditary but not saturated subset of E^0. Note, however, that $I(P_c(E)) = I(\overline{P_c(E)})$, by Lemma 2.4.1.

Lemma 2.7.1 *Let E be an arbitrary graph and K any field. Let $v \in P_c(E)$, and let c be the cycle without exits such that $s(c) = v$. Let Λ_v denote the (possibly infinite) set of paths in E which end at v, but which do not contain all the edges of c. Then*

$$I(c^0) = I(v) \cong \mathrm{M}_{\Lambda_v}(K[x, x^{-1}]).$$

Proof That $I(c^0) = I(v)$ is clear for any cycle c containing the vertex v: obviously $I(c^0) \supseteq I(v)$, and the reverse containment holds since there is a path p (a portion of the cycle) from v to any vertex $w \in c^0$, so $w = p^*vp \in I(v)$.

Consider the family

$$\mathscr{B} := \{\mu c^k \eta^* \mid \mu, \eta \in \Lambda_v, k \in \mathbb{Z}\},$$

where as usual c^0 denotes v and c^k denotes $(c^*)^{-k}$ for $k < 0$. By Corollary 1.5.12, $\mathscr{B}$ is a K-linearly independent set.

By Lemma 2.4.1 we have that every element in $I(v)$ is a K-linear combination of elements of the form $\alpha\beta^*$, where $r(\alpha) = r(\beta) \in T(v)$. But $T(v)$ consists precisely of the vertices in c, as c has no exits. So $\alpha = \mu c^\ell$ and $\beta = \eta c^m$ for some $\mu, \eta \in \Lambda_v$, and $\ell, m \geq 0$. This shows that $\mathscr{B}$ generates $I(v)$, so that $\mathscr{B}$ is a K-basis for $I(v)$.

We define $\varphi : I(v) \to \mathrm{M}_{\Lambda_v}(K[x, x^{-1}])$ by setting $\varphi(\mu c^k \eta^*) = x^k e_{\mu,\eta}$ for each $\mu c^k \eta^* \in \mathscr{B}$ (where $x^k e_{\mu,\eta}$ denotes the element of $\mathrm{M}_{\Lambda_v}(K[x, x^{-1}])$ which is x^k in the (μ, η) entry, and zero otherwise). Then one easily checks that φ is a K-algebra isomorphism. □

We record a consequence of Lemma 2.7.1 which is analogous to a previously noted consequence of Lemma 2.6.4.

Corollary 2.7.2 *Let K be any field. For any set Λ let E^c_Λ denote the graph with*

$$(E^c_\Lambda)^0 = \{v\} \cup \{u_\lambda \mid \lambda \in \Lambda\} \quad and \quad (E^c_\Lambda)^1 = \{f_c\} \cup \{f_\lambda \mid \lambda \in \Lambda\},$$

where $s(f_\lambda) = u_\lambda$ and $r(f_\lambda) = v$ for all $\lambda \in \Lambda$, and f_c is a loop based at v. Then $L_K(E^c_\Lambda) \cong \mathrm{M}_{\Lambda \cup \{v\}}(K[x, x^{-1}])$. In particular, by taking disjoint unions of graphs of this form, any direct sum of full matrix rings over $K[x, x^{-1}]$ arises as the Leavitt path algebra of a graph.

Now using Proposition 2.4.7 together with Lemma 2.7.1, we have achieved the following.

Theorem 2.7.3 *Let E be an arbitrary graph and K any field. Then*

$$I(P_c(E)) \cong \oplus_{i \in \Upsilon} \mathrm{M}_{\Lambda_{v_i}}(K[x, x^{-1}]),$$

where $\{c_i\}_{i\in\Upsilon}$ is the set of distinct cycles without exits in E (i.e., for which $c_i^0 \neq c_j^0$ for $i \neq j$), and, for each $i \in \Upsilon$, Λ_{v_i} is the set of paths in E which end at the base v_i of the cycle c_i, but do not contain all the edges of c_i.

For the following corollary, we will need to consider vertices for which its tree does not contain infinite bifurcations.

Definition 2.7.4 Let E be an arbitrary graph.

Denote by $P_{b^\infty}(E)$ the set of all vertices v in E^0 such that $T(v)$ contains either infinitely many distinct bifurcation vertices, or at least one infinite emitter.

Denote by $P_{ne}(E)$ the set of vertices whose tree does not contain cycles with exits.

Corollary 2.7.5 *Let E be an arbitrary graph and K any field. Denote by H the set $P_l(E) \cup P_c(E) \subseteq E^0$.*

(i) *There is an isomorphism of K-algebras*

$$I(H) \cong \left(\oplus_{i\in\Upsilon_1} \mathrm{M}_{\Lambda_{v_i}}(K)\right) \oplus \left(\oplus_{i\in\Upsilon_2} \mathrm{M}_{\Lambda_{v_i}}(K[x,x^{-1}])\right),$$

where Υ_1 is the set of equivalence classes of line points, and Υ_2 is the set of distinct cycles without exits.

(ii) *For every $v \in P_{ne}(E) \setminus P_{b^\infty}(E)$ for which every path starting at v connects to H, there is an isomorphism of K-algebras*

$$I(v) \cong \left(\oplus_{i\in\Upsilon_1'} \mathrm{M}_{\Lambda_{v_i}}(K)\right) \oplus \left(\oplus_{i\in\Upsilon_2'} \mathrm{M}_{\Lambda_{v_i}}(K[x,x^{-1}])\right),$$

where $\Upsilon_j' \subseteq \Upsilon_j$ for $j = 1, 2$.

Proof (i) It is clear that $P_l(E)$ and $P_c(E)$ are disjoint hereditary subsets of E^0. By Proposition 2.4.7 we have that $I(H) = I(P_l(E)) \oplus I(P_c(E))$. Now apply Theorems 2.6.14 and 2.7.3 to establish the result.

(ii) By definition, every vertex in the tree of v is a finite emitter and there are only a finite number of bifurcations in $T(v)$. We are going to show that $v \in \overline{H}$ by induction on the number of bifurcations in $T(v)$. If there are no bifurcations in the tree of v, then $v \in H$. Assume that the result is true for vertices whose tree has less than t bifurcation vertices, where $t \geq 1$, and let v be a vertex as in the statement whose tree has exactly t bifurcation vertices. Clearly, we can assume that v itself is a bifurcation vertex. Using (CK2) we may write $v = \sum_{e\in s^{-1}(v)} ee^*$, and now each $r(e)$ for $e \in s^{-1}(v)$ has less than t bifurcations in its tree, and has the property that each path starting at $r(e)$ connects to H. So, by the induction hypothesis, $r(s^{-1}(v)) \subseteq \overline{H}$. Since $\overline{H}$ is saturated, we get that $v \in \overline{H}$. This implies that $I(v) \subseteq I(\overline{H}) = I(H)$. By (i) this last ideal is isomorphic to $\left(\oplus_{i\in\Upsilon_1} \mathrm{M}_{\Lambda_{v_i}}(K)\right) \oplus \left(\oplus_{i\in\Upsilon_2} \mathrm{M}_{\Lambda_{v_i}}(K[x,x^{-1}])\right)$. The grading of $I(H)$ corresponds to a certain grading in the latter algebra, in such a way that, for each factor $\mathrm{M}_{\Lambda_{v_i}}(K[x,x^{-1}])$, the degree of $xe_{\gamma,\gamma}$ is strictly positive, for each diagonal matrix unit $e_{\gamma,\gamma}$. This is enough to show that these factors are graded-simple. Now, using this fact and that $I(v)$ is a graded ideal (as it is generated by an element of zero degree), we get the result. □

Corollary 2.7.6 *Let E be a finite graph and K any field. Let $v \in P_{ne}(E)$. Then there exist positive integers m, n, r_i, and s_i for which*

$$I(v) \cong \left(\oplus_{i=1}^{m} \mathrm{M}_{r_i}(K)\right) \oplus \left(\oplus_{i=1}^{n} \mathrm{M}_{s_i}(K[x, x^{-1}])\right).$$

In particular, $I(v)$ is a noetherian K-subalgebra of $L_K(E)$.

Proof Use Corollary 2.7.5(ii) with the fact that E finite implies $L_K(E)$ is unital to get that all Υ_j and Λ_{v_j} must be finite, for $j = 1, 2$. The second statement then follows immediately. □

The ideal we have described in Theorem 2.7.3 will play an important role in a Leavitt path algebra because as we now show, it captures all those ideals in the Leavitt path algebra which do not contain vertices.

Lemma 2.7.7 *Let E be an arbitrary graph and K any field. Let J be a nonzero ideal of $L_K(E)$ such that $J \cap E^0 = \emptyset$. Then $\{0\} \neq J \cap KE \subseteq I(P_c(E))$.*

Proof We first show that $\{0\} \neq J \cap KE$. Let y be a nonzero element in J. By the Reduction Theorem 2.2.11, either there exist $\alpha, \beta \in \mathrm{Path}(E)$ such that $\alpha^* y \beta = ku$ for some $u \in E^0$ and $k \in K^\times$, or $\alpha^* y \beta$ is a nonzero polynomial in a cycle without exits. Since J does not contain vertices, the first case cannot happen, and by multiplying by a power of the cycle without exits (if necessary), we produce a nonzero element in $J \cap KE$.

For such a nonzero element $x \in J \cap KE$, write $x = \sum_{u \in U} xu$, where $U = U(x)$ is the finite family of vertices of E such that $xu \neq 0$. Fix $u \in U$, and write $xu = \sum_{i=1}^{r} k_i \alpha_i$, with $k_i \in K^\times$, $\alpha_i = \alpha_i u \in \mathrm{Path}(E)$ for every i and $\alpha_i \neq \alpha_j$ for every $i \neq j$, and in such a way that $\deg(\alpha_i) \leq \deg(\alpha_{i+1})$ for every $i = 1, \ldots, r-1$.

We will prove that $xu \in I(P_c(E))$ by induction on the number r of summands. Note that $r \neq 1$ as otherwise we would have $xu = k_1 \alpha_1$, so $k_1^{-1} \alpha_1^* xu = u \in J$, a contradiction to the hypothesis. So the base case for the induction is $r = 2$.

Suppose first that $\deg(\alpha_1) = \deg(\alpha_2)$. In this case, since $\alpha_1 \neq \alpha_2$, we get $\alpha_1^* \alpha_2 = 0$ so that $k_1^{-1} \alpha_1^* xu = u \in J$, a contradiction again. This gives $\deg(\alpha_1) < \deg(\alpha_2)$, and then $\alpha_1^* xu = k_1 u + k_2 e_1 \cdots e_t$ for some $e_1, \cdots, e_t \in E^1$. By multiplying the left and right-hand sides by u we get

$$y_1 := u \alpha_1^* xu = k_1 u + k_2 u e_1 \cdots e_t u \in J \cap KE.$$

Observe that u and $e_1 \cdots e_t$ have different degrees, so since $k_1 u \neq 0$ we obtain that $y_1 \neq 0$. Moreover, as J does not contain vertices we have that $c := u e_1 \cdots e_t u \neq 0$, and thus c is a closed path based at u. We will prove that c does not have exits. Suppose on the contrary that there exist $w \in T(u)$ and $e, f \in E^1$ such that $e \neq f$, $s(e) = s(f) = w$, $c = aweb = aeb$ for some $a, b \in \mathrm{Path}(E)$. Then $\tau = af$ satisfies $\tau^* c = f^* a^* aeb = f^* eb = 0$ so that $\tau^* y_1 \tau = k_1 r(\tau) \in J$, again a contradiction.

Thus by definition $u \in P_c(E)$, so that, in particular, $xu \in I(P_c(E))$. So the base case $r = 2$ for the induction has been established.

We now assume the result holds for $r \geq 2$ and prove it for $r + 1$. Assume then that $xu = \sum_{i=1}^{r+1} k_i\alpha_i$; we distinguish two situations.

For the first case, suppose $\deg(\alpha_j) = \deg(\alpha_{j+1})$ for some $1 \leq j \leq r$. The element $\alpha_j^* xu\alpha_j = \alpha_j^* xu\alpha_j u \in J$ is nonzero, as follows: clearly each monomial remains with positive degree as $\deg(\alpha_j^*\alpha_i\alpha_j) = \deg(\alpha_i) \geq 0$. Moreover, at least $\alpha_j = \alpha_j^*\alpha_j\alpha_j$ appears in the expression for $\alpha_j^* xu\alpha_j$ because if we had $\alpha_j = \alpha_j^*\alpha_i\alpha_j$ for some $i \neq j$, then $\deg(\alpha_i) = \deg(\alpha_j)$, which implies $\alpha_j^*\alpha_i = 0$ and therefore $\alpha_j = 0$, a contradiction. This shows that $\alpha_j^* xu\alpha_j$ has at least one nonzero monomial summand, and because distinct paths of E are linearly independent (see Corollary 1.5.15), then $\alpha_j^* xu\alpha_j \neq 0$. Now, this element has at most r summands because $\alpha_j^*\alpha_{j+1}\alpha_j = 0$ and it satisfies the induction hypothesis, so that $u \in P_c(E)$.

The second case is when $\deg(\alpha_i) < \deg(\alpha_{i+1})$ for every $i = 1, \ldots, r$. Then $0 \neq \alpha_1^* xu = k_1 u + \sum_{i=2}^{r+1} k_i\beta_i$ with $\beta_i u = \beta_i \in \mathrm{Path}(E)$. Multiply again as follows:

$$y_2 := u\beta_{r+1}^* u\alpha_1^* xu\beta_{r+1}u = k_1 u + \sum_{i=2}^{r+1} k_i u\beta_{r+1}^* u\beta_i u\beta_{r+1}u \in J.$$

A similar argument to the one used above shows that y_2 is nonzero so that, if some monomial summand of y_2 becomes zero, then y_2 satisfies the induction hypothesis, therefore $u \in P_c(E)$. If this is not the case, since β_{r+1} has maximum degree among the β_i, then

$$y_2 = k_1 u + k_2\gamma_1 + k_3\gamma_1\gamma_2 + \ldots + k_{r+1}\gamma_1 \cdots \gamma_r,$$

where γ_i are closed paths based at u. We focus on γ_1. Proceeding in a similar fashion as before, we can conclude that γ_1 cannot have exits, as otherwise there would exist a path δ with $s(\delta) = u$ and $\delta^*\gamma_1 = 0$, which in turn would give $0 \neq \delta^* y_2 \delta = k_1 r(\delta) \in J$, a contradiction. Thus γ_1 is a closed path without exits, so that $r(\gamma) = u \in P_c(E)$, and finally $x = xu \in I(P_c(E))$.

Since this holds for every $u \in U$ we get $x = \sum_{u \in U} xu \in I(P_c(E))$. □

Prior to achieving our main result about $I(P_c(E))$, we need a general result about path algebras.

Lemma 2.7.8 *Let E be an arbitrary graph and K any field.*

(i) *Let $w \in E^0$, let $\mu \in \mathrm{Path}(E)$ with $r(\mu) = w$, and let $x \in KE$ for which $wx = x$. If $\mu x = 0$ in KE, then $x = 0$.*
(ii) *Let $v \in E^0$, let $\gamma \in \mathrm{Path}(E)$ with $s(\gamma) = v$, and let $y \in KE$ for which $yv = y$. If $y\gamma = 0$ in KE, then $y = 0$.*

Proof (i) Write $x = \sum_{i=1}^n k_i\mu_i \in KE$, where $k_i \in K^\times$, and the μ_i are distinct. Since $wx = x$, we may assume that $s(\mu_i) = w$ for all $1 \leq i \leq n$. In particular, each expression $\mu\mu_i$ is a path in E. Then from $\mu x = 0$ we get $\sum_{i=1}^n k_i\mu\mu_i = 0$, and since all the paths in the set $\{\mu\mu_i\}_{i=1}^n$ are distinct, they are K-linearly independent in KE (see Remark 1.2.4). Therefore $k_i = 0$ for all $1 \le i \le n$, and so $x = 0$.

Statement (ii) can be established analogously. □

Proposition 2.7.9 *Let E be an arbitrary graph and K any field. Let J be an ideal of $L_K(E)$ such that $J \cap E^0 = \emptyset$. Then $J \subseteq I(\overline{P_c(E)})$.*

Proof We may assume that $J \neq 0$. Let $0 \neq x \in J$, and write $x = \sum_{i=1}^n xu_i$ for the finite set of vertices $\{u_i \mid 1 \le i \le n\}$ for which $0 \neq xu_i$. As J is an ideal, $0 \neq xu_i \in J$, so that we can assume without loss of generality that $0 \neq x = xu$ for some $u \in E^0$.

We will show, by induction on the degree in ghost edges (recall Definitions 2.2.9), that if $xu \in J$, with $u \in E^0$, then $xu \in I(\overline{P_c(E)})$. If $\mathrm{gdeg}(xu) = 0$, the result follows by Lemma 2.7.7. Suppose the result is true for elements having degree in ghost edges strictly less than $\mathrm{gdeg}(xu)$, and show it for $\mathrm{gdeg}(xu)$.

Write $x = \sum_{i=1}^r \beta_i e_i^* + \beta$, with $\beta_i \in L_K(E)$, $\beta = \beta u \in KE$ and $e_i \in E^1$, with $e_i \neq e_j$ for every $i \neq j$. Then $xue_i = \beta_i + \beta e_i \in J$; since $\mathrm{gdeg}(xue_i) < \mathrm{gdeg}(xu)$, by the induction hypothesis $\beta_i + \beta e_i \in I(\overline{P_c(E)})$, for every $i \in \{1, \dots, r\}$.

Suppose first that u is a finite emitter. If $u = \sum_{i=1}^r e_i e_i^*$, then $xu = \sum_{i=1}^r \beta_i e_i^* + \sum_{i=1}^r \beta e_i e_i^* = \sum_{i=1}^r (\beta_i + \beta e_i) e_i^* \in I(\overline{P_c(E)})$, and we have finished. If $u = \sum_{i=1}^r e_i e_i^* + \sum_{j=1}^s f_j f_j^*$ (where $f_j \in E^1$), then $xuf_j = \beta f_j \in J \cap KE$. By Lemma 2.7.7, $\beta f_j \in I(\overline{P_c(E)})$ for every $j \in \{1, \dots, s\}$, hence $xu = \sum_{i=1}^r (\beta_i + \beta e_i) e_i^* + \sum_{j=1}^s \beta f_j f_j^* \in I(\overline{P_c(E)})$.

On the other hand, suppose that u is an infinite emitter. If $\beta = 0$ then for every j we have $xue_j = \beta_j \in I(\overline{P_c(E)})$, by the induction hypothesis, and so $xu \in I(\overline{P_c(E)})$. Now we are going to show by contradiction that the case $\beta \neq 0$ cannot happen, and thereby will complete the proof.

So suppose $\beta \neq 0$, and write $\beta = \sum_{i=1}^s k_i \beta_i'$, with $k_i \in K^\times$, and $\beta_i' \in \mathrm{Path}(E)$ distinct paths such that $|\beta_1'| \le \cdots \le |\beta_s'|$. Note that, as u is an infinite emitter, u is not in $I(\overline{P_c(E)})$. Since $\beta_i' = \beta_i' u$ then β_i' is not in $I(\overline{P_c(E)})$ for any i. (Because $P_c(E)$ contains no infinite emitters (by definition), then neither does $\overline{P_c(E)}$, and so neither does $I(\overline{P_c(E)})$.) Let $f \in s^{-1}(u)$ such that $f \neq e_j$ for every j. By Lemma 2.7.8(ii) we have $\beta f \neq 0$; since $\beta f = xf$, by the induction hypothesis $\beta f \in I(\overline{P_c(E)})$, therefore $0 \neq xf = \beta f \in I(\overline{P_c(E)})$.

We shall see that $r(f) \in \overline{P_c(E)}$. Consider the algebra $L_K(E)/I(\overline{P_c(E)})$ and denote by $\overline{x}$ the class of an element x of $L_K(E)$ in this quotient. Note that $0 = \overline{\beta f} = \sum_{i=1}^s k_i \overline{\beta_i' f}$, hence, by Theorem 2.4.15 we have $\overline{\beta_i' f} = 0$, i.e., $\beta_i' f \in I(\overline{P_c(E)})$ for every i and so $r(f) = f^*(\beta_i')^* \beta_i' f \in I(\overline{P_c(E)}) \cap E^0 = \overline{P_c(E)}$ by Corollary 2.4.16(i). Then $f^*(\beta_1')^* \beta f = k_1 r(f) + \sum_{i=2}^s k_i f^*(\beta_1')^* \beta_i' f$. Note that the second summand must be zero because otherwise for some $j \in \{2, \cdots, s\}$ we would have $\beta_j' = \beta_1' f \gamma$ for some $\gamma \in \mathrm{Path}(E)$, which is not possible because we know $\beta_j' \notin I(\overline{P_c(E)})$. Therefore $0 \neq k_1 r(f) \in J$, a contradiction again. Thus $\beta = 0$, which completes the proof of the result. □

We finish the section by utilizing Lemma 2.7.8 to give a graph-theoretic description of when an ideal $I(H)$ is an essential ideal of $L_K(E)$.

Proposition 2.7.10 *Let E be an arbitrary graph and K any field. Let H be a hereditary subset of E. Then $I(H)$ is an essential (left / right / two-sided) ideal of $L_K(E)$ if and only if every vertex of E connects to a vertex in H (i.e., $T(v) \cap H \neq \emptyset$ for all $v \in E^0$).*

Proof Since $L_K(E)$ is semiprime (Proposition 2.3.1), we may invoke [108, (14.1) Proposition] to conclude that $I(H)$ is essential as a left or right ideal if and only if it is essential as an ideal. Moreover, as $I(H)$ is a graded ideal, by Năstăsescu and van Oystaeyen [119, 2.3.5 Proposition] we have that essentiality and graded-essentiality (i.e., essentiality with respect to graded ideals) of $I(H)$ are equivalent. Hence, it suffices to show that $I(H)$ is a graded-essential ideal if and only if every vertex of E^0 connects to a vertex in H.

Suppose first that $I(H)$ is a graded essential ideal of $L_K(E)$. Let $v \in E^0$. If $H \cap T(v) = \emptyset$, then Proposition 2.5.25(ii) would imply $I(H) \cap I(T(v)) = 0$, but this cannot happen as $I(H)$ is a graded essential ideal. Hence $H \cap T(v) \neq \emptyset$. This implies that v connects to a vertex in H.

Conversely, suppose $H \cap T(v) \neq \emptyset$ for each $v \in E^0$. Let J be a nonzero graded ideal and pick a nonzero homogeneous element $x = uxv \in J$, where $u, v \in E^0$. By Corollary 2.2.12(ii), there exists a $\mu \in \text{Path}(E)$ such that $0 \neq x\mu \in KE$. Denote $r(\mu)$ by w. By hypothesis w connects to a vertex in H, hence there exists a $\lambda \in \text{Path}(E)$ such that $w = s(\lambda)$ and $r(\lambda) \in H$. But $x\mu\lambda \neq 0$ by Lemma 2.7.8(i), hence $0 \neq x\mu\lambda \in I(H) \cap J$, which establishes the result. □

2.8 The Structure Theorem for Ideals, and the Internal Structure of Ideals

Now that we have in hand an explicit description of the lattice of graded ideals of a Leavitt path algebra (the Structure Theorem for Graded Ideals 2.5.8), we turn our attention to explicitly describing the lattice of *all* ideals in a Leavitt path algebra. Although the structure of the field K played no role in the description of the graded ideals, the field will indeed play a pivotal role in this more general setting. The intuition which lies at the heart of this description is as follows. The prototypical example of a Leavitt path algebra which contains non-graded ideals is $L_K(R_1) \cong K[x, x^{-1}]$. The only graded ideals of $L_K(R_1)$, namely, $\{0\}$ and $L_K(R_1)$ itself, correspond to the two distinct hereditary saturated subsets of R_1. On the other hand, the non-graded ideals correspond to various polynomial expressions in the cycle c of R_1, specifically, are in bijective correspondence with polynomials of the form $1 + k_1x + \cdots + k_nx^n \in K[x]$, for $n > 0$ and $k_n \neq 0$. We will show in the main result of this section (the Structure Theorem for Ideals, Theorem 2.8.10) that such a bijection, one which associates hereditary saturated subsets of E^0 (possibly also with breaking vertices of such subsets) together with various cycles in E and polynomials in $K[x]$ on the one hand, with ideals of $L_K(E)$ on the other, may be established for arbitrary graphs E and fields K as well. To achieve this general result we will rely heavily on our previously completed analysis of the graded ideal structure of $L_K(E)$, together with the structure of the ideal $I(P_c)$ investigated in Sect. 2.7. It is not coincidental in this context that the loop in R_1 is the only closed simple path based at the vertex of R_1. Indeed, in general $L_K(E)$ will contain non-graded ideals only when E fails to satisfy Condition (K).

We remind the reader that when we talk about a *cycle based at a vertex* (say, v), then we mean a specific path $c = e_1 \cdots e_n$ in E (one for which $s(c) = r(c) = v$); on the other hand, when we speak about a *cycle*, we mean a collection of paths based at the different vertices of the path c (see Definitions 1.2.2).

Notation 2.8.1 Let E be an arbitrary graph. We define

$$C_u(E) = \{c \mid c \text{ is a cycle in } E \text{ for which } |CSP(v)| = 1 \text{ for every } v \in c^0\}, \quad \text{and}$$

$$C_{ne}(E) = \{c \mid c \text{ is a cycle in } E \text{ for which } c \text{ has no exits in } E\}.$$

Observe that $C_{ne}(E) \subseteq C_u(E)$ for any graph E, but not necessarily conversely: in the Toeplitz graph E_T, the unique cycle has an exit, but there is exactly one closed simple path at the vertex of that cycle.

Notation 2.8.2 Let E be an arbitrary graph. Let $H \in \mathscr{H}_E$. Denote by C_H the set

$$C_H = \{c \mid c \text{ is a cycle in } E \text{ such that } c^0 \cap H = \emptyset, \text{ and for which } r(e) \in H \text{ for every exit } e \text{ of } c\}.$$

We note that C_H corresponds precisely to the set of cycles without exits in the quotient graph E/H.

Lemma 2.8.3 *Let E be an arbitrary graph, and $H \in \mathscr{H}_E$. Then $C_H \subseteq C_u(E)$.*

Proof Let $c \in C_H$. We must show that $c \in C_u(E)$, i.e., that $|CSP(v)| = 1$ for every $v \in c^0$. But this holds because for every exit e of c the vertices in $T(r(e))$ are in H (since H is hereditary), and because $c^0 \cap H = \emptyset$. □

Recall the preorder $\leq$ in E^0: given $v, w \in E^0$, $v \leq w$ if and only if there is a path $\mu \in \text{Path}(E)$ such that $s(\mu) = w$ and $r(\mu) = v$.

Notation 2.8.4 Let E be an arbitrary graph. For $u, v \in E^0$ we write $u << v$ if $u \leq v$ but $v \not\leq u$. For a cycle c in E, we define:

$$c^{<<} := \{w \in E^0 \mid w << v \text{ for every } v \in c^0\}.$$

Roughly speaking, $c^{<<}$ is the tree of the set of vertices which are ranges of exits for the cycle c, but for which there are no paths from such vertices which return back to the cycle c. For instance, for the Toeplitz graph E_T of Example 1.3.6, we have $c^{<<} = \{v\}$.

Proposition 2.8.5 *Let E be an arbitrary graph and K any field. Let I be an ideal of E. Define $H := I \cap E^0$ and $S := \{v \in B_H \mid v^H \in I\}$. Let J denote $I/I(H \cup S^H)$; using Theorem 2.4.15, we view J as an ideal of the Leavitt path algebra of the quotient graph $L_K(E/(H,S))$. Then:*

(i) $J \subseteq I(P_c(E/(H,S)))$.
(ii) *There exists a set $C \subseteq C_H$ and a set $P = \{p_c(x) \in K[x] \mid c \in C\}$ such that each $p_c(x)$ is a polynomial of the form $1 + k_1x + \ldots + k_nx^n$, with $n > 0$ and $k_n \neq 0$, in such a way that $J = \oplus_{c\in C} I(p_c(c))$. (Note that C is empty precisely when I is graded, which happens precisely when $J = \{0\}$.)*
(iii) *The sets C and P are uniquely determined by I.*

Proof (i). Consider the ideal $J = I/I(H \cup S^H)$ of $L_K(E/(H,S))$. Recall that the vertices in $E/(H,S)$ are $(E^0 \setminus H) \cup \{v' \mid v \in B_H \setminus S\}$, and observe that vertices v' with $v \in B_H \setminus S$ correspond to the classes of the elements v^H through the isomorphism $L_K(E/(H,S)) \cong L_K(E)/I(H \cup S^H)$. It is clear from this that J does not contain vertices in the graph $E/(H,S)$. Now (i) follows by Proposition 2.7.9.

(ii) and (iii). By Theorem 2.7.3 we have an isomorphism

$$I(P_c(E/(H,S))) \cong \bigoplus_{i\in \Upsilon} \mathrm{M}_{\Lambda_i}(K[x,x^{-1}]),$$

where Υ is the set of cycles without exits in $E/(H,S)$. As observed previously, we may identify this set with C_H. We recall now these two well-known facts: first, that the ideals of a direct sum of matrix rings are direct sums of matrix rings over ideals of the base rings, and, second, that the Laurent polynomial ring $K[x,x^{-1}]$ is a principal ideal domain. Applying these two facts, along with (i) and the displayed isomorphism, we get that there exists a subset C of C_H and a set of polynomials P as in the statement, uniquely determined by J, for which

$$J \cong \bigoplus_{c\in C} \mathrm{M}_{\Lambda_c}(p_c(x)K[x,x^{-1}]) \cong \bigoplus_{c\in C} I(p_c(c)),$$

as desired. □

The main result of this section is Theorem 2.8.10, which shows that there is a lattice isomorphism between ideals in the Leavitt path algebra $L_K(E)$ on the one hand, and triples consisting of elements in $\mathscr{T}_E$ (see Definition 2.5.3), certain subsets of cycles in E, and families of polynomials in $K[x]$ on the other. We now describe such triples.

Definition 2.8.6 Let E be an arbitrary graph and K any field. For every pair $(H,S) \in \mathscr{T}_E$, consider a subset C of C_H; for every element $c \in C$, take an arbitrary polynomial $p_c(x) = 1 + k_1x + \cdots + k_nx^n \in K[x]$, where $n > 0$ and $k_n \neq 0$, and write $P = \{p_c(x) \mid c \in C\}$. We define $\mathscr{Q}_E$ as the set of triples:

$$\mathscr{Q}_E = \{((H,S),C,P)\}.$$

To show that there is a bijection between $\mathcal{L}_{id}(L_K(E))$ and $\mathcal{Q}_E$ we will assign to every triple $((H,S),C,P)$ the ideal generated by $H \cup S^H \cup P_C$, where for $P = \{p_c(x) \mid c \in C\}$, P_C denotes the subset $\{p_c(c) \mid c \in C\}$ of $L_K(E)$.

Definition 2.8.7 Let E be an arbitrary graph and K any field. We define a relation $\leq$ on $\mathcal{Q}_E$ as follows. For elements $((H_1,S_1),C_1,P_1)$ and $((H_2,S_2),C_2,P_2)$ of $\mathcal{Q}_E$, we set $P_i := \{p_c^{(i)} \mid c \in C_i\}$ for $i = 1,2$. We then define

$$((H_1,S_1),C_1,P_1) \leq ((H_2,S_2),C_2,P_2) \quad \text{if:}$$

$$(H_1,S_1) \leq (H_2,S_2), \quad C_1^0 \subseteq H_2 \cup C_2^0, \quad \text{and} \quad p_c^{(2)} \mid p_c^{(1)} \text{ in } K[x] \text{ for every } c \in C_1 \cap C_2.$$

Proposition 2.8.8 *Let E be an arbitrary graph and K any field. Then the relation $\leq$ defined on $\mathcal{Q}_E$ in Definition 2.8.7 is a partial order. Furthermore, using this relation, $\mathcal{Q}_E$ is a lattice, in which the supremum and infimum operators are described as follows.*

For the supremum $\vee$ of two elements, we have

$$((H_1,S_1),C_1,P_1) \;\vee\; ((H_2,S_2),C_2,P_2)$$

$$= \; ((\overline{H_1 \cup H_2 \cup C^0}^{S_1 \cup S_2}, (S_1 \cup S_2) \setminus \overline{H_1 \cup H_2 \cup C^0}^{S_1 \cup S_2}), C_1 \vee C_2, \; \{\text{g.c.d.}(p_c^{(1)}, p_c^{(2)})\}_{c \in C_1 \vee C_2}),$$

where

$$C = \{c \in C_1 \cap C_2 \mid \text{g.c.d.}(p_c^{(1)}, p_c^{(2)}) = 1\}, \quad \textit{and}$$

$$C_1 \vee C_2 = C_1 \cup C_2 \setminus \{c \in C_1 \cup C_2 \mid c^0 \subseteq \overline{H_1 \cup H_2 \cup C^0}^{S_1 \cup S_2}\}.$$

(We interpret $p_c^{(i)}$ as 0 if $c \notin C_i$ for $i = 1$ or 2.)

For the infimum $\wedge$ of two elements, we have

$$((H_1,S_1),C_1,P_1) \;\wedge\; ((H_2,S_2),C_2,P_2)$$

$$= \; ((H_1,S_1) \wedge (H_2,S_2), \; C_1 \wedge C_2, \; \{\text{l.c.m.}(p_c^{(1)}, p_c^{(2)})\}_{c \in C_1 \wedge C_2}),$$

where

$$C_1 \wedge C_2 = (C_1 \cap C_2) \cup C_1^{H_2} \cup C_2^{H_1},$$

$$\textit{with } C_1^{H_2} := \{c \in C_1 \mid c^0 \subseteq H_2\} \; \textit{and} \; C_2^{H_1} := \{c \in C_2 \mid c^0 \subseteq H_1\}.$$

(We interpret $p_c^{(i)}$ as 1 if $c \notin C_i$ for $i = 1$ or 2.)

Proof It is immediate to see that $\leq$ is reflexive. To show the antisymmetric property we use the antisymmetric property of $\leq$ on $\mathscr{T}_E$ (see Proposition 2.5.6) and the fact that for $((H,S),C,P) \in \mathscr{Q}_E$ we have $C^0 \cap H = \emptyset$ (because $C \subseteq C_H$).

To prove the transitivity, take three triples in $\mathscr{Q}_E$ such that $((H_1,S_1),C_1,P_1) \leq ((H_2,S_2),C_2,P_2)$ and $((H_2,S_2),C_2,P_2) \leq ((H_3,S_3),C_3,P_3)$. Since $(H_1,S_1) \leq (H_2,S_2)$ and $(H_2,S_2) \leq (H_3,S_3)$, it follows that $(H_1,S_1) \leq (H_3,S_3)$. In addition, $C_1^0 \subseteq H_2 \cup C_2^0$ and $C_2^0 \subseteq H_3 \cup C_3^0$ implies $C_1^0 \subseteq H_2 \cup C_2^0 \subseteq H_3 \cup C_3^0$. Finally, let $c \in C_1 \cap C_3$. Note that $c \in C_3$ implies $c^0 \cap H_3 = \emptyset$, hence $c \in C_2$ because otherwise $c^0 \subseteq H_2 \cup C_2^0$ would imply $c^0 \subseteq H_2 \subseteq H_3$, a contradiction. Therefore $c \in C_1 \cap C_2 \cap C_3$, and from the relations $p_c^{(2)} \mid p_c^{(1)}$ and $p_c^{(3)} \mid p_c^{(2)}$ in $K[x]$ we get $p_c^{(3)} \mid p_c^{(1)}$ in $K[x]$. Hence $((H_1,S_1),C_1,P_1) \leq ((H_3,S_3),C_3,P_3)$.

Now we check that the formula given in the statement corresponds to the supremum. To this end, let $((H_1,S_1),C_1,P_1), ((H_2,S_2),C_2,P_2) \in \mathscr{Q}_E$. Denote the element

$$((\overline{H_1 \cup H_2 \cup C^0}^{S_1 \cup S_2}, (S_1 \cup S_2) \setminus \overline{H_1 \cup H_2 \cup C^0}^{S_1 \cup S_2}),\ C_1 \vee C_2,\ \{\text{g.c.d.}(p_c^{(1)}, p_c^{(2)})\}_{c \in C_1 \vee C_2})$$

by $((\widetilde{H},\widetilde{S}),\widetilde{C},\widetilde{P})$. It is not difficult to show that $((H_i,S_i),C_i,P_i) \leq ((\widetilde{H},\widetilde{S}),\widetilde{C},\widetilde{P})$ for $i = 1,2$.

Now take $((H',S'),C',P') \in \mathscr{Q}_E$ such that $((H_i,S_i),C_i,P_i) \leq ((H',S'),C',P')$ for $i = 1,2$. First we prove $(\widetilde{H},\widetilde{S}) \leq (H',S')$. Note that $H_1 \cup H_2 \subseteq H'$. Now we want to show that $C^0 \subseteq H'$. We start by showing that $C \cap C' = \emptyset$. Assume $c \in C \cap C'$. Then $c \in C_1 \cap C_2$ and $\text{g.c.d.}(p_c^{(1)}, p_c^{(2)}) = 1$ (recall the definition of C). Since $((H_i,S_i),C_i,P_i) \leq ((H',S'),C',P')$ and $c \in C_i \cap C'$ we have $p'_c | p_c^{(i)}$, for $i = 1,2$, where $P' = \{p'_c \mid c \in C'\}$. Hence $p'_c = 1$, contradicting the choice of p'_c (which, by definition, is a non-invertible polynomial in $K[x,x^{-1}]$). Using that $C^0 \subseteq C_1^0 \subseteq H' \cup C'^0$, and taking into account that $C^0 \cap C'^0 = \emptyset$, we get $C^0 \subseteq H'$. This shows $H_1 \cup H_2 \cup C^0 \subseteq H'$. Since $S_1 \cup S_2 \subseteq H' \cup S'$ the same argument as in Proposition 2.5.6 shows

$$\overline{H_1 \cup H_2 \cup C^0}^{S_1 \cup S_2} \subseteq H'.$$

It is immediate that $S_1 \cup S_2 \setminus \overline{H_1 \cup H_2 \cup C^0}^{S_1 \cup S_2} \subseteq H' \cup S'$, and that $(C_1 \vee C_2)^0 \subseteq C_1^0 \cup C_2^0 \subseteq H' \cup (C')^0$.

Finally, note that for $c \in (C_1 \vee C_2) \cap C'$ we have that $p'_c | p_c^{(i)}$ for $i = 1,2$. Hence $p'_c \mid \text{g.c.d.}(p_c^{(1)}, p_c^{(2)})$. This concludes the proof of the formula for the supremum.

We leave to the reader the verification of the formula for the infimum. □

Lemma 2.8.9 *Let E be an arbitrary graph and K any field. For any ideal I of $L_K(E)$, let $H = I \cap E^0$, and $S^H = \{v \in B_H \mid v^H \in I\}$ (see Definitions 2.4.4). Then the largest graded ideal of $L_K(E)$ contained in I is precisely $I(H \cup S^H)$.*

Proof Clearly $I(H \cup S^H) \subseteq I$. Now let J be any other graded ideal contained in I. Then by the Structure Theorem for Graded Ideals 2.5.8, $J = I(H' \cup S^{H'})$ for $H' = J \cap E^0 \subseteq I \cap E^0 = H$, and $S^{H'} = \{v \in B_{H'} \mid v^{H'} \in J\} \subseteq S^H$. □

We now have all the tools in place to achieve the main result of this section, namely, a description of the collection of all two-sided ideals of $L_K(E)$. Recall that $\mathscr{L}_{id}(L_K(E))$ denotes the lattice of two-sided ideals of $L_K(E)$, under the usual order given by inclusion, and usual lattice operations given by $+$ and $\cap$.

Theorem 2.8.10 (The Structure Theorem for Ideals) *Let E be an arbitrary graph and K any field. Then the following map is a lattice isomorphism:*

$$\begin{aligned} \varphi : \quad \mathscr{Q}_E \quad &\longrightarrow \quad \mathscr{L}_{id}(L_K(E)) \\ ((H,S),C,P) &\mapsto I(H \cup S^H \cup P_C) \end{aligned}$$

with inverse given by

$$\begin{aligned} \varphi' : \mathscr{L}_{id}(L_K(E)) &\longrightarrow \quad \mathscr{Q}_E \\ I \quad &\mapsto ((H,S),C,P) \end{aligned}$$

where $H = I \cap E^0$, $S = \{v \in B_H \mid v^H \in I\}$, and C and P are as described in Proposition 2.8.5.

Proof We start by showing that $\varphi' \circ \varphi$ is the identity on $\mathscr{Q}_E$. Take $((H,S),C,P) \in \mathscr{Q}_E$, and denote by I its image under φ, that is, $I = I(H \cup S^H \cup P_C)$. We show that $I \cap E^0 = H$.

Clearly, $H \subseteq I \cap E^0 \subseteq I$. To see the reverse containment, consider $I/I(H \cup S^H) = I(\overline{P_C})$, where for any subset $X \subseteq L_K(E)$, $\overline{X}$ denotes the image of X under the epimorphism $\overline{\Psi} : L_K(E) \to L_K(E/(H,S))$ described in Theorems 2.4.12 and 2.4.15. Observe that for all $c \in C$ we have $\overline{c} \in C_{ne}(E/(H,S))$ and that $I/I(H \cup S^H)$ is an ideal of $L_K(E)/I(H \cup S^H)$ contained in $I(P_c(E/(H,S)))$. Concretely, we have

$$I/I(H,S) \cong \bigoplus_{\overline{c} \in \overline{C}} M_{\Lambda_{\overline{c}}}(p_c(x)K[x,x^{-1}]),$$

using the notation of Theorem 2.7.3. We want to see that there are no nonzero idempotents in $I/I(H \cup S^H)$. If e is an idempotent in $I/I(H \cup S^H)$, then the ideal J of $L_K(E)/I(H \cup S^H)$ generated by e is an idempotent ideal, contained in $I/I(H \cup S^H)$. However, by the structure of the ideal generated by $P_c(E/(H,S))$ (see Theorem 2.7.3), the only idempotent ideals of $I(P_c(E/(H,S)))$ are the direct sums of some subset of the ideals $\mathrm{M}_{\Lambda_i}(K[x,x^{-1}])$ appearing in the decomposition of $I(P_c(E/(H,S)))$ given by Theorem 2.7.3. Since all the polynomials p_c, for $c \in C$, are not invertible in $K[x,x^{-1}]$, we conclude that $J = 0$ and therefore that $e = 0$. Hence $I \cap E^0 \subseteq H$ by Corollary 2.4.16(i), and we have shown our claim.

We denote the set $\{v \in B_H \mid v^H \in I\}$ by S'. Then for $v \in S'$ we have that $\overline{v}$ is an idempotent in $I/I(H \cup S^H)$; apply again that this ideal has no nonzero idempotents to get $v^H \in I(H \cup S^H)$. Now apply Corollary 2.4.16 (ii) to obtain that $v \in S$.

By the proof of Proposition 2.8.5 we see that the sets of cycles and of polynomials associated to the ideal $I = I(H \cup S^H) + I(P_C)$ are precisely the sets C and P. Therefore $\varphi' \circ \varphi(\, ((H, S), C, P)\,) = ((H, S), C, P)$.

Now we establish that the composition $\varphi \circ \varphi'$ is the identity on $\mathscr{L}_{id}(L_K(E))$. To this end, consider $I \in \mathscr{L}_{id}(L_K(E))$. Recall from Proposition 2.8.5 that $\varphi'(I) = ((H, S), C, P)$, where $H = I \cap E^0$, $S = \{v \in B_H \mid v^H \in I\}$, and $C \subseteq C_H$ and $P = \{p_c\}_{c \in C}$ satisfy

$$I/I(H \cup S^H) = \bigoplus_{c \in C} I(p_c(\overline{c})).$$

Write $J = \varphi(\varphi'(I)) = I(H \cup S^H) + I(P_C)$ (where $P_C = \{p_c(c) \mid c \in C\}$). Since $I/I(H \cup S^H) = J/I(H \cup S^H)$, we get $I = J$ as desired. By Lemma 2.8.9, $I(H \cup S^H)$ is the largest graded ideal of $L_K(E)$ contained in I.

To finish the proof we check that both isomorphisms preserve the partial orders. First, assume that $((H_1, S_1), C_1, P_1) \leq ((H_2, S_2), C_2, P_2)$. Since $(H_1, S_1) \leq (H_2, S_2)$, we get that $I(H_1 \cup S_1^{H_1}) \subseteq I(H_2 \cup S_2^{H_2})$ by Theorem 2.5.8.

Now we want to show $I((P_1)_{C_1}) \subseteq I(H_2 \cup S_2^{H_2} \cup (P_2)_{C_2})$. Take $c \in C_1$. If $c \in C_2$ then $p_c^{(2)} | p_c^{(1)}$ and so $p_c^{(1)}(c) \in I((P_2)_{C_2})$. If $c \notin C_2$, then since $C_1^0 \subseteq H_2 \cup C_2^0$ we have $c^0 \subseteq H_2$ and so $p_c^1(c) \in I(H_2)$. This shows that φ preserves the order.

In what follows we will prove that the map φ' also preserves the order. So let I and J be in $\mathscr{L}_{id}(L_K(E))$ such that $I \subseteq J$. Again using Proposition 2.8.5 we have that $\varphi'(I) = ((H_1, S_1), C_1, P_1)$ and $\varphi'(J) = ((H_2, S_2), C_2, P_2)$, where H_i, S_i, C_i, P_i, for $i = 1, 2$ are as defined before. Again using Lemma 2.8.9, we have that the largest graded ideal $I(H_1 \cup S_1^{H_1})$ of I is contained in the largest graded ideal $I(H_1 \cup S_2^{H_2})$ of J. Hence, by Theorem 2.5.8, $(H_1, S_1) \leq (H_2, S_2)$.

To finish, we must prove $C_1^0 \subseteq C_2^0 \cup H_2$ and $p_c^{(2)} | p_c^{(1)}$ for every $c \in C_1 \cap C_2$. First, we claim $C_{H_1} \subseteq C_{H_2} \cup H_2$. Consider $c \in C_{H_1}$. By definition, $c^0 \cap H_1 = \emptyset$ and $r(e) \in H_1$ for every exit e of c. If $c^0 \cap H_2 \neq \emptyset$, then we have finished. If $c^0 \cap H_2 = \emptyset$, we get $c \in C_{H_2}$ as $r(e) \in H_2$. Note that $I/I(H_1 \cup S_1^{H_1}) = \bigoplus_{c \in C_1} I(P_c^{(1)}(\overline{c}))$.

Denote by π the canonical homomorphism: $\pi : L_K(E)/I(H_1 \cup S_1^{H_1}) \longrightarrow L_K(E)/I(H_2 \cup S_2^{H_2})$. Recall that

$$I(P_c(E/(H_1, S_1))) = \bigoplus_{c \in C_{H_1}} \mathrm{M}_{\Lambda_c}(K[\overline{c}, \overline{c}^{-1}]) \cong \bigoplus_{c \in C_{H_1}} \mathrm{M}_{\Lambda_c}(K[x, x^{-1}])$$

by Theorem 2.7.3 (where $\overline{c}$ denotes the class of c in $L_K(E)/I(H_1 \cup S_1^{H_1})$), and thus

$$\mathrm{Ker}(\pi) \cap I(P_c(E/(H_1, S_1))) = \bigoplus_{\{c \in C_{H_1} \mid c^0 \subseteq H_2\}} \mathrm{M}_{\Lambda_c}(K[\overline{c}, \overline{c}^{-1}]).$$

Let $\tilde{c}$ denote the class of c in $L_K(E)/I(H_2 \cup S_2^{H_2})$. Then, by the above,

$$\pi\left(I/I(H_1 \cup S_1^{H_1})\right) = \bigoplus_{\{c \in C_1 \mid c^0 \cap H_2 = \emptyset\}} I(p_c^{(1)}(\tilde{c})) \subseteq \pi\left(J/I(H_1 \cup S_1^{H_1})\right) = J/I(H_2 \cup S_2^{H_2}) = \bigoplus_{c \in C_2} I(p_c^{(2)}(\tilde{c})).$$

Therefore we have $\{c \in C_1 \mid c^0 \cap H_2 = \emptyset\} \subseteq C_2$ and thus $c_1^0 \subseteq H_2 \cup C_2^0$. Finally we observe that for every $c \in C_1 \cap C_2$ we have $p_c^{(2)} | p_c^{(1)}$ since $I(p_c^{(1)}(\tilde{c})) \subseteq I(p_c^{(2)}(\tilde{c}))$. This implies $((H_1, S_1), C_1, P_1) \leq ((H_2, S_2), C_2, P_2)$, and thereby establishes the result. □

We note that much of the information contained in the Structure Theorem for Ideals 2.8.10 was, independently, obtained in [132].

As we did with the Structure Theorem for Graded Ideals 2.5.8, we now record the Structure Theorem for Ideals in the case when E is row-finite.

Proposition 2.8.11 *Let E be a row-finite graph and K any field. Then every ideal I of $L_K(E)$ is of the form $I(H \cup P_C)$, where $H = I \cap E^0$, and C and P are as described in Proposition 2.8.5.*

Here is an example of how Theorem 2.8.10 allows us to explicitly describe all the ideals of an important Leavitt path algebra.

Example 2.8.12 Let K be any field, and let E_T be the Toeplitz graph $c \circlearrowright \bullet^u \xrightarrow{f} \bullet^v$. We easily see that

$$\mathcal{H}_{E_T} = \{\emptyset, \{v\}, \{u, v\}\} \quad \text{and} \quad C_u(E_T) = \{c\}.$$

Clearly there are no sets of breaking vertices in E_T. So by the Structure Theorem for Ideals 2.8.10, the complete set of ideals of $L_K(E_T)$ is given by:

$$I(\emptyset) = \{0\}, \quad I(\{v\}), \quad I(\{u, v\}) = L_K(E_T), \quad \text{and}$$

$$\{I(\{v\} \cup \{p(c)\}) \mid p(x) = 1 + k_1 x + \ldots + k_n x^n \in K[x], \text{ with } k_n \neq 0 \text{ and } n \geq 1\}.$$

Remark 2.8.13 Let E be an arbitrary graph and K any field. Then there exist natural embeddings of lattices:

$$\begin{array}{ccccc} \mathcal{H}_E & \longrightarrow & \mathcal{T}_E & \longrightarrow & \mathcal{Q}_E \\ H & \mapsto & (H, \emptyset) & & \\ & & (H, S) & \mapsto & ((H, S), \emptyset, \emptyset)\,. \end{array}$$

We conclude the section by presenting just one general result which follows directly from the explicit description of the lattice of all two-sided ideals of $L_K(E)$ given in the Structure Theorem for Ideals 2.8.10. We will present numerous additional such results in Sect. 2.9. First, we introduce a binary operation $\cdot$ on $\mathcal{Q}_E$, under which $\mathcal{Q}_E$ becomes a commutative monoid.

Definition 2.8.14 Let E be an arbitrary graph and K any field. We define a binary operation $\cdot$ on $\mathcal{Q}_E$ as follows. For any $q_1 = ((H_1, S_1), C_1, P_1)$ and $q_2 = ((H_2, S_2), C_2, P_2) \in \mathcal{Q}_E$, set

$$q_1 \cdot q_2 = ((H_1, S_1) \wedge (H_2, S_2),\ \ C_1 \wedge C_2,\ \ \{p_c^{(1)} p_c^{(2)}\}_{c \in C_1 \wedge C_2}),$$

where

$$C_1 \wedge C_2 = (C_1 \cap C_2) \cup C_1^{H_2} \cup C_2^{H_1},$$

$$\text{with}\quad C_1^{H_2} = \{c \in C_1 \mid c^0 \subseteq H_2\} \quad\text{and}\quad C_2^{H_1} = \{c \in C_2 \mid c^0 \subseteq H_1\}.$$

(We interpret $p_c^{(i)}$ as 1 if $c \notin C_i$ for $i = 1$ or 2.)

Clearly this operation is associative and commutative, and the neutral element is $((E^0, \emptyset), \emptyset, \emptyset)$.

Remark 2.8.15 We note that the set of idempotent elements of $\mathcal{Q}_E$ is precisely $\mathcal{T}_E$.

Using the explicit description of the lattice isomorphism φ given in the proof of the Structure Theorem for Ideals 2.8.10, we get

Proposition 2.8.16 *Let $\varphi : \mathcal{L}_{id}(L_K(E)) \to \mathcal{Q}_E$ be the isomorphism of Theorem 2.8.10, and let I and J be elements of $\mathcal{L}_{id}(L_K(E))$. Then $\varphi(IJ) = \varphi(I) \cdot \varphi(J)$.*

Using Proposition 2.8.16, the fact that the map φ therein is a lattice isomorphism, and the obvious commutativity of the operation $\cdot$ on $\mathcal{Q}_E$, we achieve the following consequence. This result is perhaps surprising, in that $L_K(E)$ is of course in general far from commutative.

Corollary 2.8.17 *Let E be an arbitrary graph and K any field. If I and J are arbitrary ideals of $L_K(E)$, then $IJ = JI$.*

2.9 Additional Consequences of the Structure Theorem for Ideals. The Simplicity Theorem

The Structure Theorem for Ideals 2.8.10 allows us great insight into various ring-theoretic properties of Leavitt path algebras. We record a number of these results in this section.

Consistent with our presentation of various consequences of the Structure Theorem for Graded Ideals, we begin by presenting the (non-graded) versions of results analogous to Proposition 2.5.13 and Corollary 2.5.15, namely, results about the simplicity and two-sided chain conditions of Leavitt path algebras.

Recall that an algebra A is said to be *simple* if $A^2 \neq 0$ and the only two-sided ideals of A are $\{0\}$ and A.

Theorem 2.9.1 (The Simplicity Theorem) *Let E be an arbitrary graph and K any field. Then the Leavitt path algebra $L_K(E)$ is simple if and only if E satisfies the following conditions:*

(i) $\mathscr{H}_E = \{\emptyset, E^0\}$ *(i.e., the only hereditary saturated subsets of E^0 are $\emptyset$ and E^0), and*

(ii) *E satisfies Condition (L) (i.e., every cycle in E has an exit).*

Proof The Structure Theorem for Ideals 2.8.10 provides a lattice isomorphism φ from the lattice $\mathscr{Q}_E$ to the lattice of all two-sided ideals of $L_K(E)$. In particular, we see immediately that if H is a hereditary saturated subset of E^0 not equal to $\emptyset$ or E^0, then $\varphi(((H,\emptyset),\emptyset,\emptyset))$ is a nontrivial ideal of $L_K(E)$. Similarly, if c is a cycle in E without an exit, then $c \in C_\emptyset$ (see Notation 2.8.3), and then $\varphi(((\emptyset,\emptyset),\{c\},1+x))$ gives a nontrivial ideal of $L_K(E)$. Thus the two conditions on E are necessary for the simplicity of $L_K(E)$.

Conversely, suppose E satisfies the two properties. First, as noted subsequent to Definition 2.4.4, we have that both $B_\emptyset = \emptyset$ and $B_{E^0} = \emptyset$. Additionally, $C_{E^0} = \emptyset$, and the hypothesis that every cycle in E has an exit yields that $C_\emptyset = \emptyset$ as well. Thus $\mathscr{Q}_E$ consists precisely of the two elements $((E^0,\emptyset),\emptyset,\emptyset)$ and $((\emptyset,\emptyset),\emptyset,\emptyset)$. The simplicity of $L_K(E)$ now follows from the Structure Theorem for Ideals. □

Example 2.9.2 Consider once again the graphs R_n consisting of one vertex and n loops. Obviously Condition (i) of the Simplicity Theorem is satisfied for R_n. When $n \geq 2$, Condition (ii) is satisfied for R_n as well. Thus $L_K(R_n)$ is simple for $n \geq 2$; i.e., the Leavitt algebra $L_K(1,n)$ is simple for $n \geq 2$. We note that Condition (ii) is not satisfied for the graph R_1, which implies that $L_K(R_1) \cong K[x,x^{-1}]$ is not simple. (Of course this last statement is well known.)

Remark 2.9.3 Note that graphs having infinite emitters may give rise to simple Leavitt path algebras: for example, the graph $R_\mathbb{N}$ having one vertex and countably many loops at that vertex satisfies the conditions of the Simplicity Theorem 2.9.1.

Due to its importance in the general theory of Leavitt path algebras, due to the importance that these attendant ideas and definitions will play later, and due to its historical significance, we offer now a second proof of the Simplicity Theorem.

Definitions 2.9.4 Let E be an arbitrary graph. By an *infinite path* in E we mean a sequence $\gamma = e_1, e_2, \ldots$ for which $r(e_i) = s(e_{i+1})$ for all $i \in \mathbb{N}$. We often denote such γ by $e_1e_2\cdots$. (We note that the terminology *infinite path* is perhaps misleading, but standard: despite its name, an infinite path in E is not an element of $\text{Path}(E)$.) By a *vertex in* an infinite path $\gamma = e_1, e_2, \ldots$ we mean a vertex of the form $s(e_i)$ for some $i \in \mathbb{N}$.

We denote by E^∞ the set of all infinite paths of E, and by $E^{\leq\infty}$ the set E^∞ together with the set of finite paths in E whose range vertex is a singular vertex.

We say that a vertex $v \in E^0$ is *cofinal* if for every $\gamma \in E^{\leq\infty}$ there is a vertex w in the path γ such that $v \geq w$. We say that a graph E is *cofinal* if every vertex in E is cofinal.

If c is a closed path in E, then c gives rise to the infinite path $ccc\cdots$ of E. Thus if E is cofinal, then in particular every vertex of E connects to every cycle in E, and to every sink in E.

Lemma 2.9.5 *Let E be a cofinal graph, and let $v \in E^0$ be a sink.*

(i) The only sink of E is v.
(ii) For every $w \in E^0$, $v \in T(w)$.
(iii) E contains no infinite paths. In particular, E is acyclic.

Proof

(i) is obvious.
(ii) Since $T(v) = \{v\}$, the result follows from the definition of $T(v)$ by considering the path $\gamma = v \in E^{\leq\infty}$.
(iii) If $\alpha \in E^{\infty}$, then there exists a $w \in \alpha^0$ such that $v \geq w$, which is impossible. Thus, in particular, E contains no closed paths. □

Lemma 2.9.6 *A graph E is cofinal if and only if $\mathscr{H}_E = \{\emptyset, E^0\}$.*

Proof Suppose E is cofinal. Let $H \in \mathscr{H}_E$ with $\emptyset \neq H \neq E^0$. We choose and fix $v \in E^0 \setminus H$, and subsequently build a path $\gamma \in E^{\leq\infty}$ such that $\gamma^0 \cap H = \emptyset$, as follows. If $v \in \mathrm{Sing}(E)$, take $\gamma = v$, and we are done. If not, then $v \in \mathrm{Reg}(E)$, so $0 < |s^{-1}(v)| < \infty$ and $r(s^{-1}(v)) \not\subseteq H$ (otherwise, H saturated implies $v \in H$). Hence, there exists an $e_1 \in s^{-1}(v)$ such that $r(e_1) \notin H$. Let $\gamma_1 = e_1$ and repeat this process with $r(e_1)$. Continuing in this way, either we reach a singular vertex, or we have an infinite path γ whose vertices are not in H, as desired. Now consider $w \in H$ (which exists, as $\emptyset \neq H$ by hypothesis). By cofinality, there exists a $z \in \gamma^0$ such that $w \geq z$, and by the hereditariness of H we get $z \in H$, contradicting the construction of γ.

Conversely, suppose that $\mathscr{H}_E = \{\emptyset, E^0\}$. Take $v \in E^0$ and $\gamma \in E^{\leq\infty}$, with $v \notin \gamma^0$ (the case $v \in \gamma^0$ is obvious). By hypothesis the hereditary saturated subset generated by v is E^0, i.e., $E^0 = \bigcup_{n\geq 0} \Lambda_n(v)$ as described in Lemma 2.0.7. Consider m, the minimum n such that $\Lambda_n(v) \cap \gamma^0 \neq \emptyset$, and let $w \in \Lambda_m(v) \cap \gamma^0$. If $m > 0$, then by minimality of m it must be that w is a regular vertex and that $r(s^{-1}(w)) \subseteq \Lambda_{m-1}(v)$. Since w is a regular vertex and $\gamma = (\gamma_n) \in E^{\leq\infty}$, there exists an $i \geq 1$ such that $s(\gamma_i) = w$ and $r(\gamma_i) = w' \in \gamma^0$, the latter meaning that $w' \in r(s^{-1}(w)) \subseteq \Lambda_{m-1}(v)$, contradicting the minimality of m. Therefore $m = 0$ and then $w \in \Lambda_0(v) = T(v)$, as we needed. □

The previous discussion allows us to re-establish the Simplicity Theorem without the need to invoke the full power of the Structure Theorem for Ideals 2.8.10.

Theorem 2.9.7 (The Simplicity Theorem, Revisited) *Let E be an arbitrary graph and K any field. Then the Leavitt path algebra $L_K(E)$ is simple if and only if* *the graph E satisfies the following conditions:*

(i) *The graph E is cofinal, and*
(ii) *E satisfies Condition (L).*

Proof We will use the characterization of cofinality given in Lemma 2.9.6. Suppose first that $L_K(E)$ is simple. By Theorem 2.4.8, $\mathcal{H}_E = \{\emptyset, E^0\}$. On the other hand, if E does not satisfy Condition (L), then there exists a cycle c in E which has no exits. This implies that $I(P_c(E))$ is a nonzero ideal of $L_K(E)$, and so by the simplicity of $L_K(E)$, we must have $I(P_c(E)) = L_K(E)$. But, by Theorem 2.7.3, the algebra $I(P_c(E))$ is not simple. This is a contradiction and, therefore, E must satisfy Condition (L).

Now, suppose that the graph E satisfies Conditions (i) and (ii) in the statement, and let I be a nonzero ideal of $L_K(E)$. By Corollary 2.2.14, $I \cap E^0 \neq \emptyset$. Since $I \cap E^0 \in \mathcal{H}_E$ (by Lemma 2.4.3), the cofinality of E with Lemma 2.9.6 imply $I \cap E^0 = E^0$ or, in other words, $E^0 \subseteq I$. This immediately gives $I = L_K(E)$. □

We now record the two-sided chain condition results for Leavitt path algebras. Since the verifications of these results follow from the Structure Theorem for Ideals, using arguments similar to those presented in Theorem 2.9.1 and Lemma 2.5.12, we omit the proofs. We note, however, that with the Structure Theorem for Ideals in hand, such proofs are significantly shorter than those offered originally in [9, Theorems 3.6 and 3.9].

Proposition 2.9.8 *Let E be an arbitrary graph and K any field.*

(i) *$L_K(E)$ is two-sided artinian if and only if E satisfies Condition (K), $\mathcal{H}_E$ satisfies the descending chain condition with respect to inclusion, and, for each $H \in \mathcal{H}_E$, the set B_H of breaking vertices is finite.*
(ii) *$L_K(E)$ is two-sided noetherian if and only if $\mathcal{H}_E$ satisfies the ascending chain condition with respect to inclusion, and, for each $H \in \mathcal{H}_E$, the set B_H of breaking vertices is finite.*

We comment that, by Proposition 2.5.13(ii), $L_K(E)$ is noetherian if and only if $L_K(E)$ is graded noetherian (as the two graph-theoretic conditions on E are identical). The same cannot be said for the artinian condition: for instance, $K[x, x^{-1}] \cong L_K(R_1)$ is graded artinian, but is well known to not be artinian. In addition, we note that if E does not satisfy Condition (K), then there is some hereditary saturated subset H of E^0 for which the quotient graph E/H contains a cycle without an exit; this is how Condition (K) becomes incorporated into the Structure Theorem for Ideals.

For the next consequence of the Structure Theorem for Ideals, we record the previously promised result regarding a characterization of Condition (K) in terms of the graded ideals of $L_K(E)$.

Proposition 2.9.9 *Let E be an arbitrary graph and K any field. Then every ideal of $L_K(E)$ is graded if and only if E satisfies Condition* (K).

Proof If E satisfies Condition (K), then $C_u(E) = \emptyset$ and so, by the Structure Theorem for Ideals 2.8.10, every ideal of $L_K(E)$ is of the form $I(H \cup S^H)$, and hence is graded.

Conversely, suppose that E does not satisfy Condition (K). Then there exists a cycle c in $C_u(E)$. Let H denote the saturated closure of the tree of the ranges of the exits of c. Then $H \in \mathscr{H}_E$, $c^0 \cap H = \emptyset$, and the range of every exit of c belongs to H. Therefore $c \in C_H$ and so, choosing for example $p(x) = 1 + x \in K[x]$, we have that $\varphi(((H, \emptyset), \{c\}, \{p(x)\})) = I(H \cup \{1 + c\})$ is a nongraded ideal of $L_K(E)$. □

Example 2.9.10 As one specific consequence of Proposition 2.9.9, we conclude that the list of graded ideals of the Leavitt path algebra of the infinite clock graph $C_{\mathbb{N}}$, presented in Example 2.5.10, indeed represents the list of *all* ideals of $L_K(C_{\mathbb{N}})$.

Yet another immediate application of the Structure Theorem for Ideals is the following result, in which we present (among other things) the converse of Corollary 2.5.23 regarding the structure of graded ideals in $L_K(E)$.

Corollary 2.9.11 *Let E be an arbitrary graph and K any field. For an ideal I of the Leavitt path algebra $L_K(E)$, the following are equivalent.*

(1) *I is a graded ideal.*
(2) *I is generated by idempotents.*
(3) $I = I^2$.
(4) *I is K-algebra isomorphic to a Leavitt path K-algebra.*

In particular, by Proposition 2.9.9, E satisfies Condition (K) *if and only if every ideal of $L_K(E)$ is generated by idempotents.*

Proof (1) $\Longrightarrow$ (2) follows by Theorem 2.4.8.
(2) $\Longrightarrow$ (3) is trivial.
(3) $\Longrightarrow$ (1) follows from the observation made in Remark 2.8.15.
(1) $\Longrightarrow$ (4) is Corollary 2.5.23.
(4) $\Longrightarrow$ (3) follows because any Leavitt path algebra has local units (Lemma 1.2.12). □

Corollary 2.9.12 *Let E be an arbitrary graph and K any field. If J is an ideal of a graded ideal I of $L_K(E)$, then J is an ideal of $L_K(E)$.*

Proof Let $a \in L_K(E)$ and $y \in J \subseteq I$. By Corollary 2.9.11(4) and Lemma 1.2.12(v) there exists an $x \in I$ such that $y = xy$. Then $ay = (ax)y \in IJ \subseteq J$. □

We finish Chap. 2 by presenting a result which serves as an appropriate bridge to Chap. 3, in that this result relates an ideal structure property to a property of idempotents. Rings for which every nonzero one-sided ideal contains a nonzero idempotent were studied in [122].

Proposition 2.9.13 *Let E be an arbitrary graph and K any field. The following are equivalent.*

(1) *E satisfies Condition* (L).
(2) *Every nonzero two-sided ideal of $L_K(E)$ contains a vertex.*
(3) *Every nonzero one-sided ideal of $L_K(E)$ contains a nonzero idempotent.*

Proof (1) ⇒ (3). Let a be a nonzero element in a left ideal I of $L_K(E)$. By Condition (L), an application of the Reduction Theorem 2.2.11 gives the existence of $\mu, \nu \in \text{Path}(E)$, $v \in E^0$ and $k \in K^\times$ such that $0 \neq \mu^* a \nu = kv$. Define $e = k^{-1}\nu\mu^* a$. Then $e \in I$, e is nonzero (because $0 \neq v = v^2 = k^{-2}\mu^* a(\nu\mu^* a)\nu$), and e is an idempotent, as $(k^{-1}\nu\mu^* a)(k^{-1}\nu\mu^* a) = k^{-1}\nu v\mu^* a = k^{-1}\nu\mu^* a$. An analogous proof, or an appeal to Corollary 2.0.9, establishes the result for right ideals as well.

(3) ⇒ (1). If E does not satisfy Condition (L), then there exists a cycle without exits c in E. Denote by I the (graded) ideal of $L_K(E)$ generated by the vertices of c. Lemma 2.7.1 implies that I is isomorphic to $\mathrm{M}_\Lambda(K[x, x^{-1}])$ for some set Λ. Since the ideals of I are ideals of $L_K(E)$ by Corollary 2.9.12, the hypothesis implies that every nonzero ideal of $\mathrm{M}_\Lambda(K[x, x^{-1}])$ contains a nonzero idempotent, which is not true. This proves our claim.

An argument similar to the one given in the previous paragraph also establishes (2) ⇒ (1), while (1) ⇒ (2) is Corollary 2.2.14. □

Chapter 3
Idempotents, and Finitely Generated Projective Modules

In this chapter we consider various topics related to the structure of the idempotents in $L_K(E)$. We start with a discussion of the purely infinite simplicity of a Leavitt path algebra, a topic which has fueled much of the investigative effort in the subject. In the subsequent section we analyze the structure of the monoid $\mathscr{V}(L_K(E))$ of isomorphism classes of finitely generated projective modules over $L_K(E)$. This will allow us to more fully describe Bergman's construction (presented earlier in Sect. 1.4), which was essential to the genesis of the subject. In Sect. 3.3 we remind the reader of the definition of an exchange ring, and subsequently show that the exchange Leavitt path algebras are exactly those arising from graphs which satisfy Condition (K). Von Neumann regularity is taken up in Sect. 3.4; in addition to showing that the von Neumann regular Leavitt path algebras are precisely those arising from acyclic graphs, we identify the set of vertices in E which generate the largest von Neumann regular ideal of $L_K(E)$. We continue our discussion of the idempotents in $L_K(E)$ in Sect. 3.5 by identifying the collection of primitive idempotents which are not minimal.

We consider in Sect. 3.6 the monoid-theoretic structure of $\mathscr{V}(L_K(E))$. While the monoid $\mathscr{V}(R)$ for a general ring R necessarily satisfies certain properties (e.g., $\mathscr{V}(R)$ is conical), we will show that when E is a row-finite graph and $R = L_K(E)$ then $\mathscr{V}(R)$ enjoys many additional properties, including refinement and separativity. In the subsequent Sect. 3.7 we consider the *extreme* cycles in a graph, and show that the ideal of $L_K(E)$ generated by the vertices in such cycles may be appropriately viewed as the "purely infinite socle" of $L_K(E)$. We conclude the chapter with Sect. 3.8, in which we remind the reader of the general notion of a purely infinite (but not necessarily simple) ring, and then identify those graphs E for which $L_K(E)$ is purely infinite.

We start by presenting an easily established but fundamental result regarding isomorphisms between various left $L_K(E)$-modules. This result expands on the idea presented in Lemma 2.6.10.

G. Abrams et al., *Leavitt Path Algebras*, Lecture Notes in Mathematics 2191,
DOI 10.1007/978-1-4471-7344-1_3

Proposition 3.0.1 *Let E be an arbitrary graph and K any field. Let $\mu \in \text{Path}(E)$ for which $s(\mu) = v$ and $r(\mu) = w$.*

(i) *There is a direct sum decomposition*

$$L_K(E)v = L_K(E)\mu\mu^* \oplus L_K(E)(v - \mu\mu^*)$$

as left ideals of $L_K(E)$.

(ii) *There is an isomorphism of left $L_K(E)$-modules*

$$L_K(E)w \cong L_K(E)\mu\mu^*.$$

Consequently, there is an isomorphism $L_K(E)v \cong L_K(E)w \oplus T$ for some left ideal T of $L_K(E)$.

Proof (i) Since $\mu\mu^*$ is an idempotent which commutes with v, we have that $v-\mu\mu^*$ is also an idempotent. But $\mu\mu^*(v - \mu\mu^*) = \mu\mu^* - \mu\mu^* = 0 = (v - \mu\mu^*)\mu\mu^*$, which gives easily that $L_K(E)v = L_K(E)\mu\mu^* \oplus L_K(E)(v - \mu\mu^*)$ as left $L_K(E)$-modules. (We note that in general the second summand might be $\{0\}$.)

(ii) We define $\varphi = \rho_{\mu^*} : L_K(E)w \to L_K(E)\mu\mu^*$ to be the 'right multiplication by μ^*' map, so $(rw)\varphi = rw\mu^* = r\mu^*$. The observation that $\mu^*\mu\mu^* = \mu^*$ shows that φ indeed maps into $L_K(E)\mu\mu^*$. Now define $\psi = \rho_\mu : L_K(E)\mu\mu^* \to L_K(E)w$ to be the 'right multiplication by μ' map, so $(r\mu\mu^*)\psi = r\mu\mu^*\mu = r\mu$. Using that $\mu^*\mu = w$ and that $\mu\mu^*\mu = \mu$ shows that φ and ψ are inverses. The second part of the statement now follows from (i). □

3.1 Purely Infinite Simplicity, and the Dichotomy Principle

In Sect. 2.9 we identified the simple Leavitt path algebras. Intuitively speaking, such algebras can be partitioned into two types: those which behave much like full matrix rings over K, and those which behave much like the Leavitt algebras $L_K(1, n)$. The goal of this section is to make this dichotomy precise.

Definitions 3.1.1 (See e.g. [29, Definitions 1.2]) Let R be a ring. An idempotent e in R is said to be *infinite* if there exist orthogonal idempotents $f, g \in R$ such that $e = f + g$, $g \neq 0$, and $Re \cong Rf$ as left R-modules. Rephrased, the idempotent e is infinite if Re is isomorphic to a proper direct summand of itself. In such a situation we say Re is a *directly infinite module*.

Remark 3.1.2 We note that if e is an infinite idempotent in a ring R, then the left R-module Re cannot satisfy either the ascending or the descending chain condition on submodules. In particular, a Noetherian ring contains no infinite idempotents.

Example 3.1.3 In our context, the quintessential example of an infinite idempotent is provided in the Leavitt algebra $R = L_K(R_2) \cong L_K(1,2)$. We show that 1_R is an infinite idempotent. If e, f are the loops based at v in R_2, then by (CK2) we have $v = 1_R = ee^* + ff^*$. By Proposition 3.0.1(i) we get $L_K(R_2) = L_K(R_2)1_R = L_K(R_2)ee^* \oplus L_K(R_2)(v - ee^*) = L_K(R_2)ee^* \oplus L_K(R_2)ff^*$ (where each of the two summands is clearly nonzero), and by Proposition 3.0.1(ii) we have that $L_K(R_2)1_R \cong L_K(R_2)ee^*$. A similar conclusion can be drawn in any of the Leavitt algebras $L_K(1,n)$. (Indeed, we will show in Example 3.2.6 that *every* nonzero idempotent of $L_K(1,n)$ is infinite.)

Remark 3.1.4 Suppose e is an infinite idempotent in a ring R, and suppose that g is an idempotent of R such that $Rg \cong Re \oplus Q$ for some left R-module Q. Then g is infinite as well. This is easy to see, as by hypothesis, $Re \cong Re \oplus P$ for some nonzero left R-module P, so that $Rg \cong Re \oplus Q \cong (Re \oplus P) \oplus Q \cong (Re \oplus Q) \oplus P \cong Rg \oplus P$.

There is a strong connection between infinite idempotents in $L_K(E)$ and cycles having exits in E.

Lemma 3.1.5 *Let E be an arbitrary graph and K any field. Suppose c is a cycle based at w, and suppose e is an exit for c with $s(e) = w$. Then $L_K(E)w = P \oplus Q$, where P and Q are nonzero left ideals of $L_K(E)$, and $L_K(E)w \cong P$ as left $L_K(E)$-modules. In particular, w is an infinite idempotent of $L_K(E)$.*

Proof By Proposition 3.0.1(i), we get a decomposition $L_K(E)w = L_K(E)cc^* \oplus L_K(E)(w - cc^*)$. But since $r(c) = w$, we get by Proposition 3.0.1(ii) that $L_K(E)w \cong L_K(E)cc^*$. Since e is an exit for c we have $c^*e = 0$ (by (CK1)). This yields that $w - cc^* \neq 0$, since, if otherwise $w - cc^* = 0$, then multiplying on the right by e would give $e = 0$ in $L_K(E)$, violating Corollary 1.5.13. Thus $P = L_K(E)cc^*$ and $Q = L_K(E)(w - cc^*)$ give the desired result. □

We now identify those vertices of E which are infinite idempotents of $L_K(E)$.

Proposition 3.1.6 *Let E be an arbitrary graph and K any field. Let $v \in E^0$. Then v is an infinite idempotent in $L_K(E)$ if and only if v connects to a cycle with exits in E.*

Proof Suppose first that v connects to a cycle with exits. Specifically, suppose there exists a cycle c in E with an exit e to which v connects. Let w denote $s(e)$. Since v connects to c, there exists a $\mu \in \text{Path}(E)$ with $s(\mu) = v$ and $r(\mu) = w$. By Proposition 3.0.1(i) we have $L_K(E)v \cong L_K(E)w \oplus T$ for some left ideal T of $L_K(E)$. But $L_K(E)w$ is infinite by Lemma 3.1.5, so that Remark 3.1.4 yields the result.

Conversely, assume that $T(v)$ does not contain any cycle with exits. By Theorem 1.6.10, it suffices to consider the case of a finite graph E. (Observe that if F is a finite complete subgraph of E containing a cycle c which has no exits in E, then c is also a cycle without exits in the graph $F(\text{Reg}(E) \cap \text{Reg}(F))$ built in Definition 1.5.16, because the vertices in c are regular both in E and in F.)

Now, by Corollary 2.7.6, we have

$$I(v) \cong \mathrm{M}_{r_1}(K) \oplus \cdots \oplus \mathrm{M}_{r_k}(K) \oplus \mathrm{M}_{s_1}(K[x, x^{-1}]) \oplus \cdots \oplus \mathrm{M}_{s_\ell}(K[x, x^{-1}]),$$

and by Remark 3.1.2 this ring contains no infinite idempotents. □

We now utilize a result which we will discuss in further detail in Sect. 3.8 below.

Proposition 3.1.7 *Let R be a (not necessarily unital) ring. Then the following are equivalent.*

(1) *For each nonzero $x \in R$ there exist elements $s, t \in R$ such that sxt is an infinite idempotent.*
(2) *Every nonzero one-sided ideal of R contains an infinite idempotent.*

Proof (1) $\Rightarrow$ (2). Let a be a nonzero element of R. By (1) there are $s, t \in R$ such that $e := sat$ is an infinite idempotent. Observe that we can assume that $s = es$ and $t = te$. It then follows that $a(ts)$ is an infinite idempotent in aR, because $(ats)R \cong (sat)R$. The proof for left ideals is similar.

(2) $\Rightarrow$ (1). Let x be a nonzero element in R. Then, for some $t \in R$ we have that $e := xt$ is an infinite idempotent. Hence $e = ext$ is an infinite idempotent. □

Definition 3.1.8 A simple ring R which satisfies the equivalent conditions of Proposition 3.1.7 is called a *purely infinite simple* ring.

Remark 3.1.9 We will show below that for simple unital rings, the conditions of Proposition 3.1.7 are equivalent to: R is not a division ring, and for every nonzero $x \in R$ there exists elements $s, t \in R$ with $sxt = 1_R$. It is of historical importance to note that the proof given by Leavitt of the simplicity of $L_K(1, n)$ for each $n \geq 2$ [113, Theorem 2] in fact demonstrates that $L_K(1, n)$ has this property, and thus is purely infinite simple.

We now have all the tools necessary to characterize the purely infinite simple Leavitt path algebras in terms of properties of the associated graph.

Theorem 3.1.10 (The Purely Infinite Simplicity Theorem) *Let E be an arbitrary graph and K any field. Then the Leavitt path algebra $L_K(E)$ is purely infinite simple if and only if E satisfies the following conditions:*

(i) $\mathscr{H}_E = \{\emptyset, E^0\}$,
(ii) *E satisfies Condition* (L), *and*
(iii) *every vertex in E^0 connects to a cycle.*

Equivalently, (iii) *may be replaced by:*

(iii') *E contains at least one cycle.*

Proof Suppose first that conditions (i), (ii) and (iii) are satisfied. By the Simplicity Theorem 2.9.1, (i) and (ii) together imply that $L_K(E)$ is a simple ring. Note that (ii) and (iii) together give that every vertex connects to a cycle with exits. So by Proposition 3.1.6 we get that all the vertices of E are infinite idempotents in $L_K(E)$. Now let $0 \neq \alpha \in L_K(E)$. Since E satisfies Condition (L), by the Reduction Theorem 2.2.11 there exist $\mu, \kappa \in \text{Path}(E)$ and $k \in K^{\times}$ with $k^{-1}\mu^*\alpha\kappa = v$ for some vertex v. Since v is an infinite idempotent by the previous paragraph, we see from Proposition 3.1.7(1) that $L_K(E)$ is purely infinite.

Conversely, suppose that $L_K(E)$ is purely infinite simple. Again invoking the Simplicity Theorem 2.9.1, the graph E satisfies conditions (i) and (ii) in the statement. Now we will show that condition (iii) holds as well. By Proposition 3.1.6, it suffices to show that every vertex v of E is an infinite idempotent in $L_K(E)$. By hypothesis (using Proposition 3.1.7(2)), the nonzero left ideal $L_K(E)v$ contains an infinite idempotent y; write $y = rv$ for some $r \in L_K(E)$. As y is infinite, necessarily $y \neq 0$. Then, since $rv \cdot rv = rv$, it is easy to show that $x = vrv$ is an idempotent as well; moreover, $x \neq 0$, as otherwise $x = 0$ would give $rx = 0$, which would give $rvrv = rv = 0$, contrary to the choice of $y = rv$. Thus x is a nonzero idempotent in $L_K(E)v$ which commutes with v, and so $L_K(E)v = L_K(E)x \oplus L_K(E)(v - x)$. But $L_K(E)vrv = L_K(E)rv$; the inclusion $\subseteq$ is clear, while $\supseteq$ follows from $rv = rvrv$. Rephrased, $L_K(E)x = L_K(E)y$. Thus $L_K(E)v = L_K(E)y \oplus L_K(E)(v - x)$. As y is infinite, we get that v must be infinite as well, using Remark 3.1.4.

We finish by showing that conditions (iii) and (iii') are equivalent in the presence of conditions (i) and (ii). By Theorem 2.9.7, condition (i) may be replaced by the condition that E is cofinal. In particular, every vertex of E must connect to every cycle of E (as each cycle gives rise to an infinite path in E). So the existence of at least one cycle suffices to give (iii), and conversely. □

With both the Simplicity Theorem 2.9.7 and Purely Infinite Simplicity Theorem 3.1.10 now established, Proposition 2.6.20 immediately yields the following.

Theorem 3.1.11 (The Dichotomy Principle for Simple Leavitt Path Algebras) *Let E be an arbitrary graph and K any field. If $L_K(E)$ is simple, then either $L_K(E)$ is locally matricial or $L_K(E)$ is purely infinite simple.*

Example 3.1.12 Any algebra of the form $M_\Lambda(K)$ (for any set Λ) is an example of a locally matricial simple Leavitt path algebra (see Corollary 2.6.6). Additional such examples exist as well, for instance, let E denote the "doubly infinite line graph"

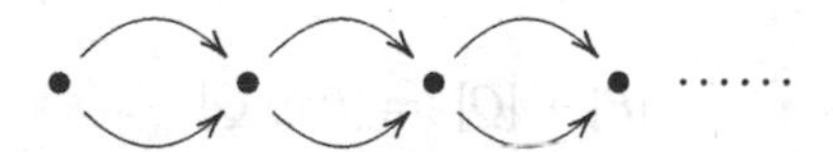

The corresponding Leavitt path algebra $L_K(E)$ is simple, but is not isomorphic to $M_\Lambda(K)$ for any set Λ, as $\mathrm{Soc}(L_K(E)) = \{0\}$ by Theorem 2.6.14.

Remark 3.1.13 We note that as a result of condition (iii) in Theorem 3.1.10, if E is a graph for which $L_K(E)$ is purely infinite simple, then necessarily E contains no sinks.

Indeed, the cofinality condition yields a version of the Dichotomy Principle with respect to graded simplicity.

Proposition 3.1.14 (The Trichotomy Principle for Graded Simple Leavitt Path Algebras) *Let E be an arbitrary graph and K any field. If $L_K(E)$ is graded simple, then exactly one of the following occurs:*

(i) *$L_K(E)$ is locally matricial, or*
(ii) *$L_K(E) \cong M_\Lambda(K[x, x^{-1}])$ for some set Λ, or*
(iii) *$L_K(E)$ is purely infinite simple.*

Proof By Corollary 2.5.15 and Lemma 2.9.6, the graded simplicity of $L_K(E)$ is equivalent to the cofinality of E. The three possibilities given in the statement correspond precisely to whether: (i) E contains no cycles; resp., (ii) contains exactly one cycle; resp., (iii) contains two or more cycles.

If E contains no cycles then (i) follows by Proposition 2.6.20. If E contains at least two cycles then by cofinality each cycle in E must connect to each of the other cycles in E. Consequently, each cycle in E has an exit, and (iii) follows by the Purely Infinite Simplicity Theorem 3.1.10. Now suppose that E contains exactly one cycle c. Then c has no exits (otherwise, if e were an exit for c then by cofinality $r(e)$ would connect to c, and would thus produce a second cycle in E). So $P_c(E)$ is nonempty, which yields that $I(P_c(E))$ is a nonzero (necessarily graded) ideal of $L_K(E)$. But then graded simplicity gives that $L_K(E) = I(P_c(E))$, from which Theorem 2.7.3 yields the desired result. □

3.2 Finitely Generated Projective Modules: The $\mathcal{V}$-Monoid

The goal of this section is to establish Theorem 1.4.3, the fundamental result which was presented (without proof) in the first chapter. This result provided one of the main springboards from which the entire subject of Leavitt path algebras was launched. We restate the result below as Theorem 3.2.5. We recall now the definitions of its two main ingredients.

Definition 3.2.1 Let R be a unital ring. We denote by $\mathcal{V}(R)$ the set of isomorphism classes (denoted using [ ]) of finitely generated projective left R-modules. We endow $\mathcal{V}(R)$ with the structure of a commutative monoid by defining

$$[P] + [Q] := [P \oplus Q]$$

for $[P], [Q] \in \mathcal{V}(R)$.

Suppose more generally that R is a not-necessarily-unital ring. We consider any unital ring S containing R as a two-sided ideal, and denote by $FP(R, S)$ the class of finitely generated projective left S-modules P for which $P = RP$. In this situation, $\mathcal{V}(R)$ is defined as the monoid of isomorphism classes of objects in $FP(R, S)$. This definition of $\mathcal{V}(R)$ does not depend on the particular unital ring S in which R sits as a two-sided ideal, as can be seen from the following alternative description: $\mathcal{V}(R)$ is the set of equivalence classes of idempotents in $\mathrm{M}_{\mathbb{N}}(R)$, where $e \sim f$ in $\mathrm{M}_{\mathbb{N}}(R)$ if and only if there are $x, y \in \mathrm{M}_{\mathbb{N}}(R)$ such that $e = xy$ and $f = yx$. (See [117, p. 296].)

For an idempotent $e \in R$ we will sometimes denote the element $[Re]$ of $\mathcal{V}(R)$ simply by $[e]$.

We note that if R is a ring with local units, then the well-studied *Grothendieck group* $K_0(R)$ of R is the universal group corresponding to the monoid $\mathcal{V}(R)$, see [117, Proposition 0.1]. We will study the Grothendieck group of Leavitt path algebras in great depth throughout Chap. 6.

For any graph E one can associate a monoid M_E; this monoid will play a central role in the topic of Leavitt path algebras. We recall the description of the monoid M_E associated to a graph given in Definition 1.4.2. Specifically, M_E is the free abelian monoid (written additively), having generating set $\{a_v \mid v \in E^0\}$, and with relations given by setting $a_v = \sum_{e \in s^{-1}(v)} a_{r(e)}$ for every $v \in \text{Reg}(E)$. For notational clarity, we often denote the zero element of M_E by z.

Examples 3.2.2 Some examples of the construction of the monoid M_E will be helpful.

(i) As noted in Sect. 1.4, if R_n is the rose with n petals graph ($n \geq 2$), then

$$M_{R_n} = \{z, a_v, 2a_v, \ldots, (n-1)a_v\}, \text{ with relation } na_v = a_v.$$

Although perhaps counterintuitive at first glance, we have that the subset $M_{R_n} \setminus \{z\}$ of M_{R_n} is not only closed under $+$ (and thereby forms a subsemigroup of M_{R_n}), but $M_{R_n} \setminus \{z\}$ is in fact a *group*, isomorphic to $\mathbb{Z}/(n-1)\mathbb{Z}$, with identity element $(n-1)a_v$.

(ii) For the graph R_1 having one vertex v and one loop, we see that M_{R_1} is the monoid $\{z, a_v, 2a_v, \ldots\} \cong \mathbb{Z}^+$.

(iii) For the oriented line graph A_n ($n \geq 1$), M_{A_n} is generated by the n elements $a_{v_1}, a_{v_2}, \ldots, a_{v_n}$, with relations $a_{v_i} = a_{v_{i+1}}$ for $1 \leq i \leq n-1$. Thus $M_{A_n} = \{z, a_{v_n}, 2a_{v_n}, \ldots\} \cong \mathbb{Z}^+$.

(iv) For the Toeplitz graph E_T of Example 1.3.6, M_{E_T} is the free abelian monoid generated by $\{a_u, a_v\}$, modulo the single relation $a_u = a_u + a_v$.

Definition 3.2.3 The category $\mathscr{RG}$ is defined to be the full subcategory of the category $\mathscr{G}$ (given in Definition 1.6.2) whose objects are the pairs $(E, \text{Reg}(E))$, where E is a row-finite graph. We identify the objects of $\mathscr{RG}$ with the row-finite graphs. Note that the morphisms between two objects E and F of $\mathscr{RG}$ are precisely the complete homomorphisms $\psi: E \to F$, that is, the graph homomorphisms $\psi: E \to F$ such that ψ^0 and ψ^1 are injective and such that, for each $v \in \text{Reg}(E)$, the map ψ^1 induces a bijection from $s_E^{-1}(v)$ onto $s_F^{-1}(\psi^0(v))$. The subcategory $\mathscr{RG}$ of $\mathscr{G}$ is closed under direct limits, and the assignment $E \mapsto L_K(E)$ $(= C_K^{\text{Reg}(E)}(E))$ extends to a continuous functor from $\mathscr{RG}$ to the category of K-algebras (cf. Proposition 1.6.4).

Lemma 3.2.4 *The assignment $E \mapsto M_E$ can be extended to a continuous functor from the category $\mathscr{RG}$ of row-finite graphs and complete graph homomorphisms to the category of abelian monoids. Moreover, this assignment commutes with direct limits. It follows that every graph monoid M_E arising from a row-finite graph E is the direct limit of graph monoids corresponding to finite graphs.*

Proof Every complete graph homomorphism $f: E \to F$ induces a natural monoid homomorphism

$$M(f): M_E \to M_F,$$

and so we get a functor M from the category $\mathcal{RG}$ to the category of abelian monoids. The fact that M commutes with direct limits is established in the same way as in Proposition 1.6.4. □

We recall that a unital ring R is called *left hereditary* if every left ideal of R is projective. We are ready to prove Theorem 1.4.3, slightly restated and expanded here.

Theorem 3.2.5 *Let E be a row-finite graph and K any field. Then there is a natural monoid isomorphism $\mathcal{V}(L_K(E)) \cong M_E$. Moreover, if E is finite, then $L_K(E)$ is hereditary.*

Proof Because of the defining relations used to build M_E, for each row-finite graph E there is a unique monoid homomorphism $\gamma_E : M_E \to \mathcal{V}(L_K(E))$ such that $\gamma_E(a_v) = [L_K(E)v]$. Clearly these homomorphisms induce a natural transformation from the functor M to the functor $\mathcal{V} \circ L_K$; that is, if $f : E \to F$ is a complete graph homomorphism, then the following diagram commutes.

$$\begin{array}{ccc} M_E & \xrightarrow{\gamma_E} & \mathcal{V}(L_K(E)) \\ {\scriptstyle M(f)}\downarrow & & \downarrow{\scriptstyle \mathcal{V}(L_K(f))} \\ M_F & \xrightarrow{\gamma_F} & \mathcal{V}(L_K(F)) \end{array}$$

We need to show that γ_E is a monoid isomorphism for every row-finite graph E. By Lemma 3.2.4 and Corollary 1.6.16, we see that it is enough to show that γ_E is an isomorphism for any finite graph E.

So let E be a finite graph, and let $\{v_1, \ldots, v_m\} = \mathrm{Reg}(E)$ (i.e., the non-sinks of E). We start by defining the algebra

$$B_0 = \prod_{v \in E^0} K.$$

In B_0 we clearly have a family $\{p_v : v \in E^0\}$ of orthogonal idempotents such that $\sum_{v \in E^0} p_v = 1$. Now we consider the two finitely generated projective left B_0-modules $P = B_0 p_{v_1}$ and $Q = \bigoplus_{\{e \in E^1 | s(e) = v_1\}} B_0 p_{r(e)}$. By a beautiful (and delicate) construction of Bergman (see [51, p. 38]), there exists an algebra $B_1 := B_0 \langle i, i^{-1} : \overline{P} \cong \overline{Q} \rangle$ which admits a universal isomorphism $i : \overline{P} := B_1 \otimes_{B_0} P \to \overline{Q} := B_1 \otimes_{B_0} Q$. By examining the construction, we see that this algebra is precisely the algebra $L_K(X_1)$, where X_1 is the graph having $X_1^0 = E^0$, and where v_1 emits the same edges as it does in E, but all other vertices do not emit any edges. More explicitly, the row $(x_e : s(e) = v_1)$ implements an isomorphism $\overline{P} = B_1 p_{v_1} \to \overline{Q} = \bigoplus_{\{e \in E^1 | s(e) = v_1\}} B_1 p_{r(e)}$, with inverse given by the column $(y_e : s(e) = v_1)^T$, which is clearly universal. By Bergman [51, Theorem 5.2], the monoid $\mathcal{V}(B_1)$ is obtained from $\mathcal{V}(B_0)$ by adjoining the relation $[P] = [Q]$. Because in our situation we have that $\mathcal{V}(B_0)$ is the free abelian monoid on generators $\{a_v \mid v \in E^0\}$, where $a_v = [p_v]$, we get that $\mathcal{V}(B_1)$ is given by generators $\{a_v \mid v \in E^0\}$ and a single relation

$$a_{v_1} = \sum_{\{e \in E^1 \mid s(e)=v_1\}} a_{r(e)}.$$

Now we proceed inductively. For $n \geq 1$, let B_n be the Leavitt path algebra $B_n = L_K(X_n)$, where X_n is the graph with the same vertices as E, but where only the first n vertices $v_1, \ldots, v_n$ emit edges, and these vertices emit the same edges as they do in E. We assume by induction that $\mathcal{V}(B_n)$ is the abelian monoid given by generators $\{a_v \mid v \in E^0\}$ and relations

$$a_{v_i} = \sum_{\{e \in E^1 \mid s(e)=v_i\}} a_{r(e)},$$

for $i = 1, \ldots, n$. Let X_{n+1} be the analogous graph, corresponding to vertices $v_1, \ldots, v_n, v_{n+1}$. Then we have $B_{n+1} = B_n\langle i, i^{-1} : \overline{P} \cong \overline{Q}\rangle$ for $P = B_n p_{v_{n+1}}$ and $Q = \bigoplus_{\{e \in E^1 \mid s(e)=v_{n+1}\}} B_n p_{r(e)}$, and so we can again apply [51, Theorem 5.2] to deduce that $\mathcal{V}(B_{n+1})$ is the monoid with the same generators as before, and with relations corresponding to those given in the displayed equations. This establishes the desired isomorphism of monoids.

It follows from a related result of Bergman [51, Theorem 6.2] that the global dimension of $L_K(E)$ is at most 1, i.e., that $L_K(E)$ is hereditary. □

Example 3.2.6 By Theorem 3.2.5 and Examples 3.2.2(i), we see that, for $n \geq 2$,

$$\mathcal{V}(L_K(R_n)) \cong \{z, a_v, 2a_v, \ldots, (n-1)a_v\}, \text{ with relation } na_v = a_v.$$

In particular, $\mathcal{V}(L_K(R_n)) \setminus \{z\}$ is isomorphic to the group $\mathbb{Z}/(n-1)\mathbb{Z}$ (with neutral element $(n-1)a_v$). We note that this conclusion regarding the explicit description of the $\mathcal{V}$-monoid of the Leavitt algebras $L_K(1,n) \cong L_K(R_n)$ is quite non-trivial; we do not know of a "direct" or "first principles" proof of this statement.

Further, this property implies that every nonzero finitely generated projective module over $L_K(1,n)$ is necessarily infinite, as the regular module $L_K(1,n)$ itself is infinite.

Of course we may also apply Theorem 3.2.5 to the graphs R_1 and A_n to get the well-known facts that the $\mathcal{V}$-monoid of each of the algebras $L_K(R_1) \cong K[x, x^{-1}]$ and $L_K(A_n) \cong \mathrm{M}_n(K)$ is isomorphic to $\mathbb{Z}^+$.

Examples 3.2.7 Let E denote the following graph.

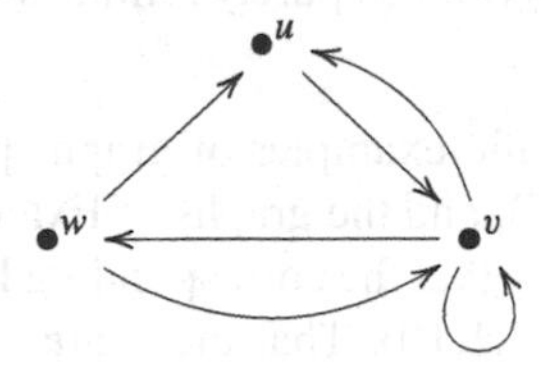

Then M_E is the monoid generated by $\{a_u, a_v, a_w\}$, modulo the relations $a_u = a_v$; $a_v = a_u + a_v + a_w$; and $a_w = a_u + a_v$. By some tedious computations, it is not hard to show that $M_E = \{z, a_u, 2a_u, 3a_u\}$. (We will give a streamlined approach to the computation of M_E in Sect. 6.3.) We note that, as was the case with the M_{R_n} examples ($n \geq 2$), this monoid M_E has the property that $M_E \setminus \{z\}$ is a group (isomorphic to $\mathbb{Z}/3\mathbb{Z}$).

Below are some additional examples of the descriptions of the $\mathscr{V}$-monoids of the Leavitt path algebras of various graphs. For each of these graphs, the Leavitt path algebra is purely infinite simple by Theorem 3.1.10. Thus, as one consequence of these examples, we see that there are many purely infinite simple Leavitt path algebras which are not isomorphic to the classical Leavitt algebras $L_K(1, n)$, because the nonzero elements of the $\mathscr{V}$-monoid of these algebras is not isomorphic to a finite cyclic group (see Example 3.2.6). (For each of these, we will also give a streamlined approach to the computation of the associated graph monoid M in Sect. 6.3.)

First, let F be the graph

Then $L_K(F)$ is (unital) purely infinite simple by Theorem 3.1.10, and $\mathscr{V}(L_K(F)) \setminus \{z\} \cong \mathbb{Z}$.

Next, let G be the following graph.

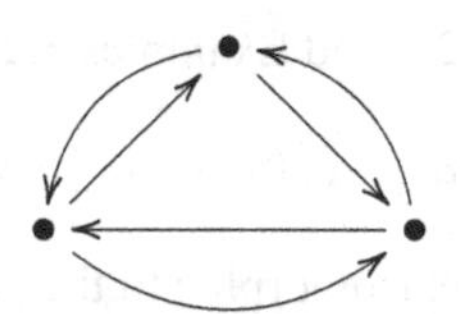

Then $L_K(G)$ is (unital) purely infinite simple, and $\mathscr{V}(L_K(G)) \setminus \{z\} \cong (\mathbb{Z}/2\mathbb{Z}) \oplus (\mathbb{Z}/2\mathbb{Z})$.

A final example is that associated with the graph H

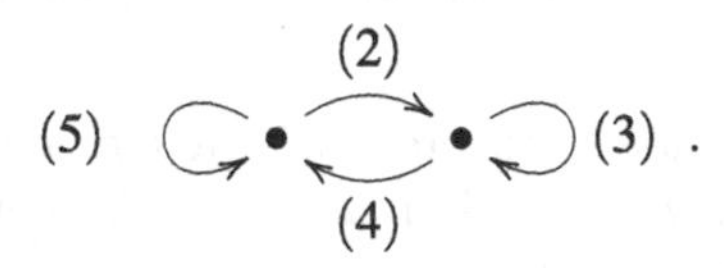

(Recall that the notation $\bullet^v \xrightarrow{(n)} \bullet^w$ indicates that $|\{e \in E^1 \mid s(e) = v, r(e) = w\}| = n$.) Here again $L_K(H)$ is (unital) purely infinite simple, and $\mathscr{V}(L_K(H)) \setminus \{z\} \cong \mathbb{Z} \oplus (\mathbb{Z}/2\mathbb{Z})$.

Remark 3.2.8 Of all the specific examples of graphs presented in this section, the R_n graphs of Examples 3.2.2(i), and the graphs of Examples 3.2.7, are precisely the graphs which have the property that the corresponding Leavitt path algebra is purely infinite simple (by Theorem 3.1.10). That these are also precisely the graphs for which $M_E \setminus \{z\}$ is a group is not coincidental, as we will show in Proposition 6.1.12 below.

Recall the category $\mathscr{G}$ presented in Definition 1.6.2, whose objects are the pairs (E,X), where E is a directed graph and X is a subset of $\mathrm{Reg}(E)$.

We now describe the monoid M_E corresponding to an arbitrary graph E. Indeed, we do more than this: we describe the monoid corresponding to any object (E,X) in the category $\mathscr{G}$ investigated in Chap. 1. As the reader likely has guessed, this assignment will be extended to a continuous functor from $\mathscr{G}$ to the category of abelian monoids. (A complete treatment in the more general framework of separated graphs appears in [27].)

Definition 3.2.9 Let (E,X) be an object of the category $\mathscr{G}$. We define the *graph monoid* $M(E,X)$ as the abelian monoid given by the set of generators

$$E^0 \sqcup \{q'_Z \mid Z \subseteq s^{-1}(v),\ v \in E^0, 0 < |Z| < \infty\},$$

modulo relations we now describe. First, for notational convenience, we denote, for each finite subset Y of E^1,

$$\mathbf{r}(Y) := \sum_{e \in Y} r(e).$$

Now impose on the indicated generators the following relations:

(i) $v = \mathbf{r}(Z) + q'_Z$ for every $v \in E^0$ and $Z \subseteq s^{-1}(v)$ for which $0 < |Z| < \infty$,
(ii) $q'_{Z_1} = \mathbf{r}(Z_2 \setminus Z_1) + q'_{Z_2}$ for every $v \in E^0$ and every pair of finite nonempty subsets Z_1 and Z_2 of $s^{-1}(v)$ for which $Z_1 \subsetneqq Z_2$, and
(iii) $q'_Z = 0$ for $Z = s^{-1}(v)$ when $v \in X$.

Informally, the elements q'_Z of $M(E,X)$ are intended to represent the equivalence classes of the idempotents $v - \sum_{e \in Z} ee^*$ in $C_K^X(E)$, for Z a finite nonempty subset of $s^{-1}(v)$, $v \in E^0$.

Clearly we see that $M(E,\mathrm{Reg}(E)) = M_E$ when E is a row-finite graph, so these monoids $M(E,X)$ generalize the monoids M_E defined above for row-finite graphs.

In order to simplify the notation, we will denote elements in the monoid $M(E,X)$ corresponding to vertices $v \in E^0$ simply by using the same symbol v. Of course, these correspond to the elements denoted by a_v in the monoid $M_E = M(E,\mathrm{Reg}(E))$. Due to the various descriptions of the generators of $M(E,X)$, we think this simplification will be helpful for the reader.

There is some redundancy among these generators and relations. In particular, we could omit the generators q'_Z for nonempty proper subsets Z of $s^{-1}(v)$ for $v \in \mathrm{Reg}(E)$, since relation (ii) gives q'_Z in terms of $q'_{s^{-1}(v)}$, and relation (i) for Z follows from the corresponding relation for $s^{-1}(v)$ in light of (ii). In general, (i) may be viewed as a form of (ii) with $Z_1 = \emptyset$, except that the notation $q'_\emptyset$ would not be well-defined.

Taking into account these comments, an alternative definition of the monoid $M(E,X)$ is as follows: the monoid $M(E,X)$ is the abelian monoid given by the set of generators

$$E^0 \sqcup \{q_v \mid v \in \mathrm{Reg}(E) \setminus X\} \sqcup \{q'_Z \mid Z \subseteq s^{-1}(v),\ v \in \mathrm{Inf}(E),\ 0 < |Z| < \infty\}$$

and the following relations:

(i′) $v = \mathbf{r}(Z) + q'_Z$ for $v \in \mathrm{Inf}(E), Z \subseteq s^{-1}(v)$, and $0 < |Z| < \infty$,
(ii′) $q'_{Z_1} = \mathbf{r}(Z_2 \setminus Z_1) + q'_{Z_2}$ for finite nonempty subsets Z_1 and Z_2 of $s^{-1}(v)$, $v \in \mathrm{Inf}(E)$, with $Z_1 \subsetneq Z_2$,
(iii′) $v = \mathbf{r}(s^{-1}(v))$ for each $v \in X$, and
(iv′) $v = \mathbf{r}(s^{-1}(v)) + q_v$ for each $v \in \mathrm{Reg}(E) \setminus X$.

Informally, the elements q_v for $v \in \mathrm{Reg}(E) \setminus X$ are intended to represent the equivalence classes of the idempotents $v - \sum_{e \in s^{-1}(v)} ee^*$ in $C^X_K(E)$, and correspond to the elements $q_{s^{-1}(v)}$ in the above notation.

Although this alternate definition might seem intuitively clearer, the reason to work instead with the first definition becomes apparent when we look for the natural definition of the morphism associated to a map in $\mathscr{G}$. Consider a morphism $\phi : (F,Y) \to (E,X)$ in $\mathscr{G}$. There is a unique monoid homomorphism $M(\phi) : M(F,Y) \to M(E,X)$ sending $v \mapsto \phi^0(v)$ for $v \in F^0$, and sending $q'_Z \mapsto q'_{\phi^1(Z)}$ for nonempty finite sets $Z \subseteq s^{-1}(v)$, $v \in E^0$. The latter assignments are well-defined because if Z is a nonempty finite subset of $s^{-1}(v)$ for some $v \in E^0$, then $\phi^1(Z)$ is a nonempty finite subset of $s^{-1}(\phi^0(v))$. Moreover, the conditions (2) and (3) in Definition 1.6.2 make clear that relation (iii) above is preserved by $M(\phi)$. The assignments $(E,X) \mapsto M(E,X)$ and $\phi \mapsto M(\phi)$ define a functor M from $\mathscr{G}$ to the category of abelian monoids. It is easily checked (just as for the functor C^X_K in Proposition 1.6.4) that M is continuous.

We denote by **Mon** the category of abelian monoids.

Theorem 3.2.10 *Let E be an arbitrary graph and K any field. Let $\mathscr{G}$ be the category presented in Definition 1.6.2. For each object (E,X) of $\mathscr{G}$, define*

$$\Gamma(E,X)\colon M(E,X) \to \mathscr{V}(C^X_K(E))$$

to be the monoid homomorphism sending $v \mapsto [v]$ for $v \in E^0$, and, for each $w \in X$, $q'_Z \mapsto [w - \sum_{e \in Z} ee^*]$ for each finite nonempty subset $Z \subseteq s^{-1}(w)$. Then $\Gamma : M \to \mathscr{V} \circ C_K$ is an isomorphism of functors $\mathscr{G} \to$ **Mon**.

Proof It is easily seen that the maps $\Gamma(E,X)$ are well-defined monoid homomorphisms, and that Γ defines a natural transformation from M to $\mathscr{V} \circ C_K$.

We have observed that M is continuous, as is $\mathscr{V} \circ C_K$ (by taking into account that $\mathscr{V}$ is continuous, and invoking Proposition 1.6.4). Thus, by Theorem 1.6.10, we see that it is sufficient to show that $\Gamma(E,X)$ is an isomorphism in the case where E is a finite graph.

We use induction on $|\mathrm{Reg}(E)|$ (i.e., the number of non-sinks in E) to establish the result for finite objects (E, X) in $\mathscr{G}$. The result is trivial if $|\mathrm{Reg}(E)| = 0$ (i.e., if there are no edges in E). Assume that $\Gamma(F, Y)$ is an isomorphism for all finite objects (F, Y) of $\mathscr{G}$ for which $|\mathrm{Reg}(F)| \leq n - 1$ for some $n \geq 1$, and let (E, X) be a finite object in $\mathscr{G}$ such that $|\mathrm{Reg}(E)| = n$. Select $v \in E^0$ such that $s^{-1}(v) \neq \emptyset$. We can apply induction to the object (F, Y) obtained from (E, X) by deleting all the edges in $s^{-1}(v)$, and leaving intact the structure corresponding to the remaining vertices (keeping $F^0 = E^0$).

Assume first that $v \in X$. Then $M(E, X)$ is obtained from $M(F, Y)$ by factoring out the relation $v = \mathbf{r}(s^{-1}(v))$. On the other hand, the algebra $C_K^X(E)$ is the Bergman algebra obtained from $C_K^Y(F)$ by adjoining a universal isomorphism between the pair of finitely generated projective modules $C_K^Y(F)v$ and $\bigoplus_{e \in s^{-1}(v)} C_K^Y(F)r(e)$. Accordingly, it follows from [51, Theorem 5.2] that $\mathscr{V}(C_K^X(E))$ is the quotient of $\mathscr{V}(C_K^Y(F))$ modulo the relation $[v] = [\mathbf{r}(s^{-1}(v))]$. Since $\Gamma(F, Y) : M(F, Y) \to \mathscr{V}(C_K^Y(F))$ is an isomorphism by the induction hypothesis, we obtain that $\Gamma(E, X)$ is an isomorphism in this case. (The proof in this case is indeed similar to the proof of Theorem 3.2.5.)

Assume now that $v \notin X$. In this case, $M(E, X)$ is obtained from $M(F, Y)$ by adjoining a new generator q_v and factoring out the relation $v = \mathbf{r}(s^{-1}(v)) + q_v$. On the K-algebra side, we shall make use of another of Bergman's constructions, namely "the creation of idempotents". Write $s^{-1}(v) = \{e_1, \dots, e_m\}$. Let R be the algebra obtained from $C_K^Y(F)$ by adjoining $m + 1$ pairwise orthogonal idempotents $g_1, \dots, g_m, q'_v$ with

$$v = g_1 + \cdots + g_m + q'_v.$$

It follows from [51, Theorem 5.1] that $\mathscr{V}(R)$ is the monoid obtained from $\mathscr{V}(C_K^Y(F))$ by adjoining $m + 1$ new generators $z_1, \dots, z_m, q''_v$, and factoring out the relation $[v] = \sum_{j=1}^m z_j + q''_v$.

It is then clear that $C_K^X(E)$ is isomorphic to the Bergman algebra obtained from R by consecutively adjoining universal isomorphisms between the left modules generated by the idempotents $r(e_i)$ and g_i, for $i = 1, \dots, m$. It follows that $\mathscr{V}(C_K^X(E))$ is the monoid obtained from $\mathscr{V}(C_K^Y(F))$ by adjoining a new generator q''_v and factoring out the relation $[v] = [\mathbf{r}(s^{-1}(v))] + q''_v$. Therefore, applying the induction hypothesis to (F, Y), we again conclude that $\Gamma(E, X)$ is an isomorphism. □

We can now obtain the description of $\mathscr{V}(L_K(E))$ for an arbitrary graph E. To match the notation utilized in the row-finite case, we set $M_E := M(E, \mathrm{Reg}(E))$. From Definition 3.2.9 we see that M_E is the abelian monoid given by the set of generators

$$E^0 \sqcup \{q'_Z \mid Z \subseteq s^{-1}(v),\ v \in \mathrm{Inf}(E),\ 0 < |Z| < \infty\},$$

and the following relations:

(i) $v = \mathbf{r}(Z) + q'_Z$ for $v \in \mathrm{Inf}(E)$, $Z \subseteq s^{-1}(v)$, and $0 < |Z| < \infty$,
(ii) $q'_{Z_1} = \mathbf{r}(Z_2 \setminus Z_1) + q'_{Z_2}$ for finite nonempty subsets Z_1 and Z_2 of $s^{-1}(v)$, $v \in \mathrm{Inf}(E)$, with $Z_1 \subsetneqq Z_2$, and
(iii) $v = \mathbf{r}(s^{-1}(v))$ for each $v \in \mathrm{Reg}(E)$.

Consequently, Theorem 3.2.10 yields the following.

Corollary 3.2.11 *Let E be an arbitrary graph and K any field. Then $\mathscr{V}(L_K(E)) \cong M_E$.*

3.3 The Exchange Property

Our next excursion into the idempotent structure of Leavitt path algebras brings us to the notion of an *exchange ring*. The exchange property for modules was introduced by Crawley and Jónsson in [67]. Roughly speaking, it is the suitable condition which yields a version of the Krull–Schmidt Theorem even in situations where the modules do not decompose as direct sums of indecomposables. Following [154], the (unital) ring R is an *exchange ring* if ${}_RR$ has the property that for every left R-module M and any two decompositions of M as $M = M' \oplus N$ and $M = \bigoplus_{i=1}^n M_i$, for which $M' \cong {}_RR$, then there exist submodules $M'_i \subseteq M_i$ such that $M = M' \oplus \left(\bigoplus_{i=1}^n M'_i\right)$.

A multiplicative characterization of unital exchange rings was obtained independently by Goodearl [88] and by Nicholson [121]. Concretely, R is an exchange ring if and only if for every element $a \in R$ there exists an idempotent $e \in R$ such that $e \in Ra$ and $1 - e \in R(1 - a)$. The appropriate generalization of the notion of exchange ring to not-necessarily-unital rings was provided in [15]: R is exchange if there is a unital ring S containing R as an ideal, for which, for every $x \in R$, there exists an idempotent $e \in R$ for which $e - x \in S(x - x^2)$.

Many classes of rings are exchange rings. In the following three results we identify how the exchange property plays out for the three primary colors of Leavitt path algebras.

Because the exchange property in a ring can be formulated as the existence of a solution to a specific type of equation in the ring, and because it is easy to show that any finite-dimensional matrix algebra $\mathrm{M}_n(K)$ is an exchange ring, we get the following.

Proposition 3.3.1 *The direct limit of exchange rings is an exchange ring. In particular, let K be a field. Then any locally matricial K-algebra is an exchange ring. Specifically, $\mathrm{M}_\Lambda(K)$ is an exchange ring for any set Λ.*

In the current context, the most important class of exchange rings is the following.

Theorem 3.3.2 ([17, Corollary 1.2]) *Let R be a purely infinite simple ring. Then R is an exchange ring.*

On the other hand, the K-algebra $R = K[x, x^{-1}]$ is not an exchange ring, as follows. Since the only idempotents in R are 0 and 1, and $a = 1 + x + x^2$ is not invertible in R, and $1-a = -x-x^2$ is also not invertible in R, the exchange condition fails for the element a. More generally,

Proposition 3.3.3 *For any field K, and for any set Λ, the matrix algebra $\mathrm{M}_\Lambda(K[x, x^{-1}])$ is not an exchange ring.*

We will need the following additional property of exchange rings (which we state here in less than its full generality).

Theorem 3.3.4 ([15, Lemma 3.1(a) and Theorem 2.2]) *Let R be a ring and let I be an ideal of R. Then R is an exchange ring if and only if I and R/I are exchange rings, and the natural map $\mathscr{V}(R) \mapsto \mathscr{V}(R/I)$ is surjective.*

Having given this background information, we now focus on our goal of identifying those Leavitt path algebras $L_K(E)$ which are exchange rings. Recall that for $X \subseteq E^0$, we denote by $\overline{X}$ the hereditary saturated closure of X.

Proposition 3.3.5 *Let E be a graph and suppose that c is a cycle with exits such that, for every $v \in c^0$, c is the only cycle based at v. Let $v \in c^0$, and consider the set*

$$X = \{w \in E^0 \mid v \geq w \text{ and } w \ngeq v\}.$$

Then X is a hereditary subset of E^0 and $H := \overline{X}$ is a hereditary saturated subset of E^0 for which $c^0 \cap H = \emptyset$. In particular, c is a cycle without exits in the quotient graph E/H.

Proof Clearly X is a hereditary subset of E^0 with $X \cap c^0 = \emptyset$. Since the hypotheses yield that no vertex in $\overline{X} \setminus X$ can be contained in a cycle, we see that $\overline{X} \cap c^0 = \emptyset$ as well. □

Lemma 3.3.6 *Let E be an arbitrary graph. If E does not satisfy Condition* (K)*, then there exists a hereditary saturated subset H in E^0 such that E/H does not satisfy Condition* (L).

Proof Since E does not satisfy Condition (K), there exists a $u \in E^0$ which is the base of a unique closed simple path, hence of a unique cycle; denote it by c. As in Proposition 3.3.5, the hereditary set $X = \{w \in E^0 \mid v \geq w, w \ngeq v\}$ has the property that $\overline{X} \cap c^0 = \emptyset$. Set $H := \overline{X}$. Then c is a cycle without exits in E/H, so that E/H does not satisfy Condition (L). □

Lemma 3.3.7 *Let E be an arbitrary graph and K any field. If $L_K(E)$ is an exchange ring, then E satisfies Condition (L).*

Proof Suppose on the contrary that E does not satisfy Condition (L). Then there exists a cycle c in E which has no exits. Denote by I the ideal of $L_K(E)$ generated by c^0. Then Lemma 2.7.1 gives that I is isomorphic to $\mathrm{M}_\Lambda(K[x, x^{-1}])$ for some set Λ, which is not an exchange ring by Proposition 3.3.3. But every ideal of an exchange ring is exchange (Theorem 3.3.4), so I must be exchange, a contradiction. □

Lemma 3.3.7, together with relationships between Condition (K) and Condition (L), will help us reach the main goal in this section, namely, to show that the exchange Leavitt path algebras are precisely those arising from graphs having Condition (K). One of the fundamental steps in the proof of this result is the following graph theoretic property.

Lemma 3.3.8 *Let E be a graph satisfying Condition (K), and let X be a finite subgraph of E. Then there is a finite complete subgraph F of E, containing X, such that F satisfies Condition (K).*

Proof By Theorem 1.6.10 there is a finite complete subgraph G of E such that $X \subseteq G$. The goal is to embed G in a finite complete subgraph F of E such that F satisfies Condition (K). Let $\sim_E$ be the symmetric closure of the relation $\geq$ on E^0: that is, for $v, w \in E^0$, $v \sim_E w$ if either $v = w$, or there is a closed path in E containing both v and w.

We claim that if $v \sim_E w$ then $|CSP_E(v)| > 1$ if and only if $|CSP_E(w)| > 1$. Indeed, it suffices to show one of the implications. Assume that $|CSP_E(v)| > 1$ and that $v \neq w$ and $v \sim_E w$. Since $v \sim_E w$, one can easily show that there is a closed simple path $e_1 e_2 \cdots e_n \in CSP_E(v)$ such that $s(e_i) = w$ for exactly one i with $1 < i \leq n$. By hypothesis, there is a distinct path $\gamma = f_1 f_2 \cdots f_m$ in $CSP_E(v)$. If γ^0 does not contain w, then $e_i e_{i+1} \cdots e_n e_1 e_2 \cdots e_{i-1}$ and $e_i e_{i+1} \cdots e_n \gamma e_1 e_2 \cdots e_{i-1}$ are distinct elements of $CSP_E(w)$. If γ^0 contains w, and $e_1 e_2 \cdots e_{i-1} \neq f_1 f_2 \cdots f_{i-1}$, then taking j such that $s(f_j) = w$, we obtain that $e_i e_{i+1} \cdots e_n f_1 f_2 \cdots f_{j-1}$ and $e_i e_{i+1} \cdots e_n e_1 e_2 \cdots e_{i-1}$ are distinct elements of $CSP_E(w)$. Similarly, if γ^0 contains w, $f_{m-(n-i)} \cdots f_m \neq e_i \cdots e_n$, and j is as above, then $f_j \cdots f_m e_1 \cdots e_{i-1}$ and $e_i \cdots e_n e_1 \cdots e_{i-1}$ are distinct elements of $CSP_E(w)$. Finally, if both $e_1 e_2 \cdots e_{i-1} = f_1 f_2 \cdots f_{i-1}$ and $f_{m-(n-i)} \cdots f_m = e_i \cdots e_n$, then $e_i e_{i+1} \cdots e_n e_1 e_2 \cdots e_{i-1}$ and $f_i f_{i+1} \cdots f_{j-1}$ are two different elements of $CSP_E(w)$, where j is the first index for which $j > i$ and $s(f_j) = w$. This establishes the claim.

There is a finite number of cycles $c_1, \ldots, c_r$ in G, based at $v_1, \ldots, v_r$ respectively, for which $|CSP_G(v_i)| = 1$ for all i. We form a new graph G' by adding to G the vertices and edges in a closed simple path $\gamma_i \neq c_i$ based at v_i, for $i = 1, \ldots, r$. Let F be the completion of G' in E, so that F is formed by adding the edges departing from vertices $v \in (G')^0$ such that $v \in \mathrm{Reg}(E)$ and $s_{G'}^{-1}(v) \neq \emptyset$, together with the corresponding range vertices (in case these edges were not already in G').

We show now that F satisfies Condition (K). First, we see that for $v \in (G')^0$, either $|CSP_{G'}(v)| \geq 2$ or $|CSP_{G'}(v)| = 0$, as follows. If $v \in G^0$ and $|CSP_G(v)| = 1$ then $v \in \cup_{i=1}^r c_i^0$ and thus $|CSP_{G'}(v)| \geq 2$. If $v \in \gamma_i^0$ for some i then $v \sim_{G'} v_i$ and so $|CSP_{G'}(v)| \geq 2$, because $|CSP_{G'}(v_i)| \geq 2$, using the observation above. Finally, if $v \in G^0$, $|CSP_G(v)| = 0$ and $|CSP_{G'}(v)| \neq 0$, then $v \sim_{G'} v_i$ for some i, and so $|CSP_{G'}(v)| \geq 2$.

Since all vertices in $F^0 \setminus (G')^0$ are sinks in F, it therefore suffices to show that $|CSP_F(w)| \neq 1$ for all $w \in (G')^0$ having $|CSP_{G'}(w)| = 0$. Suppose that there is a cycle $c = e_1 e_2 \cdots e_m$ based at w in F and that $|CSP_{G'}(w)| = 0$. If $w \notin G^0$, then $w \in \gamma_i^0$ for some i, and so $|CSP_{G'}(w)| \geq 2$, because $w \sim_{G'} v_i$. Therefore, $w \in G^0$. Let p be the smallest index with $e_p \notin G^1$. Then we have $s(e_p) \in G^0$. Since G is complete, the vertex $s(e_p)$ is a sink in G, and is not a sink in G'. It follows that $s(e_p) \in \gamma_i^0$ for some i, and so $|CSP_{G'}(s(e_p))| \geq 2$ as before. Hence

$$|CSP_F(s(e_p))| \geq |CSP_{G'}(s(e_p))| \geq 2.$$

Since $w \sim_F s(e_p)$, we get that $|CSP_F(w)| \geq 2$, as desired. □

Lemma 3.3.9 *Let E be a graph and K a field for which the ideal lattice $\mathscr{L}_{id}(L_K(E))$ of $L_K(E)$ is finite. Then E satisfies Condition* (K).

Proof By Lemma 3.3.6, it suffices to show that the quotient graph E/H satisfies Condition (L) for every $H \in \mathscr{H}_E$. Suppose on the contrary that there exists a hereditary saturated subset H of E^0 such that E/H does not satisfy Condition (L). This means that E/H contains a cycle without exits, say c. Since $L_K(E/H) \cong L_K(E)/I(H \cup B_H^H)$ (see Theorem 2.4.15) thus has a finite number of ideals, we may assume that $H = \emptyset$.

Denote by I the ideal of $L_K(E)$ generated by c^0. By Lemma 2.7.1 the ideal I is isomorphic to $\mathrm{M}_\Lambda(K[x,x^{-1}])$ for some set Λ, so that I has infinitely many ideals. Since I is a graded ideal, the ideals of I are also ideals of $L_K(E)$ (by Lemma 2.9.12), so $L_K(E)$ has infinitely many ideals, a contradiction. □

We note that the converse of Lemma 3.3.9 is clearly not true, with any graph having infinitely many vertices and no edges providing a counterexample.

Although at first glance the following result might seem to be quite limited in its scope, it will indeed provide the basis of the key theorem of this section.

Proposition 3.3.10 *Let E be a row-finite graph for which the ideal lattice $\mathscr{L}_{id}(L_K(E))$ is finite. Then $L_K(E)$ is an exchange ring.*

Proof Observe first that Lemma 3.3.9 implies that the graph E satisfies Condition (K). Since $\mathscr{L}_{id}(L_K(E))$ is finite, we can build an ascending chain of ideals

$$0 = I_0 \subseteq I_1 \subseteq \cdots \subseteq I_n = L_K(E)$$

such that, for every $i \in \{1, \ldots, n-1\}$, the ideal I_i is maximal among the ideals of $L_K(E)$ contained in I_{i+1}. Now we prove the result by induction on n.

If $n = 1$, then $L_K(E)$ is a simple ring. By the Dichotomy Principle 3.1.11, $L_K(E)$ is either locally matricial or purely infinite simple. But then Proposition 3.3.1 together with Theorem 3.3.2 imply that $L_K(E)$ is an exchange ring.

Now suppose the result holds for any Leavitt path algebra in which there are a finite number of ideals, and a maximal chain of two-sided ideals has length $k < n$. Since the graph satisfies Condition (K) (by Lemma 3.3.9), Proposition 2.9.9 can be applied to get that every ideal of $L_K(E)$ is graded. Since E is row-finite, by Theorem 2.5.9 there exist $H_i \in \mathscr{H}_E$, for $i \in \{1, \ldots, n\}$, such that:

(i) $I_i = I(H_i)$ for every $1 \leq i \leq n$,
(ii) $H_i \subsetneq H_{i+1}$ for every $i \in \{1, \ldots, n-1\}$, and

(iii) for every $i \in \{1, \ldots, n-1\}$, there is no hereditary and saturated set T such that $H_i \subsetneq T \subsetneq H_{i+1}$.

At this point we may apply the induction hypothesis to I_{n-1}, which is the Leavitt path algebra of a row-finite graph by Proposition 2.5.19, and has finitely many ideals by Corollary 2.9.12. Thus we have that I_{n-1} is an exchange ring. But $L_K(E)/I_{n-1} \cong L_K(E/H_{n-1})$ (as E is row-finite, so we may invoke Corollary 2.4.13(i)), and thus is a simple Leavitt path algebra (by the maximality of I_{n-1} inside $L_K(E)$). By the first step of the induction, $L_K(E)/I_{n-1} \cong L_K(E/H_{n-1})$ is an exchange ring. Since $\mathscr{V}(L_K(E/H_{n-1}))$ is generated by the isomorphism classes arising from its vertices (by Theorem 3.2.5), we obviously have that the natural map $\mathscr{V}(L_K(E)) \to \mathscr{V}(L_K(E)/I_{n-1})$ is surjective. So Theorem 3.3.4 can be applied, which finishes the proof. □

We are now in position to present the main result of the section.

Theorem 3.3.11 *Let E be an arbitrary graph and K any field. Then the following are equivalent.*

(1) *$L_K(E)$ is an exchange ring.*
(2) *E/H satisfies Condition* (L) *for every hereditary saturated subset H of E^0.*
(3) *E satisfies Condition* (K).
(4) *$\mathscr{L}_{gr}(L_K(E)) = \mathscr{L}_{id}(L_K(E))$; that is, every two-sided ideal of $L_K(E)$ is graded.*
(5) *The graphs E_H and E/H both satisfy Condition* (K) *for every hereditary saturated subset H of E^0.*
(6) *The graphs E_H and E/H both satisfy Condition* (K) *for some hereditary saturated subset H of E^0.*

Proof (1) ⇒ (2). Consider a hereditary saturated subset $H \in \mathscr{H}_E$. By Corollary 2.4.13(ii) we have that $L_K(E)/I(H \cup B_H^H)$ is isomorphic to the Leavitt path algebra $L_K(E/H)$. Since the quotient of an exchange ring by an ideal is an exchange ring (Theorem 3.3.4), Lemma 3.3.7 applies to give (2).

(2) ⇒ (3) is Lemma 3.3.6.

(3) ⇒ (1). By Lemma 3.3.8 and Theorem 1.6.10, we can write

$$L_K(E) \cong \varinjlim_{F \in \mathscr{F}} C_K^{X_F}(F) \cong \varinjlim_{F \in \mathscr{F}} L_K(F(X_F)),$$

where $\mathscr{F}$ is the family of finite complete subgraphs of E satisfying Condition (K), and $F(X_F)$ is the finite graph obtained from F by applying Theorem 1.5.18. Recalling Definition 1.5.16, we see that the graph $F(X_F)$ satisfies Condition (K) if F satisfies Condition (K), because both graphs contain the same closed paths, and the new vertices added to F in order to form $F(X_F)$ are sinks. Since the class of exchange rings is closed under direct limits (Proposition 3.3.1), it suffices to prove the result for finite graphs.

Let E be a finite graph with Condition (K). Then all the ideals of $L_K(E)$ are graded by Proposition 2.9.9, and so, by Theorem 2.5.9, the lattice of ideals of $L_K(E)$ is finite. The result follows therefore from Proposition 3.3.10.

(3) $\Leftrightarrow$ (4) is Proposition 2.9.9.

(3) $\Leftrightarrow$ (5) $\Leftrightarrow$ (6). It is easy to see that for every $H \in \mathscr{H}_E$ we have $\mathrm{CSP}_E(v) = \mathrm{CSP}_{E_H}(v)$ for all $v \in H$, and $\mathrm{CSP}_E(w) = \mathrm{CSP}_{E/H}(w)$ for all $w \in E^0 \setminus H$. This gives the result. □

We close the section by giving another characterization of the exchange Leavitt path algebras. Recall that an ideal I of $L_K(E)$ is self-adjoint if $\alpha^* \in I$ for every $\alpha \in I$.

Proposition 3.3.12 *Let E be an arbitrary graph and K any field. Then E satisfies Condition* (K) *if and only if every two-sided ideal of $L_K(E)$ is self-adjoint.*

Proof Suppose E satisfies Condition (K). Then by Theorem 3.3.11 every ideal of $L_K(E)$ is graded, and by Corollary 2.4.10 every such ideal is self-adjoint.

Conversely, suppose every ideal of $L_K(E)$ is self-adjoint. Let H be a hereditary saturated subset of E^0. We will show that E/H satisfies Condition (L), and thus Theorem 3.3.11 will yield the desired result. On the contrary, if E/H does not satisfy Condition (L), then there exists a cycle without exits c in E/H. By Lemma 2.7.1 the ideal I of $L_K(E/H)$ generated by c^0 is isomorphic to $\mathrm{M}_\Lambda(K[x,x^{-1}])$ for some set Λ. By Corollary 2.4.10 $I(H \cup B_H^H)$ is a self-adjoint ideal, hence the hypothesis implies that every ideal of $L_K(E)/I(H \cup B_H^H)$ is self-adjoint. Since $L_K(E/H) \cong L_K(E)/I(H \cup B_H^H)$ (by Corollary 2.4.13(ii)), we get that every ideal of $L_K(E/H)$ is self-adjoint. But every ideal of I is an ideal of $L_K(E/H)$ (by Lemma 2.9.12), hence every ideal of I, and consequently of $\mathrm{M}_\Lambda(K[x,x^{-1}])$, is self-adjoint. But this is a contradiction, as can easily be seen by using the same ideas as presented in the $|\Lambda| = 1$ case given prior to Corollary 2.4.10. □

3.4 Von Neumann Regularity

In this section we will show that the Leavitt path algebras arising from acyclic graphs are precisely the von Neumann regular Leavitt path algebras. Subsequently, we will give an explicit description of the largest von Neumann regular ideal of a Leavitt path algebra.

Recall that an element a in a ring R is said to be *von Neumann regular* if there exists a $b \in R$ such that $aba = a$. The ring R is called a *von Neumann regular ring* if every element in R is von Neumann regular. Note that in this situation the element $x = ba$ is idempotent. Indeed, von Neumann regular rings are characterized as those rings for which every finitely generated left ideal is generated by an idempotent, so that the topic of von Neumann regularity fits well with the theme of this chapter.

Theorem 3.4.1 *Let E be an arbitrary graph and K any field. Then the following are equivalent.*

(1) *$L_K(E)$ is von Neumann regular.*
(2) *E is acyclic.*
(3) *$L_K(E)$ is locally K-matricial.*

Proof (1) $\Rightarrow$ (2). Suppose that there exists a cycle c in E; denote $s(c)$ by v. We will prove that the element $v-c$ cannot be von Neumann regular. Suppose otherwise that there exists an element $\beta \in L_K(E)$ such that $(v-c)\beta(v-c) = (v-c)$. Replacing β by $v\beta v$ if necessary, there is no loss of generality in assuming that $\beta = v\beta v$. We write β as a sum of homogeneous elements $\beta = \sum_{i=m}^{n} \beta_i$, where $m, n \in \mathbb{Z}$, $\beta_m \neq 0$, $\beta_n \neq 0$, and $\deg(\beta_i) = i$ for all nonzero β_i with $m \leq i \leq n$. Since $\deg(v) = 0$, we have $v\beta_i v = \beta_i$ for all i. Then

$$v - c = (v-c)\left(\sum_{i=m}^{n} \beta_i\right)(v-c).$$

Equating the lowest degree terms on both sides, we get $\beta_m = v$. Since $\deg(v) = 0$, we conclude that $m = 0$, and that $\beta_0 = v$. Thus $\beta = \sum_{i=0}^{n} \beta_i$. Suppose $\deg(c) = s > 0$. By again equating terms of like degree in the displayed equation, we see that $\beta_i = 0$ whenever i is nonzero and not a multiple of s, so that

$$\sum_{i=m}^{n} \beta_i = v + \sum_{t=1}^{n/s} \beta_{ts}.$$

So upon rewriting the equation above, we have

$$v-c = (v-c)v(v-c)+(v-c)\left(\sum_{t=1}^{k} \beta_{ts}\right)(v-c), \text{ which gives } 0 = -c+c^2+(v-c)\left(\sum_{t=1}^{k} \beta_{ts}\right)(v-c).$$

By equating the degree s components on both sides we obtain $\beta_s = c$. Similarly, by equating the degree $2s$ components, we get $0 = c^2 - c\beta_s - \beta_s c + \beta_{2s}$. But substituting $\beta_s = c$ yields $\beta_{2s} = c^2$, and continuing in this manner we get $\beta_{ts} = c^t$, for every $t \in \mathbb{N}$. But this is not possible, as $\beta_{ts} = 0$ for $t > n/s$.

(2) $\Rightarrow$ (3) is Proposition 2.6.20.

(3) $\Rightarrow$ (1). It is well known that every matricial K-algebra is a von Neumann regular ring, and hence easily so too is any direct union of such algebras. □

Every ring R contains a largest von Neumann regular ideal (see e.g., [86, Proposition 1.5]), which we denote here by $U(R)$. Specifically, $U(R)$ is an ideal of R, which is von Neumann regular as a ring, with the property that if J is any ideal of R which is von Neumann regular as a ring, then $J \subseteq U(R)$. This ideal is often called the *Brown–McCoy radical* of R. It is not hard to show that $R/U(R)$ contains no nonzero von Neumann regular ideals.

Remark 3.4.2 It is clear that if R is matricial, then $U(R) = R$. On the other hand, using an idea which amounts to a special case of the idea used in the proof of Theorem 3.4.1, it is easy to show that $U(K[x, x^{-1}]) = \{0\}$. This in turn can be used to show that $U(R) = \{0\}$ for any K-algebra R of the form which arises in Theorem 2.7.3. In particular, $U(I(P_c(E)) = \{0\}$, where $P_c(E)$ is the set of vertices in E^0 which lie in a cycle without exits (cf. Notation 2.2.4).

We begin by showing that every von Neumann regular ideal of a Leavitt path algebra is graded.

Lemma 3.4.3 *Let E be an arbitrary graph and K any field. Then every von Neumann regular ideal of $L_K(E)$ is a graded ideal.*

Proof Clearly the result holds for the zero ideal, so let I be a nonzero von Neumann regular ideal of $L_K(E)$. By the Structure Theorem for Ideals 2.8.10 we have that $I = I(H \cup S^H \cup P_C)$, where H, S^H and P_C are as described therein. If I were not graded, then necessarily $P_C \neq \emptyset$, and the ideal $I/I(H \cup S^H)$ of the Leavitt path algebra $L_K(E)/I(H \cup S^H)$ would be isomorphic to $\bigoplus_{\overline{c} \in \overline{C}} M_{\Lambda_{\overline{c}}}(p_c(x)K[x, x^{-1}])$. But algebras of the latter form contain no nonzero von Neumann regular elements, which contradicts the von Neumann regularity of $I/I(H \cup S^H)$ (which is a consequence of it being the quotient of the von Neumann regular ring I). Therefore I must be graded, as required. □

In the context of Leavitt path algebras, we are able to describe the Brown–McCoy radical of $L_K(E)$ in terms of a specific subset of E^0.

Definition 3.4.4 For a graph E, we denote by $P_{nc}(E)$ the set of all vertices in E^0 which do not connect to any cycle in E.

It is clear from the definition that $P_{nc}(E)$ is both hereditary and saturated.

Proposition 3.4.5 *Let E be an arbitrary graph and K any field. Let H denote $P_{nc}(E)$. Then $U(L_K(E)) = I(H \cup B_H^H)$.*

Proof We first establish that $I(H \cup B_H^H)$ is a von Neumann regular ideal of $L_K(E)$. Indeed, by Theorem 2.5.22, this ideal (viewed as a ring) is isomorphic to the Leavitt path algebra of the graph ${}_{(H,B_H)}E$. Since none of the vertices in H connects to a cycle in E then it is straightforward from the definition of ${}_{(H,B_H)}E$ that this graph is necessarily acyclic. So by Theorem 3.4.1, $L_K({}_{(H,B_H)}E)$, and hence $I(H \cup B_H^H)$, is a von Neumann regular ring. Thus $I(H \cup B_H^H) \subseteq U(L_K(E))$.

To establish the reverse inclusion, we first invoke Lemma 3.4.3 to get that $U(L_K(E))$ is a graded ideal. So by the Structure Theorem for Graded Ideals 2.5.8 we have $U(L_K(E)) = I(H' \cup S^{H'})$ for some $S \subseteq B_{H'}$, where $H' = U(L_K(E)) \cap E^0$. We claim that $H' \subseteq H$; to establish the claim, we consider the ideal $I(H')$. By Theorem 2.5.19, $I(H')$ is isomorphic to $L_K({}_{H'}E)$. On the other hand, $I(H') \subseteq I(H' \cup S^{H'}) = U(L_K(E))$, so that $I(H')$ is von Neumann regular (as it is an ideal of the von Neumann regular ring $U(L_K(E))$). Thus Theorem 3.4.1 applies to yield that ${}_{H'}E$ is acyclic and, consequently, that H' has no cycles. By definition, this gives that $H' \subseteq P_{nc}(E) = H$, which establishes the claim.

Now, use that $H' \subseteq H$ implies $B_{H'} \subseteq H \cup B_H$ and, consequently, that $S^{H'} \subseteq H \cup B_H^H$, to get $U(L_K(E)) = I(H' \cup S^{H'}) \subseteq I(H \cup B_H^H)$. □

Remark 3.4.6 Since $P_{nc}(E) \in \mathscr{H}_E$, we have $I(P_{nc}(E)) = L_K(E)$ if and only if $P_{nc}(E) = E^0$. But by definition, the latter statement is equivalent to E being acyclic. So Proposition 3.4.5 can be viewed as a generalization of Theorem 3.4.1.

We recall the following subset of E^0 given in Definitions 2.6.1: the set of line points of E, denoted $P_l(E)$, is the set of those vertices of E which connect neither to bifurcations nor to cycles. In particular, $P_l(E)$ contains all the sinks of E. Additionally, by definition we have $P_l(E) \subseteq P_{nc}(E)$, so that $I(P_l(E)) \subseteq I(P_{nc}(E))$ for any graph E.

Corollary 3.4.7 *Let E be a finite graph and K any field. Then* $\mathrm{Soc}(L_K(E)) = U(L_K(E))$*; that is, the socle coincides with the Brown–McCoy radical for the Leavitt path algebra of a finite graph.*

Proof Using Theorem 2.6.14 and Proposition 3.4.5, we need only show that $I(P_l(E)) = I(P_{nc}(E))$. As noted immediately above, the containment $I(P_l(E)) \subseteq I(P_{nc}(E))$ holds for any graph E. Conversely, recall that for a finite graph E, each vertex connects either to a cycle or to a sink. So $v \in P_{nc}(E)$ and the finiteness of E implies that there is an integer N for which every path starting at v ends in a sink in at most N steps. But then using the (CK2) relation as many times as necessary at each of these N steps (together with the finiteness of the graph), we see that v is in the saturated closure of the sinks of E, and hence $v \in I(P_l(E))$. So $I(P_{nc}(E)) \subseteq I(P_l(E))$, completing the proof. □

Example 3.4.8 In the particular case of the Toeplitz algebra $\mathscr{T}_K = L_K(E_T)$ (see Example 1.3.6), the largest von Neumann regular ideal $U(\mathscr{T}_K)$ is the ideal generated by the sink, which by Corollary 3.4.7 is precisely $\mathrm{Soc}(\mathscr{T}_K)$.

Remark 3.4.9 Corollary 3.4.7 does not extend to infinite graphs, not even to infinite acyclic graphs. This can already be seen in Example 3.1.12, in which the graph E given there is acyclic (so that $U(L_K(E)) = L_K(E)$ by Theorem 3.4.1), but has zero socle. As an additional example (in which the socle of the Leavitt path algebra is nonzero), let F denote the graph

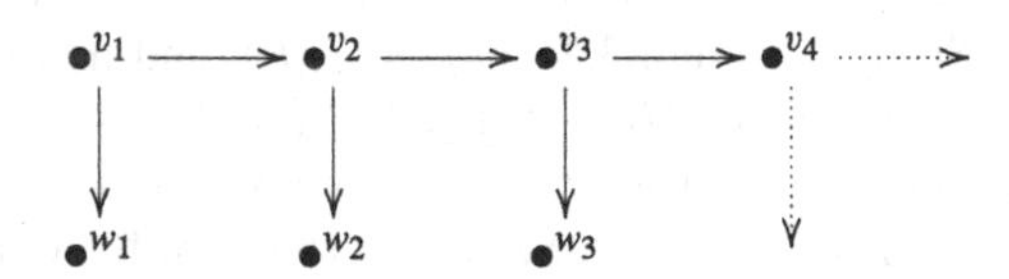

Then $P_l(F) = \{w_n\}_{n \in \mathbb{N}}$. It is easy to see that $I(P_l(F))$ is not all of $L_K(F)$ (since $v_i \notin I(P_l(F))$ for all $i \in \mathbb{N}$), so that by Theorem 2.6.14 we have $\mathrm{Soc}(L_K(F)) \neq L_K(F)$. But as above we have that $L_K(F)$ is von Neumann regular, so that $U(L_K(F)) = L_K(F)$.

Remark 3.4.10 There are a number of additional general ring-theoretic properties which are related to von Neumann regularity, including π-regularity and strong π-regularity, to name just two. It was established in [12, Theorem 1] that in the context of Leavitt path algebras, the three properties von Neumann regularity, π-regularity, and strong π-regularity are equivalent.

3.5 Primitive Non-Minimal Idempotents

We continue our presentation of idempotent-related topics by considering the primitive, non-minimal idempotents of $L_K(E)$. We focus first on the ideal generated by these elements; this ideal will play a role similar to that played by $\mathrm{Soc}(L_K(E))$, but with respect to the vertices which lie on cycles without exits. We will utilize the following general ring-theoretic result.

Proposition 3.5.1 ([108, Proposition 21.8]) *Let e be an idempotent in a (not-necessarily-unital) ring R. The following are equivalent.*

(1) *Re is an indecomposable left R-module (equivalently, eR is an indecomposable right R-module).*
(2) *eRe is a ring without nontrivial idempotents.*
(3) *e cannot be decomposed as $a + b$, where a, b are nonzero orthogonal idempotents in R.*

A nonzero idempotent of R which satisfies these conditions is called a primitive *idempotent.*

Clearly (by (1)) any minimal idempotent of R (Definitions 2.6.7) is necessarily primitive.

Proposition 3.5.2 *Let E be an arbitrary graph and K any field. Let $v \in E^0$. Then v is a primitive idempotent of $L_K(E)$ if and only if $T(v)$ has no bifurcations.*

Proof Suppose that $T(v)$ has bifurcations; say $T(v)$ has its first bifurcation at w, with μ being the shortest path which connects v to w. Since there are no bifurcations in μ, the (CK2) relation at each non-final vertex of μ yields $\mu\mu^* = v$. Hence we get $L_K(E)v = L_K(E)\mu\mu^*$. Let e and f be two different edges emitted by w; then $ee^* \neq w$ (as otherwise $w - ee^* = 0$, which on right multiplication by f would give $f = 0$), and so by Proposition 3.0.1(i) we get $L_K(E)w = L_K(E)ee^* \oplus L_K(E)(w - ee^*)$ is a decomposition of the desired type.

Conversely, suppose that $T(v)$ has no bifurcations. Two cases can occur. First, suppose $T(v)$ does not contain vertices in cycles. In this case, $v \in P_l(E)$, which means that v is minimal by Proposition 2.6.11, and so necessarily primitive. On the other hand, suppose $T(v) \cap P_c(E) \neq \emptyset$. Since $T(v)$ has no bifurcations, there can be only one cycle $c \in L_K(E)$ such that $T(v) \cap c^0 \neq \emptyset$, which in addition can have no exits. Furthermore, every vertex of $T(v)$ is either in c^0 or connects to a vertex w in c^0 via a path μ, where there are no bifurcations at any of the vertices of μ. Since then $\mu\mu^* = v$, we get $L_K(E)v \cong L_K(E)w$ as left $L_K(E)$-modules by Proposition 3.0.1(ii). Since w is in a cycle without exits, by Proposition 2.2.7 we have $wL_K(E)w \cong K[x, x^{-1}]$, which is a ring without nontrivial idempotents. Now Proposition 3.5.1 gives that w and v are both primitive, and completes the proof. □

Remark 3.5.3 If $vL_K(E)v$ is a ring with no nontrivial idempotents, then v is a primitive idempotent and, as a consequence of the proof of Proposition 3.5.2, we have either $vL_K(E)v \cong K$ (if v is minimal) or $vL_K(E)v \cong K[x, x^{-1}]$ (if v is not minimal).

We have found a graphical relationship between the primitive and the minimal vertices of the Leavitt path algebra of any graph: the minimal vertices are those whose trees do not contain bifurcations nor connect to cycles, while the primitive vertices see this second condition suppressed. In particular,

Remark 3.5.4 A vertex $v \in E^0$ is a primitive non-minimal idempotent of $L_K(E)$ if and only if $vL_K(E)v \cong K[x, x^{-1}]$. In particular, the vertices in $P_c(E)$ are primitive non-minimal.

Proposition 3.5.2 provides us with a tool to distinguish between those cycles with exits and those cycles without exits in a graph, giving us a characterization of Condition (L) in terms of primitive vertices.

Corollary 3.5.5 *Let E be an arbitrary graph and K any field. Then E satisfies Condition* (L) *if and only if every primitive vertex in $L_K(E)$ is minimal.*

In particular, if every vertex in $L_K(E)$ is infinite, then E satisfies Condition (L).

Proof By Proposition 3.5.2 and Remark 3.5.4, $L_K(E)$ contains a primitive non-minimal vertex if and only if E contains a cycle without exits. The additional statement follows vacuously. □

In Theorem 3.5.7 we extend Corollary 3.5.5 from the primitive non-minimal vertices to the primitive non-minimal idempotents of a Leavitt path algebra. As one consequence, this will show (Corollary 3.5.8) that Condition (L) is a ring isomorphism invariant of Leavitt path algebras.

Proposition 3.5.6 *Let E be an arbitrary graph and K any field. If $z \in L_K(E)$ is a primitive idempotent and we can write $\alpha z \beta = kv$ for $\alpha, \beta \in L_K(E)$, $k \in K^\times$, and $v \in E^0$, then $L_K(E)z \cong L_K(E)v$. If, moreover, z is primitive non-minimal, then $zL_K(E)z \cong K[x, x^{-1}]$.*

Proof We may assume $\alpha = v\alpha$ and $\beta = \beta v$. Define $a = k^{-1}\alpha z$ and $b = z\beta$. Then $ab = v$, and $e := ba = k^{-1}z\beta\alpha z$ is in $zL_K(E)z$. Moreover, $e^2 = baba = bva = ba = e$ and thus $L_K(E)e \cong L_K(E)v$ as left ideals of $L_K(E)$ by a standard ring theory result. (The maps $\rho_b : L_K(E)e \to L_K(E)v$ and $\rho_a : L_K(E)v \to L_K(E)e$ give the isomorphisms.) Note in particular that this implies $L_K(E)e \neq \{0\}$. Since z is a primitive idempotent, $zL_K(E)z$ is a ring without nontrivial idempotents, so that $e \in \{0, z\}$; since $e \neq 0$, we have $z = e$, so that $L_K(E)z \cong L_K(E)v$, as desired. If in addition z is primitive non-minimal, then necessarily so is v, and hence $zL_K(E)z \cong vL_K(E)v \cong K[x, x^{-1}]$ by Remark 3.5.4. □

We are now in position to establish a result similar to Corollary 3.5.5, but with respect to *all* idempotents in $L_K(E)$.

Theorem 3.5.7 *Let E be an arbitrary graph and K any field. Then E satisfies Condition* (L) *if and only if every primitive idempotent in $L_K(E)$ is minimal.*

Proof If $L_K(E)$ has no primitive non-minimal idempotents, in particular it has no primitive non-minimal vertices, so that by Corollary 3.5.5, E satisfies Condition (L).

Now suppose E satisfies Condition (L), and let x be a primitive non-minimal idempotent of $L_K(E)$. By the Reduction Theorem 2.2.11 there exist $v \in E^0$, $k \in K^\times$, and $\mu, \kappa \in \text{Path}(E)$ such that $\mu^* x \kappa = kv$. Note that, by Corollary 3.5.5, v cannot be primitive non-minimal. But this is a contradiction since by Proposition 3.5.6, $L_K(E)v \cong L_K(E)x$. □

Because Theorem 3.5.7 yields a characterization of Condition (L) in E as a ring-theoretic condition on $L_K(E)$, we immediately get the next result (which can also be derived from Proposition 2.9.13 as well).

Corollary 3.5.8 *Let E, F be arbitrary graphs and K any field, and suppose $L_K(E) \cong L_K(F)$ as rings. Then E satisfies Condition (L) if and only if F satisfies Condition (L).*

The tools developed above will allow us to reformulate, in terms of idempotents, the Simplicity and Purely Infinite Simplicity Theorems. By the Simplicity Theorem 2.9.1, $L_K(E)$ is simple if and only if $\mathscr{H}_E = \{\emptyset, E^0\}$, and E satisfies Condition (L). The condition $\mathscr{H}_E = \{\emptyset, E^0\}$ is equivalent to the nonexistence of nontrivial two-sided ideals of $L_K(E)$ generated by idempotents (see Theorem 2.5.8 and Corollary 2.9.11). So Theorem 3.5.7 yields the following.

Corollary 3.5.9 *Let E be an arbitrary graph and K any field. Then $L_K(E)$ is simple if and only if every primitive idempotent in $L_K(E)$ is minimal, and $L_K(E)$ contains no nontrivial two-sided ideals generated by idempotents.*

By the Purely Infinite Simplicity Theorem 3.1.10, $L_K(E)$ is purely infinite simple if and only if $L_K(E)$ is simple, and every vertex of E connects to a cycle. If E is finite, then the latter condition may be replaced by the condition that there are no minimal idempotents in $L_K(E)$, as follows. On the one hand, if every vertex connects to a cycle (necessarily with an exit), then there are no minimal vertices in E (indeed, by Proposition 3.1.6, every vertex is infinite in this case). On the other hand, if there are no minimal vertices then there are no sinks, and since E is finite, this yields that every vertex must connect to a cycle. But $\text{Soc}(L_K(E)) = I(P_l(E))$ (Theorem 2.6.14), and $P_l(E) = \emptyset$ (because E is finite and there are no sinks), so that $\text{Soc}(L_K(E)) = \{0\}$. Specifically, there are no minimal idempotents in $L_K(E)$. So we have established

Corollary 3.5.10 *Let E be a finite graph and K any field. Then $L_K(E)$ is purely infinite simple if and only if $L_K(E)$ contains no primitive idempotents and no nontrivial two-sided ideals generated by idempotents.*

3.6 Structural Properties of the $\mathscr{V}$-Monoid

For a ring R with enough idempotents, the monoid $\mathscr{V}(R)$ of isomorphism classes of finitely generated projective left R-modules was discussed in Sect. 3.2. The monoid $\mathscr{V}(R)$ is clearly *conical*; that is, if $p, q \in \mathscr{V}(R)$ have $p + q = 0$, then $p = q = 0$. In the specific case of a Leavitt path algebra $L_K(E)$, we show in this section that the monoid $\mathscr{V}(L_K(E))$ satisfies some additional monoid-theoretic properties (properties which, unlike the conical property, fail for some monoids of the form $\mathscr{V}(S)$ for some rings S). These properties arise in various contexts associated with decomposition and cancellation properties among finitely generated projective left $L_K(E)$-modules.

Definitions 3.6.1 Let $(M, +)$ denote an abelian monoid.

(i) M is called a *refinement monoid* if whenever $a + b = c + d$ in M, there exist $x, y, z, t \in M$ such that $a = x + y$ and $b = z + t$, while $c = x + z$ and $d = y + t$.
(ii) There is a canonical preorder on any abelian monoid M (the *algebraic preorder*), defined by setting $x \leq y$ if and only if there exists an $m \in M$ such that $y = x + m$. Following [28], M is called a *separative monoid* if it satisfies the following condition: if $a, b, c \in M$ satisfy $a + c = b + c$, and $c \leq na$ and $c \leq nb$ for some $n \in \mathbb{N}$, then $a = b$.

There are analogous definitions from a ring-theoretic point of view.

Definitions 3.6.2 Let R be a ring with enough idempotents. The class of finitely generated projective left R-modules is denoted by $FP(R)$.

(i) We say that $FP(R)$ satisfies the *refinement property* if whenever $A_1, A_2, B_1, B_2 \in FP(R)$ satisfy $A_1 \oplus A_2 \cong B_1 \oplus B_2$, then there exist decompositions $A_i = A_{i1} \oplus A_{i2}$ for $i = 1, 2$ such that $A_{1j} \oplus A_{2j} \cong B_j$ for $j = 1, 2$.
(ii) We say that R is *separative* if whenever $A, B, C \in FP(R)$ satisfy $A \oplus C \cong B \oplus C$ and C is isomorphic to direct summands of both nA and nB for some $n \in \mathbb{N}$, then $A \cong B$.

Remark 3.6.3 We note that, while the monoid $\mathscr{V}(R)$ of *isomorphism classes* of finitely generated projective left R-modules has been, and will continue to be, a key player in the subject of Leavitt path algebras, it is more common in the literature to focus on the class of *all* finitely generated projective left R-modules in a discussion of the properties of R presented in Definitions 3.6.2.

The following is then clear.

Proposition 3.6.4 *Let R be a ring with enough idempotents.*

(i) *$\mathscr{V}(R)$ is a refinement monoid if and only if $FP(R)$ satisfies the refinement property.*
(ii) *$\mathscr{V}(R)$ is separative if and only if R is separative.*

We will show in this section that $\mathscr{V}(L_K(E))$ is both separative and a refinement monoid for every graph E and field K. The approach will be to first establish these results for row-finite graphs, and subsequently invoke appropriate direct limit theorems from Chap. 1. For context, we note that it has been shown [28, Proposition 1.2] that every exchange ring satisfies the refinement property. On the other hand, as of 2017 it is an outstanding open question to determine whether every exchange ring is separative.

We recall again the definition of the monoid M_E (Definition 1.4.2), but here stated in the context of row-finite graphs: M_E denotes the abelian monoid given by the generators $\{a_v \mid v \in E^0\}$, with the relations:

$$a_v = \sum_{\{e \in E^1 \mid s(e)=v\}} a_{r(e)} \quad \text{for every } v \in E^0 \text{ that emits edges.} \tag{M}$$

We introduce some helpful notation. Let E be a row-finite graph, and let $\mathbb{F}_E$ (or simply $\mathbb{F}$ when E is clear) be the free abelian monoid on the set E^0. Each of the nonzero elements of $\mathbb{F}_E$ can be written in a unique form (up to permutation) as $\sum_{i=1}^n x_i$, where $x_i \in E^0$ (and repeats are allowed). Now we will give a description of the congruence on $\mathbb{F}_E$ generated by the relations (M). For $x \in \mathrm{Reg}(E)$, write

$$\mathbf{r}(x) := \sum_{\{e \in E^1 \mid s(e)=x\}} r(e) \in \mathbb{F}.$$

(This notation is consistent with that given in Definition 3.2.9; in the current notation, the expression $\mathbf{r}(x)$ is being used to more efficiently denote the set $\mathbf{r}(r(s^{-1}(x)))$.) With this notation, the relations (M) are expressed more efficiently as $x = \mathbf{r}(x)$ for every $x \in \mathrm{Reg}(E)$.

Definition 3.6.5 Let $\mathbb{F} = \mathbb{F}_E$ be the free abelian monoid on the set of vertices E^0 of a row-finite graph E. Define a binary relation $\to_1$ on $\mathbb{F} \setminus \{0\}$ as follows. Let $\sum_{i=1}^n x_i$ be an element in $\mathbb{F} \setminus \{0\}$ as above and let $j \in \{1, \dots, n\}$ be an index such that x_j emits edges. In this situation we write

$$\sum_{i=1}^n x_i \to_1 \sum_{i \neq j} x_i + \mathbf{r}(x_j).$$

Let $\to$ be the transitive and reflexive closure of $\to_1$ on $\mathbb{F} \setminus \{0\}$, that is, $\alpha \to \beta$ if and only if there is a finite sequence $\alpha = \alpha_0 \to_1 \alpha_1 \to_1 \cdots \to_1 \alpha_t = \beta$. Let $\sim$ be the congruence on $\mathbb{F} \setminus \{0\}$ generated by the relation $\to_1$ (or, equivalently, by the relation $\to$). Namely $\alpha \sim \alpha$ for all $\alpha \in \mathbb{F} \setminus \{0\}$ and, for $\alpha, \beta \neq 0$, we have $\alpha \sim \beta$ if and only if there is a finite sequence $\alpha = \alpha_0, \alpha_1, \dots, \alpha_n = \beta$, such that, for each $i = 0, \dots, n-1$, either $\alpha_i \to_1 \alpha_{i+1}$ or $\alpha_{i+1} \to_1 \alpha_i$. The number n above will be called the *length* of the sequence. The congruence $\sim$ on $\mathbb{F} \setminus \{0\}$ is extended to $\mathbb{F}$ by adding the single pair $0 \sim 0$.

It is clear that $\sim$ is the congruence on $\mathbb{F}$ generated by relations (M), and so $M_E = \mathbb{F}/\sim$.

The *support* of an element γ in $\mathbb{F}$, denoted $\mathrm{supp}(\gamma) \subseteq E^0$, is the set of basis elements appearing in the canonical expression of γ.

Lemma 3.6.6 *Let $\to$ be the binary relation on $\mathbb{F}$ given in Definition 3.6.5. Suppose $\alpha, \beta, \alpha_1, \beta_1 \in \mathbb{F} \setminus \{0\}$ with $\alpha = \alpha_1 + \alpha_2$ and $\alpha \to \beta$. Then β can be written as $\beta = \beta_1 + \beta_2$ with $\alpha_1 \to \beta_1$ and $\alpha_2 \to \beta_2$.*

Proof By induction, it is enough to prove the result in the case where $\alpha \to_1 \beta$. In this situation, there is an element x in the support of α such that $\beta = (\alpha - x) + \mathbf{r}(x)$. The element x belongs either to the support of α_1 or to the support of α_2. Assume, for instance, that the element x belongs to the support of α_1. Then we set $\beta_1 = (\alpha_1 - x) + \mathbf{r}(x)$ and $\beta_2 = \alpha_2$. The case where x is in the support of α_2 is similar. □

Note that the elements β_1 and β_2 in Lemma 3.6.6 are not uniquely determined by α_1 and α_2 in general, because the element $x \in E^0$ considered in the proof could belong to both the support of α_1 and the support of α_2.

The following lemma gives the important "confluence" property of the congruence $\sim$ on the free abelian monoid $\mathbb{F}_E$.

Lemma 3.6.7 (The Confluence Lemma) *Let α and β be nonzero elements in $\mathbb{F}_E$. Then $\alpha \sim \beta$ if and only if there is a $\gamma \in \mathbb{F}_E \setminus \{0\}$ such that $\alpha \to \gamma$ and $\beta \to \gamma$.*

Proof Assume that $\alpha \sim \beta$. Then there exists a finite sequence $\alpha = \alpha_0, \alpha_1, \dots, \alpha_n = \beta$, such that, for each $i = 0, \dots, n-1$, either $\alpha_i \to_1 \alpha_{i+1}$ or $\alpha_{i+1} \to_1 \alpha_i$. We proceed by induction on n. If $n = 0$, then $\alpha = \beta$ and there is nothing to prove. Assume the result is true for sequences of length $n-1$, and let $\alpha = \alpha_0, \alpha_1, \dots, \alpha_n = \beta$ be a sequence of length n. By the induction hypothesis, there is a $\lambda \in \mathbb{F}$ such that $\alpha \to \lambda$ and $\alpha_{n-1} \to \lambda$. Now there are two cases to consider. If $\beta \to_1 \alpha_{n-1}$, then $\beta \to \lambda$ and we are done. Assume that $\alpha_{n-1} \to_1 \beta$. By definition of $\to_1$, there is a basis element $x \in E^0$ in the support of α_{n-1} such that $\alpha_{n-1} = x + \alpha'_{n-1}$ and $\beta = \mathbf{r}(x) + \alpha'_{n-1}$. By Lemma 3.6.6, we have $\lambda = \lambda(x) + \lambda'$, where $x \to \lambda(x)$ and $\alpha'_{n-1} \to \lambda'$. If the length of the sequence from x to $\lambda(x)$ is positive, then we have $\mathbf{r}(x) \to \lambda(x)$ and so $\beta = \mathbf{r}(x) + \alpha'_{n-1} \to \lambda(x) + \lambda' = \lambda$. On the other hand, if $x = \lambda(x)$, we define $\gamma = \mathbf{r}(x) + \lambda'$. Then $\lambda \to_1 \gamma$ and so $\alpha \to \gamma$, and also $\beta = \mathbf{r}(x) + \alpha'_{n-1} \to \mathbf{r}(x) + \lambda' = \gamma$. This concludes the proof. □

We are now ready to show the refinement property of M_E.

Proposition 3.6.8 *The monoid M_E associated with any row-finite graph E is a refinement monoid.*

Proof We use the identification $M_E = \mathbb{F}/\sim$. Let $\alpha = \alpha_1 + \alpha_2 \sim \beta = \beta_1 + \beta_2$, with $\alpha_1, \alpha_2, \beta_1, \beta_2 \in \mathbb{F}$. By the Confluence Lemma 3.6.7, there is a $\gamma \in \mathbb{F}$ such that $\alpha \to \gamma$ and $\beta \to \gamma$. By Lemma 3.6.6, we can write $\gamma = \alpha'_1 + \alpha'_2 = \beta'_1 + \beta'_2$, with $\alpha_i \to \alpha'_i$ and $\beta_i \to \beta'_i$ for $i = 1, 2$. Since $\mathbb{F}$ is a free abelian monoid, $\mathbb{F}$ has the refinement property and so there are decompositions $\alpha'_i = \gamma_{i1} + \gamma_{i2}$ for $i = 1, 2$ such that $\beta'_j = \gamma_{1j} + \gamma_{2j}$ for $j = 1, 2$. The result follows. □

Our next goal is to establish a lattice isomorphism between the lattice $\mathcal{H}_E$ of hereditary saturated subsets of E^0, and the lattice of order-ideals of the associated monoid M_E, where E is row-finite. This in turn can be interpreted as a lattice isomorphism with the graded ideals of $L_K(E)$ (Theorem 2.5.9), and thereby also an isomorphism with the lattice of the ideals of $L_K(E)$ generated by idempotents (Corollary 2.9.11).

An *order-ideal* of a monoid M is a submonoid I of M such that, for each $x, y \in M$, if $x+y \in I$ then $x \in I$ and $y \in I$. An order-ideal can also be described as a submonoid I of M which is hereditary with respect to the canonical preorder $\leq$ on M: $x \leq y$ and $y \in I$ imply $x \in I$. (Recall that the preorder $\leq$ on M is defined by setting $x \leq y$ if and only if there exists an $m \in M$ such that $y = x + m$.)

The set $\mathcal{L}(M)$ of order-ideals of M forms a (complete) lattice $\left(\mathcal{L}(M), \subseteq, \overline{\sum}, \cap\right)$. Here, for a family of order-ideals $\{I_i\}$, we denote by $\overline{\sum} I_i$ the set of elements $x \in M$ such that $x \leq y$ for some y belonging to the algebraic sum $\sum I_i$ of the order-ideals I_i. Note that $\sum I_i = \overline{\sum} I_i$ whenever M is a refinement monoid.

Recall again that $\mathbb{F}_E$ is the free abelian monoid on E^0, and $M_E = \mathbb{F}_E/\sim$. For $\gamma \in \mathbb{F}_E$ we will denote by $[\gamma]$ its class in M_E. Note that any order-ideal I of M_E is generated *as a monoid* by the set $\{[v] \mid v \in E^0\} \cap I$.

The set $\mathcal{H}_E$ of hereditary saturated subsets of E^0 is also a complete lattice $(\mathcal{H}_E, \subseteq, \overline{\cup}, \cap)$ (Remark 2.5.2).

Proposition 3.6.9 *Let E be a row-finite graph. Then there are order-preserving mutually inverse maps*

$$\varphi: \mathcal{H}_E \longrightarrow \mathcal{L}(M_E) \quad \text{and} \quad \psi: \mathcal{L}(M_E) \longrightarrow \mathcal{H}_E,$$

where, for $H \in \mathcal{H}_E$, $\varphi(H)$ is the order-ideal of M_E generated by $\{[v] \mid v \in H\}$, and, for $I \in \mathcal{L}(M_E)$, $\psi(I) = \{v \in E^0 \mid [v] \in I\}$.

Proof The maps φ and ψ are obviously order-preserving. We claim that to establish the result it suffices to show

(i) for $I \in \mathcal{L}(M_E)$, the set $\psi(I)$ is a hereditary saturated subset of E^0, and
(ii) if $H \in \mathcal{H}_E$ then $[v] \in \varphi(H)$ if and only if $v \in H$.

To see this, if (i) and (ii) hold, then ψ is well-defined by (i), and $\psi(\varphi(H)) = H$ for $H \in \mathcal{H}_E$, by (ii). On the other hand, if I is an order-ideal of M_E, then obviously $\varphi(\psi(I)) \subseteq I$, and since I is generated as a monoid by $\{[v] \mid v \in E^0\} \cap I = [\psi(I)]$, it follows that $I \subseteq \varphi(\psi(I))$.

Proof of (i) Let I be an order-ideal of M_E, and set $H := \psi(I) = \{v \in E^0 \mid [v] \in I\}$. To see that H is hereditary, we have to prove that, whenever we have $\gamma = e_1 e_2 \cdots e_n$ in Path(E) with $s(e_1) = v$ and $r(e_n) = w$ and $v \in H$, then $w \in H$. If we consider the corresponding sequence $v \to_1 \gamma_1 \to_1 \gamma_2 \to_1 \cdots \to_1 \gamma_n$ in $\mathbb{F}_E$, we see that w belongs to the support of γ_n, so that $w \leq \gamma_n$ in $\mathbb{F}_E$. This implies that $[w] \leq [\gamma_n] = [v]$, and so $[w] \in I$ because I is hereditary.

To show saturation, take a non-sink $v \in E^0$ such that $r(s^{-1}(v)) \subseteq H$. We then have $\text{supp}(\mathbf{r}(v)) \subseteq H$, so that $[\mathbf{r}(v)] \in I$ because I is a submonoid of M_E. But $[v] = [\mathbf{r}(v)]$, so that $[v] \in I$ and $v \in H$.

Proof of (ii) Let H be a hereditary saturated subset of E^0, and let $I := \varphi(H)$ be the order-ideal of M_E generated by $\{[v] \mid v \in H\}$. Clearly $[v] \in I$ if $v \in H$. Conversely, suppose that $[v] \in I$. Then $[v] \leq [\gamma]$, where $\gamma \in \mathbb{F}_E$ satisfies $\text{supp}(\gamma) \subseteq H$. Thus we can write $[\gamma] = [v] + [\delta]$ for some $\delta \in \mathbb{F}_E$. By the Confluence Lemma 3.6.7, there exists a $\beta \in \mathbb{F}_E$ such that $\gamma \to \beta$ and $v + \delta \to \beta$. Since H is hereditary and $\text{supp}(\gamma) \subseteq H$, we get $\text{supp}(\beta) \subseteq H$. By Lemma 3.6.6, we have $\beta = \beta_1 + \beta_2$, where $v \to \beta_1$ and $\delta \to \beta_2$. Observe that $\text{supp}(\beta_1) \subseteq \text{supp}(\beta) \subseteq H$. Using that H is saturated, it is a simple matter to check that, if $\alpha \to_1 \alpha'$ and $\text{supp}(\alpha') \subseteq H$, then $\text{supp}(\alpha) \subseteq H$. Using this and induction, we obtain that $v \in H$, as desired. □

We now show that the monoid M_E associated with a row-finite graph E is always a separative monoid. Recall (Definitions 3.6.1) this means that for elements $x, y, z \in M_E$, if $x + z = y + z$ and $z \leq nx$ and $z \leq ny$ for some positive integer n, then $x = y$.

The separativity of M_E follows from results of Brookfield [53] on primely generated monoids; see also [157, Chap. 6]. Indeed, the class of primely generated refinement monoids satisfies many other nice cancellation properties. We will highlight *unperforation* later, and refer the reader to [53] for further information.

Definition 3.6.10 Let M be a monoid. An element $p \in M$ is *prime* if for all $a_1, a_2 \in M$, $p \leq a_1 + a_2$ implies $p \leq a_1$ or $p \leq a_2$. A monoid is *primely generated* if each of its elements is a sum of primes.

Proposition 3.6.11 ([53, Corollary 6.8]) *Any finitely generated refinement monoid is primely generated.*

It follows from Propositions 3.6.8 and 3.6.11 that, for a finite graph E, the monoid M_E is primely generated. Note that the primely generated property does not extend in general to row-finite graphs, as is demonstrated by the following graph G:

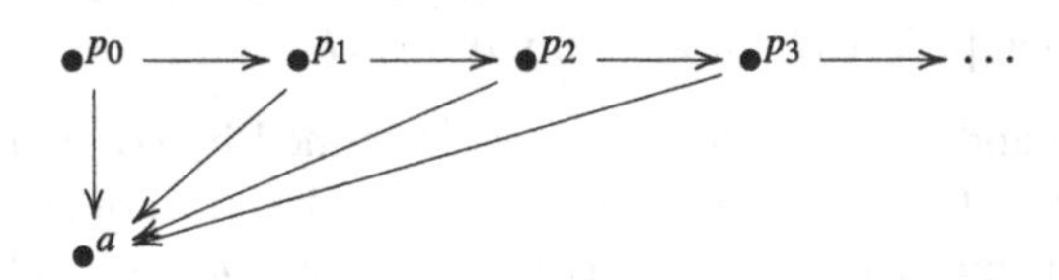

The corresponding monoid M_G has generators $a, p_0, p_1, \ldots$, and relations given by $p_i = p_{i+1} + a$ for all $i \geq 0$. One can easily see that the only prime element in M is a, so that M is not primely generated.

Theorem 3.6.12 *Let E be a row-finite graph. Then the monoid M_E is separative.*

Proof By Lemma 3.2.4, we get that M_E is the direct limit of monoids M_{X_i} corresponding to finite graphs X_i. Therefore, in order to check separativity, we can assume that the graph E is finite. In this situation, we have that M_E is generated by the finite set E^0 of vertices of E, and thus M_E is finitely generated. By

Proposition 3.6.8, M_E is a refinement monoid, so it follows from Proposition 3.6.11 that M_E is a primely generated refinement monoid. By Brookfield [53, Theorem 4.5], the monoid M_E is separative. □

As remarked previously, primely generated refinement monoids satisfy many nice cancellation properties, as shown in [53]. Some of these properties are preserved in direct limits, so they are automatically true for the graph monoids corresponding to any row-finite graph (and, as we will show below in Theorem 3.6.21, turn out to be true for arbitrary graphs). Especially important in several applications is the property of unperforation.

Definition 3.6.13 The monoid $(M, +)$ is said to be *unperforated* if, for all elements $a, b \in M$ and all positive integers n, we have $na \leq nb \implies a \leq b$.

Proposition 3.6.14 *Let E be a row-finite graph. Then the monoid M_E is unperforated.*

Proof As in the proof of Theorem 3.6.12, we can reduce to the case of a finite graph E. In this case, the result follows from [53, Corollary 5.11(5)]. □

Corollary 3.6.15 *Let E be a row-finite graph. Then $FP(L_K(E))$ satisfies the refinement property, and $L_K(E)$ is a separative ring. Moreover, the monoid $\mathcal{V}(L_K(E))$ is unperforated.*

Proof By Theorem 3.2.5, we have $\mathcal{V}(L_K(E)) \cong M_E$. So the result follows from Proposition 3.6.8, Theorem 3.6.12 and Proposition 3.6.14. □

Another useful technique to deal with graph monoids of finite graphs consists of considering composition series of order-ideals in the monoid. These composition series correspond via Proposition 3.6.9 and Theorem 2.5.9 to composition series of graded ideals in $L_K(E)$. (Using [47, Theorem 4.1(b)], they also correspond to composition series of closed gauge-invariant ideals of the graph C^*-algebra $C^*(E)$; this approach will be used in the proof of Theorem 5.3.5 below.) The composition series approach can be used to obtain a different proof of the separativity of M_E (Theorem 3.6.12), an approach we sketch in Remark 3.6.19.

Definition 3.6.16 Given an order-ideal S of a monoid M we define a congruence $\sim_S$ on M by setting $a \sim_S b$ if and only if there exist $e, f \in S$ such that $a + e = b + f$. Let M/S be the factor monoid obtained from the congruence $\sim_S$ (see e.g., [28]). We denote by $[x]_S$ the class of an element $x \in M$ in M/S.

In particular, if I is any ideal of a ring R, the monoid $\mathcal{V}(I)$ is an order-ideal of $\mathcal{V}(R)$. Using the construction of the factor monoid given in Definition 3.6.16, it can be shown that for a large class of rings R, one has $\mathcal{V}(R/I) \cong \mathcal{V}(R)/\mathcal{V}(I)$ for any ideal I of R (see e.g., [28, Proposition 1.4]). We present here some useful general facts about $\mathcal{V}$-monoids.

Proposition 3.6.17 *Let R be any ring with local units.*

(i) *Assume that $\mathcal{V}(R)$ is a refinement monoid. Then the map*

$$I \mapsto \mathcal{V}(I)$$

gives a lattice isomorphism between the lattice $\mathscr{L}_{\mathrm{idem}}(R)$ consisting of those ideals of R which are generated by idempotents, and the lattice $\mathscr{L}(\mathscr{V}(R))$ of order-ideals of $\mathscr{V}(R)$.

(ii) *If I is an ideal of R generated by idempotents, then there is a canonical injective map*

$$\omega \colon \mathscr{V}(R)/\mathscr{V}(I) \to \mathscr{V}(R/I)$$

such that $\omega([e]_{\mathscr{V}(I)}) = [e + I]$ for every idempotent e in R.

Proof (i) Since R has local units and $\mathscr{V}(R)$ is a refinement monoid, every idempotent in $M_{\mathbb{N}}(R)$ is equivalent to an idempotent of the form $e_1 \oplus \cdots \oplus e_n$ for some idempotents $e_1, \ldots, e_n$ of R. (See Definition 3.8.2 below.) It follows that the set of trace ideals considered in [27, Definition 10.9] is exactly the set of ideals of R generated by idempotents. Therefore the bijective correspondence follows from [27, Proposition 10.10] (see [78, Theorem 2.1(c)] for the unital case).

(ii) Since R has local units, the proof of [24, Proposition 5.3(c)] can be easily adapted to get that the map ω is injective. Note that ω is just the map induced by the canonical projection $\pi \colon R \to R/I$. □

Observe that, by combining Theorem 2.5.9, Corollary 2.9.11, Theorems 3.2.5 and 3.6.8, Proposition 3.6.9 can be re-established by using Proposition 3.6.17(i). A similar route can also be used to prove the following result.

Lemma 3.6.18 *Let E be a row-finite graph. For a hereditary saturated subset H of E^0, consider the order-ideal $S = \varphi(H)$ of M_E associated with H, as in Proposition 3.6.9. Let E/H be the quotient graph (recall Definition 2.4.11). Then there are natural monoid isomorphisms*

$$M_E/S \cong \mathscr{V}(L_K(E))/\mathscr{V}(I(H)) \cong \mathscr{V}(L_K(E)/I(H)) \cong \mathscr{V}(L_K(E/H)) \cong M_{E/H}.$$

Proof By Theorem 3.2.5 we have $M_E \cong \mathscr{V}(L_K(E))$. By Proposition 3.6.17(ii), the map

$$\omega \colon \mathscr{V}(L_K(E))/\mathscr{V}(I(H)) \to \mathscr{V}(L_K(E)/I(H)) \quad \text{defined by} \quad \omega([e]_{\mathscr{V}(I(H))}) = [e+I(H)]$$

is injective. Moreover, there is an isomorphism $L_K(E)/I(H) \cong L_K(E/H)$, given in Corollary 2.4.13(i). Since $\mathscr{V}(L_K(E/H)) \cong M_{E/H}$, the monoid $\mathscr{V}(L_K(E/H))$ is generated by the classes of vertices v in $E^0 \setminus H$, so we get that the map ω is surjective. The result follows. □

Remark 3.6.19 We sketch a proof of the separativity of M_E, different from the one presented in Theorem 3.6.12, using the theory of order-ideals. For a row-finite graph E, we call M_E *simple* if the only order-ideals of M_E are trivial. This corresponds by Proposition 3.6.9 to the situation where the hereditary saturated subset generated by any vertex of E is all of E^0. By Lemma 2.9.6, this happens if and only if E is cofinal (Definition 2.9.4).

As in the proof of Theorem 3.6.12, we can assume that E is a finite graph. In this case it is obvious that E^0 has a finite number of hereditary saturated subsets, so M_E has a finite number of order-ideals. Take a finite chain $0 = S_0 \leq S_1 \leq \cdots \leq S_n = M_E$ such that each S_i is an order-ideal of M_E, and all the quotient monoids S_i/S_{i-1} are simple. By Proposition 3.6.9 we have $S_i \cong M_{H_i}$ for some finite graph H_i, and by Lemma 3.6.18 we have $S_i/S_{i-1} \cong M_{G_i}$ for some cofinal finite graph G_i. By Proposition 3.6.8, S_i is a refinement monoid for all i, so the Extension Theorem for refinement monoids [28, Theorem 4.5] tells us that S_i is separative if and only if so are S_{i-1} and S_i/S_{i-1}. It follows by induction that it is enough to consider the case where E is a cofinal finite graph.

So let E be a cofinal finite graph. We distinguish three cases. First, suppose that E is acyclic. Then necessarily there is a sink v in E, and by cofinality for every vertex w of E there is a path from w to v. It follows that M_E is a free abelian monoid of rank one (i.e., isomorphic to $\mathbb{Z}^+$), generated by a_v. In particular, M_E is a separative monoid. Secondly, assume that E has a cycle without exits, and let v be any vertex in this cycle. By using the cofinality condition, it is easy to see that there are no other cycles in E, and that every vertex in E connects to v. It follows again that M_E is a free abelian monoid of rank one, generated by a_v. Finally, we consider the case where every cycle in E has an exit. By cofinality, every vertex connects to every cycle. Using this and the property that every cycle has an exit, it is easy to show that for every nonzero element x in M_E there is a nonzero element y in M_E such that $x = x + y$. It follows that $M_E \setminus \{0\}$ is a group; see for example [28, Proposition 2.4]. In particular, M_E is a separative monoid. □

Example 3.6.20 This example will be useful later on. Consider the following graph E:

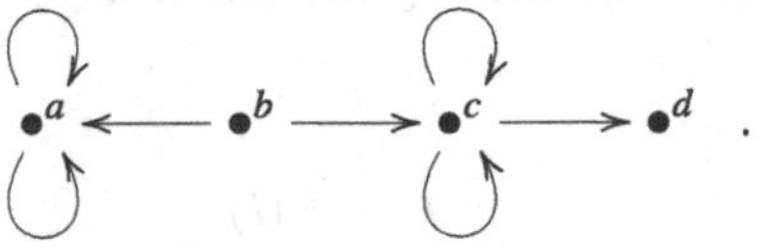

Then M_E is the monoid generated by a, b, c, d with defining relations $a = 2a$, $b = a + c$, $c = 2c + d$. A composition series of order-ideals for M_E is obtained from the graph monoids corresponding to the following chain of hereditary saturated subsets of E:

$$\emptyset \subsetneq \{d\} \subsetneq \{c, d\} \subsetneq \{a, b, c, d\} = E^0.$$

By Lemma 3.6.18, the corresponding simple quotient monoids are the graph monoids corresponding to the following graphs:

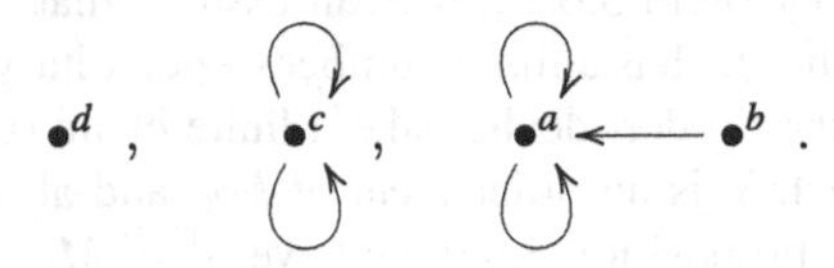

It is a relatively straightforward matter to generalize the previously established structural results about graph monoids of row-finite graphs to arbitrary graphs, using the direct limit machinery from Sect. 1.6. We complete this section by providing the details.

Theorem 3.6.21 *Let E be an arbitrary graph, let K be a field, and let X be a subset of* $\mathrm{Reg}(E)$. *Then the monoid $\mathcal{V}(C_K^X(E))$ is an unperforated, separative, refinement monoid. In particular, the monoid $\mathcal{V}(L_K(E))$ is an unperforated, separative, refinement monoid.*

Proof Since the properties in the statement are preserved under direct limits, and since the functor $\mathcal{V}$ is continuous, we see from Theorem 1.6.10 that it suffices to prove the result for finite graphs E. So suppose that E is a finite graph and that X is a finite subset of $\mathrm{Reg}(E)$. By Theorem 1.5.18, we have that $C_K^X(E) \cong L_K(E(X))$ for a certain finite graph $E(X)$. By Proposition 3.6.8, Theorem 3.6.12, Proposition 3.6.14 and Theorem 3.2.5, $\mathcal{V}(L_K(E(X))$ is an unperforated, separative, refinement monoid, and thus so is $\mathcal{V}(C_K^X(E))$. □

Remark 3.6.22 For a refinement monoid, unperforation implies separativity. This follows immediately from [59, Theorem 1], and it was noted independently in [156, Corollary 2.4].

Theorem 3.6.23 *Let E be an arbitrary graph and K any field.*

(i) *The map*

$$I \mapsto \mathcal{V}(I)$$

gives a lattice isomorphism between the lattice $\mathcal{L}_{\mathrm{gr}}(L_K(E))$ of graded ideals of $L_K(E)$ and the lattice $\mathcal{L}(\mathcal{V}(L_K(E)))$ of order-ideals of $\mathcal{V}(L_K(E))$.

(ii) *Let I be a graded ideal of $L_K(E)$. Then there is a natural monoid isomorphism*

$$\omega\colon \frac{\mathcal{V}(L_K(E))}{\mathcal{V}(I)} \longrightarrow \mathcal{V}(L_K(E)/I).$$

Proof (i) Since $\mathcal{V}(L_K(E))$ is a refinement monoid (Theorem 3.6.21) and the graded ideals of $L_K(E)$ are precisely the idempotent-generated ideals (Corollary 2.9.11), the result follows directly from Proposition 3.6.17(i).

(ii) Again by Corollary 2.9.11, we have that I is an idempotent-generated ideal, so the map ω is injective by Proposition 3.6.17(ii). Now by Theorem 2.5.8 there exist $H \in \mathscr{H}_E$ and $S \subseteq B_H$ such that $I = I(H \cup S^H)$. Therefore, by using Theorem 2.4.15 and Corollary 3.2.11, we get

$$\mathscr{V}(L_K(E)/I) = \mathscr{V}(L_K(E)/I(H \cup S^H)) \cong \mathscr{V}(L_K(E/(H,S))) \cong M_{E/(H,S)}.$$

It follows that $\mathscr{V}(L_K(E)/I)$ is generated by elements of the form $[v - \sum_{f \in Z} ff^*]$, where $v \in E^0 \setminus H$ and Z is a finite (possibly empty) subset of $s_E^{-1}(v)$ such that $r(f) \notin H$ for every $f \in Z$. Thus the map ω is surjective, and consequently a monoid isomorphism. □

3.7 Extreme Cycles

In Chap. 1 we described the three "primary colors" of Leavitt path algebras: $n \times n$ matrix rings $\mathrm{M}_n(K) \cong L_K(A_n)$, Laurent polynomials $K[x, x^{-1}] \cong L_K(R_1)$, and Leavitt algebras $L_K(1, n) \cong L_K(R_n)$ (for $n \geq 2$). In Theorem 2.6.14 we showed that the ideal of $L_K(E)$ generated by the set of line points $P_l(E)$ yields a piece of $L_K(E)$ similar in appearance to the first color, while in Theorem 2.7.3 we showed that the ideal of $L_K(E)$ generated by the vertices that lie on cycles without exits $P_c(E)$ is similar in appearance to the second color. Intuitively, in this section we complete the picture by describing the piece of $L_K(E)$ which most resembles the third color. Specifically, we identify sets of vertices which generate ideals in $L_K(E)$ which are purely infinite simple as a K-algebra.

Definitions 3.7.1 Let E be a graph and c a cycle in E. We say that c is an *extreme cycle* if c has exits and, for every path λ starting at a vertex in c^0, there exists a $\mu \in \mathrm{Path}(E)$ such that $r(\lambda) = s(\mu)$, and $r(\lambda\mu) \in c^0$. We will denote by $P_{ec}(E)$ the set of vertices which belong to extreme cycles. Intuitively, c is an extreme cycle if every path which leaves c can be lengthened in such a way that the longer path returns to c.

Let X'_{ec} be the set of all extreme cycles in a graph E. We define in X'_{ec} the following relation: given $c, d \in X'_{ec}$, we write $c \sim d$ whenever c and d are connected, that is, $T(c^0) \cap d^0 \neq \emptyset$, equivalently, $T(d^0) \cap c^0 \neq \emptyset$. It is not difficult to see that $\sim$ is an equivalence relation. The set of all $\sim$ equivalence classes is denoted by $X_{ec} = X'_{ec}/\sim$. When we want to emphasize a specific graph E under consideration we will write $X'_{ec}(E)$ and $X_{ec}(E)$ for X'_{ec} and X_{ec}, respectively.

For $c \in X'_{ec}$, we let $\tilde{c}$ denote the class of c. We write $\tilde{c}^0$ to represent the set of all vertices which are in the cycles belonging to $\tilde{c}$.

Examples 3.7.2 Consider the following graphs.

$$E = {}_e\circlearrowright \bullet^v \xrightarrow{f} \bullet^w \circlearrowleft^{g}_{h} \quad \text{and} \quad F = {}_{e'}\circlearrowright \bullet^{v'} \underset{g'}{\overset{f'}{\rightleftarrows}} \bullet^{w'} \circlearrowleft^{h'_1}_{h'_2}.$$

Then straightforward computations yield that $P_{ec}(E) = \{w\}$, $X'_{ec}(E) = \{g, h\}$, and $X_{ec}(E) = \{\tilde{g}\}$. Similarly, $P_{ec}(F) = \{v', w'\}$, $X'_{ec}(F) = \{e', f'g', g'f', h'_1, h'_2\}$, and $X_{ec}(F) = \{\tilde{e'}\}$.

Example 3.7.3 Let E_T be the Toeplitz graph ${}^e\circlearrowright \bullet^u \xrightarrow{f} \bullet^v$. Then clearly $P_{ec}(E_T) = \emptyset$.

Remark 3.7.4 Let E be an arbitrary graph. These two observations are straightforward to verify.

(i) For any $c \in X'_{ec}$, $\tilde{c}^0 = T(c^0)$. Consequently, $\tilde{c}^0$ is a hereditary subset of E^0, which in turn yields that $P_{ec}(E)$ is a hereditary subset of E^0.
(ii) Given $c, d \in X'_{ec}$, $\tilde{c} \neq \tilde{d}$ if and only if $\tilde{c}^0 \cap \tilde{d}^0 = \emptyset$.

We analyze the structure of the ideal generated by $P_{ec}(E)$. Recall the construction of the hedgehog graph ${}_HE$ given in Definition 2.5.16.

Lemma 3.7.5 *Let E be an arbitrary graph and K any field. For every cycle c such that $c \in X'_{ec}$, the ideal $I(\tilde{c}^0)$ is isomorphic to a purely infinite simple Leavitt path algebra. Concretely, $I(\tilde{c}^0) \cong L_K({}_HE)$, where $H = \tilde{c}^0$.*

Proof Observing that H is a hereditary subset of E^0, we may use Theorem 2.5.19 and Remark 2.5.21(iii) to get that $I(\tilde{c}^0)$ is isomorphic to the Leavitt path algebra $L_K({}_HE)$. We will show that this Leavitt path algebra is purely infinite simple by invoking the Purely Infinite Simplicity Theorem 3.1.10.

To show that every vertex of ${}_HE$ connects to a cycle, take $v \in {}_HE^0$. If $v \in H$ then it connects to c by the definition of $H = \tilde{c}^0$. If $v \notin H$ then there is an $f \in ({}_HE)^1$ such that $s(f) = v$ and $r(f) \in H$. Hence v connects to c in this case as well.

Next, we show that every cycle in ${}_HE$ has an exit. Pick such a cycle d; then necessarily by the definition of ${}_HE$, d is a cycle in H. Since by construction we have $\tilde{d} = \tilde{c}$, this means that d connects to c and hence it has an exit in E, which is also an exit in ${}_HE$.

Finally, to show that the only hereditary saturated subsets of $({}_HE)^0$ are $\emptyset$ and $({}_HE)^0$, let $\emptyset \neq H' \in \mathscr{H}_{{}_HE}$, and consider $v \in H'$. Note that every pair of vertices in H is connected by a path, and that $({}_HE)^0$ is the saturation of H in ${}_HE$. Hence, if $v \in H$ then $H' = ({}_HE)^0$. If $v \notin H$ then there exists an $f \in ({}_HE)^1$ such that $v = s(f)$ and $r(f) \in H$. This implies $({}_HE)^0 \subseteq H'$, as desired. □

Theorem 3.7.6 *Let E be an arbitrary graph and K any field. Then*

$$I(P_{ec}(E)) = \oplus_{\tilde{c} \in X_{ec}} I(\tilde{c}^0).$$

Furthermore, $I(\tilde{c}^0)$ is isomorphic to a purely infinite simple Leavitt path algebra for each $\tilde{c} \in X_{ec}$.

Proof The hereditary set $P_{ec}(E)$ can be partitioned as $P_{ec}(E) = \sqcup_{\tilde{c}\in X_{ec}} \tilde{c}^0$. By Remark 3.7.4(ii) and Proposition 2.4.7, $I(P_{ec}(E)) = I(\sqcup_{\tilde{c}\in X_{ec}} \tilde{c}^0) = \oplus_{\tilde{c}\in X_{ec}} I(\tilde{c}^0)$. Finally, each $I(\tilde{c}^0)$ is isomorphic to a purely infinite simple Leavitt path algebra by Lemma 3.7.5. □

Lemma 3.7.7 *Let E be an arbitrary graph and K any field. Then the hereditary sets $P_l(E)$, $P_c(E)$ and $P_{ec}(E)$ are pairwise disjoint. Consequently, the ideal of $L_K(E)$ generated by their union is $I(P_l(E)) \oplus I(P_c(E)) \oplus I(P_{ec}(E))$.*

Proof By the definition of $P_l(E)$, $P_c(E)$ and $P_{ec}(E)$, they are pairwise disjoint. To get the result, apply Proposition 2.4.7. □

The ideal described in Lemma 3.7.7 will be of use later on, so we name it here.

Definition 3.7.8 Let E be an arbitrary graph and K any field. We define the ideal I_{lce} of $L_K(E)$ by setting

$$I_{lce} := I(P_l(E)) \oplus I(P_c(E)) \oplus I(P_{ec}(E)).$$

As mentioned at the start of this section, the ideal I_{lce} captures the essential structural properties of the three primary colors of Leavitt path algebras, a statement we now make more precise.

Theorem 3.7.9 *Let E be an arbitrary graph and K any field. Consider I_{lce}, the ideal of $L_K(E)$ presented in Definition 3.7.8. Then*

$$I_{lce} \cong (\oplus_{i\in\Gamma_1} \mathrm{M}_{\Lambda_i}(K)) \oplus \left(\oplus_{j\in\Gamma_2} \mathrm{M}_{\Lambda_j}(K[x,x^{-1}])\right) \oplus \left(\oplus_{l\in\Gamma_3} I(\tilde{c}_l^0)\right)$$

where:

Γ_1 is the index set of the decomposition of $P_l(E)$ into disjoint hereditary sets (i.e., $P_l(E) = \sqcup_{i\in\Gamma_1} H_i$ as in Lemma 2.6.13), and, for every $i \in \Gamma_1$, Λ_i denotes the set $\{\mu\mu^* \mid \mu \in \mathrm{Path}(E),\ r(\mu) \in H_i\}$;

Γ_2 is the index set of the cycles without exits in E, and for every $j \in \Gamma_2$, Λ_j is the set of distinct paths ending at the basis of a cycle without exits c_j and not containing all the edges of c_j; and

Γ_3 is the index set of $X_{ec}(E)$.

Proof This follows directly from Theorems 2.6.14, 2.7.3, and 3.7.6. □

In general the ideal I_{lce} of $L_K(E)$ need not be "large" in $L_K(E)$. For example, let F denote the "doubly infinite line graph" of Example 3.1.12. Since there are no cycles in F, we get vacuously that $P_{ec}(F) = \emptyset = P_c(F)$. Since there are no line points in F, we have $P_l(F) = \emptyset$, so that, by definition, $I_{lce}(F) = \{0\}$. So we have produced an example of the desired type. However, when E^0 is finite, we show below that the ideal I_{lce} is in fact essential in $L_K(E)$. The key is the following.

Lemma 3.7.10 *Let E be a graph for which E^0 is finite. Let $v \in E^0$. Then v connects to at least one of: a sink, a cycle without exits, or an extreme cycle.*

Proof Recall the preorder $\geq$ on E^0 presented in Definition 2.0.4. Consider the partial order $\geq'$ resulting from the antisymmetric closure of $\geq$. The statement will be proved once we show that the minimal elements in $(E^0, \geq')$ are sinks, vertices in cycles without exits, and vertices in extreme cycles.

Indeed, let $v \in E^0$ be a minimal element. If v is a sink, we are done. Otherwise, there exists a $w \in E^0$ such that $v \geq w$. The minimality of v implies $w \geq' v$, hence there is a closed path c in E such that $v, w \in c^0$. If c has no exits, we are done. Otherwise, let μ be a path in E of length ≥ 1 such that the first edge appearing in μ is an exit for c. Then $v \geq s(\mu)$. Again by the minimality of v we have $s(\mu) \geq' v$. This implies that every path starting at a vertex of c^0 returns to c^0 and so c is an extreme cycle as required. □

Proposition 3.7.11 *Let E be a graph for which E^0 is finite. Then I_{lce} is an essential ideal of $L_K(E)$.*

Proof Let $v \in E^0$. Since E^0 is finite then Lemma 3.7.10 ensures that v connects to a line point, or to a cycle without exits, or to an extreme cycle. This means that every vertex of E connects to the hereditary set $P_l(E) \cup P_c(E) \cup P_{ec}(E)$ and, consequently, to its hereditary saturated closure, which we denote by H. By Proposition 2.7.10 this means that $I(H)$ is an essential ideal of $L_K(E)$, and by Lemma 3.7.7 it coincides with I_{lce}. □

We note that although I_{lce} is an essential ideal of $L_K(E)$ when E^0 is finite, I_{lce} need not equal all of $L_K(E)$. We see this behavior in $L_K(E_T)$, where E_T is the Toeplitz graph as discussed in Example 3.7.3. Here we have $P_{ec}(E_T) = \emptyset = P_c(E_T)$, and $P_l(E_T)$ is the sink v. So $I_{lce}(E_T) = I(\{v\})$; but $I(\{v\}) \neq L_K(E_T)$, since $\{v\}$ is hereditary saturated.

3.8 Purely Infinite Without Simplicity

We conclude Chap. 3 by presenting a description of the purely infinite (but not necessarily simple) Leavitt path algebras arising from row-finite graphs. As happened in the purely infinite simple case (Sect. 3.1), an in-depth analysis of the idempotent structure of $L_K(E)$ will be required. Roughly speaking, the first half of this section (through Lemma 3.8.10) will be a discussion of the purely infinite notion for general rings, while the second half will be taken up in considering this notion in the specific context of Leavitt path algebras. Many of the fundamental ideas in this section can be found in the seminal paper [39].

The general theory of purely infinite rings works smoothly for s-unital rings, defined here.

Definition 3.8.1 A ring R is said to be *s-unital* if for each $a \in R$ there exists a $b \in R$ such that $a = ab = ba$. By Ara [18, Lemma 2.2], if R is s-unital then for each finite subset $F \subseteq R$ there is an element $u \in R$ such that $ux = x = xu$ for all $x \in F$.

Of course, all rings with local units are s-unital, so that all Leavitt path algebras fall under this umbrella. For an example of an s-unital ring without nonzero idempotents, consider the algebra $C_c(\mathbb{R})$ of those continuous functions on the real line having compact support.

We start by recalling the definitions of the properties *properly purely infinite* and *purely infinite* in a general non-unital, non-simple ring, introduced in [39]. We will then specialize to the simple case.

Definition 3.8.2 Let R be a ring, and suppose x and y are square matrices over R, say $x \in \mathrm{M}_k(R)$ and $y \in \mathrm{M}_n(R)$ for $k, n \in \mathbb{N}$. We use $\oplus$ to denote block sums of matrices; thus,

$$x \oplus y = \begin{pmatrix} x & 0 \\ 0 & y \end{pmatrix} \in \mathrm{M}_{k+n}(R),$$

and similarly for block sums of more than two matrices.

We define a relation $\precsim$ on matrices over R by declaring that $x \precsim y$ if and only if there exist $\alpha \in \mathrm{M}_{kn}(R)$ and $\beta \in \mathrm{M}_{nk}(R)$ such that $x = \alpha y \beta$.

Recall that the set of idempotent elements of a ring R is endowed with a partial order $\leq$ given by $e \leq f$ if and only if $e = ef = fe$. It is not hard to show that if x and y are idempotent matrices in $\mathrm{M}_{\mathbb{N}}(R)$, then $x \precsim y$ if and only if $x \sim f$, where f is an idempotent such that $f \leq y$. (The relation $\sim$ on idempotent matrices has been defined in Sect. 3.2.)

For any ring R and element $a \in R$, the expression RaR denotes the set of all finite sums $\sum_{i=1}^{n} z_i a t_i$, where $z_i, t_i \in R$. If R is s-unital, then RaR is precisely the ideal $I(a)$ of R generated by a.

Definitions 3.8.3 Let R be any ring.

(i) We call an element $a \in R$ *properly infinite* if $a \neq 0$ and $a \oplus a \precsim a$.
(ii) We call R *purely infinite* if the following two conditions are satisfied:

  (1) no quotient of R is a division ring, and
  (2) whenever $a \in R$ and $b \in RaR$, then $b \precsim a$ (i.e., $b = xay$ for some $x, y \in R$).

(iii) We call R *properly purely infinite* if every nonzero element of R is properly infinite.

Remark 3.8.4 Suppose R is a simple unital ring. Then we easily see that R is purely infinite (simple) if and only if R is not a division ring, and for all $0 \neq a \in R$ there exist $x, y \in R$ for which $1_R = xay$.

Lemma 3.8.5 *Let R be an s-unital ring.*

(i) *If R is properly purely infinite, then R is purely infinite.*
(ii) *If $\mathrm{M}_2(R)$ is purely infinite, then R is properly purely infinite.*

Proof (i) Suppose first that R/I is a division ring for some ideal I of R. Take a nonzero element $\overline{a}$ of R/I. Then a is a nonzero element in R, and thus by hypothesis is properly infinite. So there exist elements $\alpha_1, \alpha_2, \beta_1, \beta_2 \in R$ such that

$$\begin{pmatrix} a & 0 \\ 0 & a \end{pmatrix} = \begin{pmatrix} \alpha_1 & 0 \\ \alpha_2 & 0 \end{pmatrix} \begin{pmatrix} a & 0 \\ 0 & 0 \end{pmatrix} \begin{pmatrix} \beta_1 & \beta_2 \\ 0 & 0 \end{pmatrix}.$$

But then in R/I we have that

$$\begin{pmatrix} \overline{a} & 0 \\ 0 & \overline{a} \end{pmatrix} = \begin{pmatrix} \overline{\alpha}_1\overline{a}\overline{\beta}_1 & \overline{\alpha}_1\overline{a}\overline{\beta}_2 \\ \overline{\alpha}_2\overline{a}\overline{\beta}_1 & \overline{\alpha}_2\overline{a}\overline{\beta}_2 \end{pmatrix}.$$

Since $\overline{a} \neq 0$, it follows that $\overline{\alpha}_1, \overline{\alpha}_2, \overline{\beta}_1, \overline{\beta}_2$ are all nonzero. Now, since R/I is a division ring, $\overline{\alpha}_1\overline{a}\overline{\beta}_2 = 0$ implies $\overline{a} = 0$, a contradiction. This shows that no quotient of R is a division ring, so that Condition (1) of Definitions 3.8.3(ii) holds.

By using that R is s-unital, one can easily see that $r_1 + r_2 + \cdots + r_t \precsim r_1 \oplus r_2 \oplus \cdots \oplus r_t$ for all $r_1, \ldots, r_t \in R$, cf. [39, Lemma 2.2]. Now let $a \in R$ be properly infinite and $b \in RaR$. Write $b = \sum_{i=1}^{n} x_i a y_i$ for some $x_i, y_i \in R$. We have $x_i a y_i \precsim a$ for all $1 \leq i \leq n$, whence by the above, we have

$$b = \sum_{i=1}^{n} x_i a y_i \precsim x_1 a y_1 \oplus x_2 a y_2 \oplus \cdots \oplus x_n a y_n \precsim a \oplus a \oplus \cdots \oplus a \precsim a,$$

with the final $\precsim$ being a consequence of $a \oplus a \precsim a$. This establishes Condition (2) of Definitions 3.8.3(ii), and yields the result.

(ii) As R is s-unital, given $a \in R$ there exists a $u \in R$ such that $ua = au = a$. Hence,

$$a \oplus a = \begin{pmatrix} a & 0 \\ 0 & a \end{pmatrix} = \begin{pmatrix} u & 0 \\ 0 & 0 \end{pmatrix} \begin{pmatrix} a & 0 \\ 0 & 0 \end{pmatrix} \begin{pmatrix} u & 0 \\ 0 & 0 \end{pmatrix} + \begin{pmatrix} 0 & 0 \\ u & 0 \end{pmatrix} \begin{pmatrix} a & 0 \\ 0 & 0 \end{pmatrix} \begin{pmatrix} 0 & u \\ 0 & 0 \end{pmatrix} \in \mathrm{M}_2(R) \cdot a \oplus 0 \cdot \mathrm{M}_2(R).$$

Since $\mathrm{M}_2(R)$ is assumed to be purely infinite, it follows that $a \oplus a \precsim a \oplus 0$, and so $a \oplus a \precsim a$. Therefore a is either zero or properly infinite. □

For notational convenience, we will often write various square matrix expressions as sums and products of non-square matrices; for instance, $a \oplus a = \begin{pmatrix} a \\ 0 \end{pmatrix} (1 \;\; 0) + \begin{pmatrix} 0 \\ a \end{pmatrix} (0 \;\; 1)$.

The concepts of properly purely infinite and purely infinite agree for *simple* s-unital rings. Moreover, in this case we can relate these conditions to the existence of infinite idempotents in all nonzero right (or left) ideals, see Proposition 3.8.8

below. (However, there exist simple non-s-unital rings which are purely infinite but not properly purely infinite, see e.g., [39, Example 3.5].)

We first show, in the next few lemmas, that every simple s-unital purely infinite ring contains nonzero idempotents.

Lemma 3.8.6 *Let R be a not-necessarily-unital ring, and suppose that R contains nonzero elements x, y, u, v satisfying the relations*

$$vu = uv = u, \quad yu = y, \quad vx = x, \quad v = yx. \tag{3.1}$$

Then R contains a nonzero idempotent.

Proof Let $\widetilde{R}$ denote a ring obtained by adjoining a unit to R. Then in $\widetilde{R}$ we have

$$(y+(1-v))(x+(1-u)) = yx+y(1-u)+(1-v)x+(1-v)(1-u) = v+0+0+(1-v) = 1.$$

It follows that $e = (x + (1 - u))(y + (1 - v))$ is an idempotent in $\widetilde{R}$, whence $1 - e$ is an idempotent which is easily seen to belong to R.

If $e \neq 1$, then $1 - e$ is the desired nonzero idempotent in R. If $e = 1$, then $y = yeu = y(x + (1 - u))(y + (1 - v))u = yxy \in R$, which shows that $v = yx$ is a (nonzero) idempotent in R. □

Lemma 3.8.7 *If R is s-unital, simple, and purely infinite then R contains a nonzero idempotent.*

Proof Let $0 \neq x \in R$, so (as R is s-unital) there exists an $a \in R$ with $ax = xa = x$. Then $0 \neq x = xa = xa^2$, so that $a^2 \neq 0$. Now using twice the s-unitality, we see that there are $b, c \in R$ such that $ab = ba = a$ and $bc = cb = b$. Since R is purely infinite, there exist $s, t \in R$ such that $c = sa^2t$. So we have

$$ab = ba = a, \quad bc = cb = b, \quad \text{and } c = sa^2t.$$

Define $x = at$, $y = sa$, $v = c$, and $u = b$. Then $vu = uv = u$, $yx = sa^2t = v$, $vx = cat = cbat = bat = at = x$, and $yu = sab = sa = y$. So x, y, u, v are nonzero elements of R satisfying the relations (3.1), and thus it follows from Lemma 3.8.6 that R contains a nonzero idempotent. □

We now obtain the promised characterization of purely infinite simple s-unital rings. In particular, all the conditions below are equivalent for a simple Leavitt path algebra.

Proposition 3.8.8 *Let R be a simple s-unital ring. Then the following are equivalent:*

(1) *R is properly purely infinite.*
(2) *R is purely infinite.*
(3) *For every nonzero $a \in R$ there exist $s, t \in R$ such that sat is a nonzero, infinite idempotent.*
(4) *Every nonzero one-sided ideal of R contains a nonzero infinite idempotent.*

Proof (1) $\Rightarrow$ (2) follows from Lemma 3.8.5(i).

(2) $\Rightarrow$ (3). By Lemma 3.8.7 R contains a nonzero idempotent w. By simplicity of R, $w \in RaR$, so since R is purely infinite there exist s, t in R such that $w = sat$.

We show that every nonzero idempotent in R is infinite, which will complete the argument. Let e be such. Assume first that e is a unit for R. Then, since R is not a division ring, there is a nonzero a in R such that a is not left invertible in R. Again invoking the simplicity and purely infiniteness of R, there exist $s, t \in R$ be such that $sat = e$. Then $f := tsa$ is an idempotent in R with $e \sim f$ and $f \neq e$, which implies that e is infinite. Finally, assume that e is not a unit for R. We may assume that $(1-e)x \neq 0$ for some $x \in R$, where here $1 \in \widetilde{R}$ if R is not unital. As before we can find an idempotent $f \in (1-e)xR$ such that $f \sim e$. But now $g := f(1-e)$ is an idempotent in R orthogonal to e, and equivalent to e. Since $e + g = uev$ for some $u, v \in R$, there is an idempotent $h \leq e$ such that $h \sim e + g \sim e \oplus e$, showing indeed that e is properly infinite. This completes the argument.

(3) $\Rightarrow$ (4) is contained in Proposition 3.1.7.

(4) $\Rightarrow$ (1). First observe that, as R is a simple ring, every infinite idempotent in R is indeed properly infinite. Now let a be a nonzero element in R. By assumption, there is a properly infinite idempotent e in R such that $e \precsim a$. Since R is simple there exists an $n \geq 1$ such that $a \precsim n \cdot e = e \oplus e \oplus \cdots \oplus e$. Thus we get

$$a \oplus a \precsim n \cdot e \oplus n \cdot e \precsim e \precsim a,$$

showing that a is properly infinite. □

Lemma 3.8.9 *Let I be an ideal of an arbitrary ring R.*

(i) *If R is (properly) purely infinite, then so is R/I.*
(ii) *Suppose that I is s-unital when viewed as a ring. If R is (properly) purely infinite, then so is I.*

Proof (i) It is clear that proper pure infiniteness passes from R to R/I. Now assume only that R is purely infinite. Since any quotient of R/I is also a quotient of R, no quotient of R/I is a division ring. Consider $a, b \in R$ such that $\overline{b} \in (R/I)\overline{a}(R/I)$. Then there is some $c \in RaR$ such that $\overline{c} = \overline{b}$. By hypothesis, $c = xay$ for some $x, y \in R$, and therefore $\overline{b} = \overline{c} = \overline{xay}$.

(ii) Assume first the specific case in which R is properly purely infinite, and let $0 \neq a \in I$. Then there exist $\alpha_1, \alpha_2, \beta_1, \beta_2 \in R$ such that $\begin{pmatrix} a & 0 \\ 0 & a \end{pmatrix} = \begin{pmatrix} \alpha_1 \\ \alpha_2 \end{pmatrix} a \begin{pmatrix} \beta_1 & \beta_2 \end{pmatrix}$. Since I is s-unital, we also have $a = ua = au$ for some $u \in I$. Then

$$\begin{pmatrix} a & 0 \\ 0 & a \end{pmatrix} = \begin{pmatrix} \alpha_1 u \\ \alpha_2 u \end{pmatrix} a \begin{pmatrix} u\beta_1 & u\beta_2 \end{pmatrix}$$

with $\alpha_1 u, \alpha_2 u, u\beta_1, u\beta_2 \in I$. This proves that I is properly purely infinite.

Now assume the general case, so we assume only that R is purely infinite. Suppose first that I has an ideal J such that I/J is a division ring. Since I is s-unital, J is an ideal of R. Since R/J is purely infinite by (i), it suffices to find a contradiction working in R/J. Thus, there is no loss of generality in assuming that $J = 0$. If e is the unit of I, then $I = eI = Ie$, and so $I = eR = Re$. It follows that $er = ere = re$ for all $r \in R$, whence e is a central idempotent of R. But then the annihilator of e in R is an ideal T such that $R = I \oplus T$, and $R/T \cong I$ is a division ring, contradicting the assumption that R is purely infinite. Therefore no quotient of I is a division ring.

Secondly, if $a \in I$ and $b \in IaI$, then we at least have $b = xay$ for some $x, y \in R$. Since also $a = ua = au$ for some $u \in I$, we have $b = (xu)a(uy)$ with $xu, uy \in I$. Thus I satisfies the two required conditions, and is therefore purely infinite. □

Lemma 3.8.10 *Let e be an idempotent in a ring R. If R is (properly) purely infinite, then so is eRe.*

Proof Assume first that R is properly purely infinite. Any nonzero element $a \in R$ is properly infinite in R, and so $\begin{pmatrix} a & 0 \\ 0 & a \end{pmatrix} = \begin{pmatrix} \alpha_1 \\ \alpha_2 \end{pmatrix} a \begin{pmatrix} \beta_1 & \beta_2 \end{pmatrix}$ for some $\alpha_1, \alpha_2, \beta_1, \beta_2 \in R$. Then

$$\begin{pmatrix} a & 0 \\ 0 & a \end{pmatrix} = \begin{pmatrix} e\alpha_1 e \\ e\alpha_2 e \end{pmatrix} a \begin{pmatrix} e\beta_1 e & e\beta_2 e \end{pmatrix},$$

which shows that a is properly infinite in eRe. Therefore eRe is properly purely infinite in this case.

Now assume only that R is purely infinite. We first show that a prime purely infinite ring does not contain idempotents e such that eRe is a division ring. To do so, suppose that R is a prime purely infinite ring, and we have an idempotent $e \in R$ such that eRe is a division ring. Since R is prime, eR is a simple right R-module.

If $eR = R$, then $(R(1-e))^2 = 0$ and so $R(1-e) = 0$ because R is prime. (Here we are writing $R(1-e)$ for the left ideal $\{r - re \mid r \in R\}$.) But then $R = eRe$ and R is a division ring, contradicting the hypothesis that R is purely infinite. Thus, $eR \neq R$ and so $(1-e)R \neq 0$. Now $(1-e)ReR \neq 0$ because R is prime, and hence there exists a nonzero element $a \in (1-e)Re$. Note that aR is a nonzero homomorphic image of eR, whence aR is a simple right R-module. Since R is prime, $aR = gR$ for some idempotent g, and $eg = 0$ because $ea = 0$. Observe that $g - ge$ is an idempotent which generates gR, so we can replace g by $g - ge$. Hence, there is no loss of generality in assuming that $eg = ge = 0$.

Now $f = e + g$ is an idempotent such that $fR = eR \oplus aR$, and $f \in ReR$ because $gR = aR \subseteq ReR$. Since R is purely infinite, $f = xey$ for some $x, y \in R$. But then fR is a homomorphic image of eR, implying that fR is simple or zero, which is impossible in light of $fR = eR \oplus aR$. This contradiction establishes our claim.

Suppose now that I is an ideal of eRe such that eRe/I is a division ring. In this case I is a maximal ideal of eRe. Moreover, $e \notin (eRe)I(eRe) = eRIRe$, and so $e \notin RIR$. Consequently, $\overline{e}$ is a nonzero idempotent in R/RIR, and in particular, $\overline{e}$ cannot be in the Jacobson radical of R/RIR. Hence, there exists a (left) primitive ideal P of R such that $e \notin P$ and $RIR \subseteq P$. Now $I \subseteq P \cap eRe \subsetneqq eRe$, and by maximality of I in eRe we have $I = P \cap eRe$. This yields $eRe/I = eRe/(P \cap eRe) \cong \overline{e}(R/P)\overline{e}$. But this means that the purely infinite prime ring R/P has a corner which is a division ring, contradicting the claim above. Therefore no quotient of eRe is a division ring.

Establishing the second condition is easier. Suppose that $a \in eRe$ and $b \in (eRe)a(eRe) \subseteq RaR$. Since R is purely infinite, there exist $x, y \in R$ such that $b = xay$, and hence $b = (exe)a(eye)$ with $exe, eye \in eRe$. This shows that eRe is purely infinite. □

Now that the general theory of purely infinite rings has been described, we use this information in the context of Leavitt path algebras. Our first goal is to characterize the properly infinite vertices of a Leavitt path algebra. Recall that a characterization of the infinite vertices has been given in Proposition 3.1.6.

Lemma 3.8.11 *Let E be an arbitrary graph and K any field. If $v \in E^0$ and $|CSP(v)| \geq 2$, then v is a properly infinite idempotent in $L_K(E)$.*

Proof Let $e_1 \cdots e_m$ and $f_1 \cdots f_n$ be two distinct closed simple paths in E based at v. Then there is some positive integer t such that $e_i = f_i$ for $i = 1, \ldots, t-1$ while $e_t \neq f_t$. Thus, we have at least two different edges leaving the vertex $r(e_{t-1}) = r(f_{t-1})$. We compute that

$$v = s(e_1) \gtrsim r(e_1) \gtrsim \cdots \gtrsim r(e_{t-1}) \gtrsim r(e_t) \oplus r(f_t) \gtrsim r(e_{t+1}) \oplus r(f_{t+1}) \gtrsim \cdots \gtrsim r(e_m) \oplus r(f_n) = v \oplus v.$$

Therefore v is properly infinite. □

Recall that for $X \subseteq E^0$, we denote by $\overline{X}$ the hereditary saturated closure of X.

Proposition 3.8.12 *Let E be an arbitrary graph and K any field. Let $v \in E^0$. Then v is a properly infinite idempotent in $L_K(E)$ if and only if there are vertices $w_1, \ldots, w_n$ in $T(v)$ such that $|CSP(w_i)| \geq 2$ for all i and $v \in \overline{\{w_1, \ldots, w_n\}}$.*

Proof Assume that v is properly infinite. Let W be the set of vertices w in $T(v)$ such that $|CSP(w)| \geq 2$. If $I(v) = I(W)$ then there is a finite number $w_1, \ldots, w_n$ of elements of W such that $I(v) = I(\{w_1, \ldots, w_n\})$. It then follows that $v \in \overline{\{w_1, \ldots, w_n\}}$. It suffices therefore to show that $I(v) = I(W)$. On the contrary, suppose $I(W)$ is strictly contained in $I(v)$. Then by Zorn's Lemma there exists a hereditary saturated subset H properly contained in $\overline{T(v)}$ and containing $\overline{W}$. Then $L_K(E)/I(H \cup B_H^H) \cong L_K(E/H)$, and $X := \overline{T(v)} \setminus H$ is a hereditary saturated subset of E/H not containing any non-trivial hereditary saturated subsets. By Theorem 2.5.19 we have $I(v)/I(H \cup B_H^H) \cong L_K({}_X(E/H))$, and $L_K({}_X(E/H))$ is graded simple. Moreover, v is a properly infinite idempotent in $L_K({}_X(E/H))$, and it follows from the Trichotomy Principle 3.1.14 that $L_K({}_X(E/H))$ is purely infinite simple. Therefore there exists a $w \in T_{E/H}(v)$ such that $|CSP_{E/H}(w)| \geq 2$. Thus we obtain $w \in T(v) \backslash H$ and $|CSP_E(w)| \geq 2$, so that $w \in W \backslash H$, which is a contradiction, and thereby establishes one direction.

Conversely, assume that there are distinct vertices $w_1, \ldots, w_n$ in $T(v)$ such that $|CSP(w_i)| \geq 2$ for all i and $v \in \overline{\{w_1, \ldots, w_n\}}$. By Lemma 3.8.11, $e := w_1 + w_2 + \cdots + w_n$ is a properly infinite idempotent of $L_K(E)$. We claim that $e \lesssim v$. If $w_j \in T(w_i)$ for $i \neq j$, then $w_i \oplus w_j \lesssim w_j \oplus w_j \lesssim w_j$ and so we can eliminate such w_i. Thus we may assume without loss of generality that $w_i \notin T(w_j)$ for all $i \neq j$. For each i, let $\gamma_i \in \text{Path}(E)$ with $s(\gamma_i) = v$ and $r(\gamma_i) = w_i$. Since

$w_i \notin T(w_j)$ for all i, we see that the paths $\gamma_1, \gamma_2, \cdots, \gamma_n$ are pairwise incomparable, so that $\gamma_i^* \gamma_j = 0$ if $i \neq j$, and thus

$$g := \gamma_1\gamma_1^* + \gamma_2\gamma_2^* + \cdots + \gamma_n\gamma_n^*$$

is an idempotent such that $g \leq v$, and such that

$$e = w_1 + w_2 + \cdots + w_n \sim g.$$

It follows that $w_1 + w_2 + \cdots + w_n \lesssim v$. Since $I(v) = I(\{w_1, \ldots, w_n\}) = I(w_1 + \cdots + w_n)$, we have $v \lesssim \ell \cdot (w_1 + \cdots w_n) = \ell \cdot e$ for some $\ell \in \mathbb{N}$. Finally we have

$$v \oplus v \lesssim 2\ell \cdot (w_1 + \cdots + w_n) \lesssim w_1 + \cdots + w_n \lesssim v\,,$$

which shows that v is properly infinite. □

Remark 3.8.13 It follows easily from Proposition 3.8.12 that, for a vertex v of an arbitrary graph E, if v is a properly infinite idempotent in $L_K(E)$, then $|CSP(v)|$ is either 0 or ≥ 2.

Definition 3.8.14 An element a of a ring R is said to be an *infinite element* if $a \oplus b \lesssim a$ for some nonzero element $b \in R$. Obviously, a properly infinite element of R is an infinite element of R.

Lemma 3.8.15 *Let E be an arbitrary graph and K any field. Suppose that every nonzero ideal of every quotient of $L_K(E)$ contains an infinite element. Then E satisfies Condition (K), and $B_H = \emptyset$ for every $H \in \mathscr{H}_E$.*

Proof To show that E satisfies Condition (K), we have to check that $C_H = \emptyset$ for every $H \in \mathscr{H}_E$ (see the proof of Corollary 2.9.9). If $C_H \neq \emptyset$ for some $H \in \mathscr{H}_E$, then by the Structure Theorem for Ideals 2.8.10 there is a subquotient of $L_K(E)$ isomorphic to $\mathrm{M}_\Lambda(p(x)K[x, x^{-1}])$, for some set Λ, where $p(x)$ is a polynomial of the form $1 + a_1x + \cdots + a_nx^n$, with $n > 0$ and $a_n \neq 0$. Since $K[x, x^{-1}]$ embeds into a field, rank considerations show immediately that there are no infinite elements in the ring $\mathrm{M}_\Lambda(p(x)K[x, x^{-1}])$. Therefore our hypothesis implies that $C_H = \emptyset$ for all $H \in \mathscr{H}_E$.

Now suppose that, for some $H \in \mathscr{H}_E$, we have $B_H \neq \emptyset$. Then the algebra $L_K(E)/I(H) \cong L_K(E/(H, \emptyset))$ has a nonzero socle, indeed the ideal $I(H \cup B_H^H)/I(H)$ is a nonzero ideal of $L_K(E)/I(H)$ contained in the socle of $L_K(E)/I(H)$ (see Theorem 2.4.15). Since clearly the socle (of any semiprime ring) cannot contain infinite elements, we obtain a nonzero subquotient of $L_K(E)$ with no infinite elements, contradicting our hypothesis. □

Recall that a nonzero element u of a conical monoid V is said to be *irreducible* if u cannot be written as a sum of two nonzero elements [39, Definitions 6.1]. Observe that, for an idempotent e of a ring R, we have that $[e]$ is irreducible in $\mathscr{V}(R)$ if and only if e is a primitive idempotent of R.

We are now in position to present the main result of this section, in which we characterize the purely infinite Leavitt path algebras.

Theorem 3.8.16 *Let E be an arbitrary graph and K any field. The following are equivalent.*

(1) *Every nonzero ideal of every quotient of $L_K(E)$ contains an infinite vertex, i.e., if $I \subsetneqq J$ are ideals of $L_K(E)$, then there exists a $v \in E^0$ such that $v \in J \setminus I$ and such that $v + I$ is an infinite idempotent of $L_K(E)/I$.*
(2) *Every nonzero right ideal of every quotient of $L_K(E)$ contains an infinite idempotent.*
(3) *Every nonzero left ideal of every quotient of $L_K(E)$ contains an infinite idempotent.*
(4) *$L_K(E)$ is properly purely infinite.*
(5) *$L_K(E)$ is purely infinite.*
(6) *Every vertex $v \in E^0$ is properly infinite as an idempotent in $L_K(E)$, and $B_H = \emptyset$ for all $H \in \mathscr{H}_E$.*

Proof We recall that $L_K(E)$ has local units (cf. Lemma 1.2.12(v)), so that all previously established results about s-unital rings apply here.

(1) $\Rightarrow$ (2) and (3). Observe that Lemma 3.8.15 gives that E satisfies Condition (K) and that $B_H = \emptyset$ for every $H \in \mathscr{H}_E$. Therefore, by Theorem 3.3.11 and the Structure Theorem for Ideals 2.8.10, all the ideals of $L_K(E)$ are of the form $I(H)$ for some $H \in \mathscr{H}_E$. So a nonzero quotient of $L_K(E)$ will be of the form $L_K(E/H)$. Moreover, by Theorem 3.3.11, each such E/H necessarily satisfies Condition (L).

Let v be a vertex of E/H. If v does not connect to any cycle in E/H, then $\overline{T_{E/H}(v)}$ is an acyclic graph, and thus the ideal generated by v in $L_K(E/H)$ does not contain any infinite vertex, by Proposition 3.1.6, contradicting (1). Therefore every vertex of E/H connects to a cycle with exits, and again by Proposition 3.1.6, we get that every vertex is infinite.

By Proposition 2.9.13, every nonzero one-sided ideal of $L_K(E/H)$ contains a nonzero idempotent. By Corollary 3.2.11, it only remains to show that every idempotent of the form $v - \sum_{e\in Z} ee^*$, where $v \in \text{Inf}(E/H)$ and Z is a nonempty finite subset of $s^{-1}_{E/H}(v)$, is infinite. But in this situation we can choose $f \in s^{-1}_{E/H}(v) \setminus Z$, and $ff^* \leq v - \sum_{e\in Z} ee^*$, with $ff^* \sim f^*f = r(f)$, which is an infinite idempotent in $L_K(E/H)$ by the above. It follows that every nonzero idempotent of $L_K(E/H)$ is infinite, and so every nonzero one-sided ideal of $L_K(E/H)$ contains an infinite idempotent.

(2) or (3) $\Rightarrow$ (4). This holds in any s-unital ring, see e.g., [39, Proposition 3.13].

(4) $\Rightarrow$ (5). This implication also holds in any s-unital ring, by Lemma 3.8.5(i).

(5) $\Rightarrow$ (6). Let v be a vertex in E. By Proposition 3.6.21, $\mathscr{V}(L_K(E))$ is a refinement monoid. Hence, by Aranda Pino et al. [39, Theorem 6.10], in order to show that v is properly infinite as an idempotent of $L_K(E)$, it suffices to show that $[v]$ is not irreducible in any quotient of $\mathscr{V}(L_K(E))$.

By Theorem 3.6.23(i), any order-ideal I of $\mathscr{V}(L_K(E))$ is of the form $\mathscr{V}(I(H \cup S^H))$, where H is a hereditary saturated subset of E^0 and $S \subseteq B_H$.

Moreover, it follows from Theorem 3.6.23(ii) that we have monoid isomorphisms

$$\mathcal{V}(L_K(E))/I \cong \mathcal{V}(L_K(E)/I(H \cup S^H)) \cong \mathcal{V}(L_K(E/(H,S))).$$

Since there is nothing to do if $[v] \in I$, we may assume that $v \notin H$. By Lemma 3.8.9(i), $L_K(E/(H,S)) \cong L_K(E)/I(H \cup S^H)$ is purely infinite, and so for this part of the proof we may replace $L_K(E)$ by $L_K(E/(H,S))$. Thus, we need only show that $[v]$ is not irreducible in $\mathcal{V}(L_K(E))$, or equivalently, that v is not a primitive idempotent.

By Proposition 3.5.2, if v is a primitive idempotent then there cannot be any bifurcations in $T(v)$. So either v is a line point, or there is a unique shortest path connecting v to a cycle without exits. So we get that either $vL_K(E)v \cong K$, or $vL_K(E)v \cong K[x,x^{-1}]$. In any case $vL_K(E)v$ is not properly infinite, contradicting Lemma 3.8.10.

We now show that $B_H = \emptyset$ for every $H \in \mathcal{H}_E$. Let $H \in \mathcal{H}_E$. Then $L_K(E)/I(H) \cong L_K(E/(H,\emptyset))$ is properly infinite by Lemma 3.8.9(i), so by the preceding argument every vertex of $L_K(E/(H,\emptyset))$ is properly infinite. But if $v \in B_H$ then the idempotent v' in the graph $E/(H,\emptyset)$ (which corresponds to the class of v^H) belongs to the socle of $L_K(E/(H,\emptyset))$ and so cannot be properly infinite. This shows that $B_H = \emptyset$.

(6) $\Rightarrow$ (1). By Proposition 3.8.12, for every $v \in E^0$ there exist $w_1, \dots, w_n \in T(v)$ such that $|CSP(w_i)| \geq 2$ for all i such that $I(v) = I(\{w_1, \dots, w_n\})$. It follows in particular that E satisfies Condition (L). Since the same is true for every graph E/H, where H is a hereditary saturated subset of E^0, we conclude that E satisfies Condition (K) by Theorem 3.3.11. It follows from Proposition 2.9.9 that every ideal of $L_K(E)$ is a graded ideal. Since $B_H = \emptyset$ for every $H \in \mathcal{H}_E$, it follows from the Structure Theorem for Graded Ideals 2.5.8 that every ideal of $L_K(E)$ is of the form $I(H)$ for some $H \in \mathcal{H}_E$. Thus every nonzero ideal of every quotient $L_K(E)/I(H) \cong L_K(E/H)$ of $L_K(E)$ contains a vertex (by Proposition 2.9.13), which is necessarily (properly) infinite. □

As a result of Proposition 3.8.12, Condition (6) of Theorem 3.8.16 provides a characterization of purely infinite Leavitt path algebras $L_K(E)$ which depends solely on properties of the graph E, which we record here.

Corollary 3.8.17 *Let E be an arbitrary graph and K any field. The following are equivalent.*

(1) $L_K(E)$ is purely infinite.
(2) $B_H = \emptyset$ for all $H \in \mathcal{H}_E$, and for every $v \in E^0$ there exist $w_1, \dots, w_n \in T(v)$ for which: $|CSP(w_i)| \geq 2$ $(1 \leq i \leq n)$, and $v \in \overline{\{w_1, \dots, w_n\}}$.

Example 3.8.18 We present an example of a purely infinite non-simple Leavitt path algebra. Consider the following graph E:

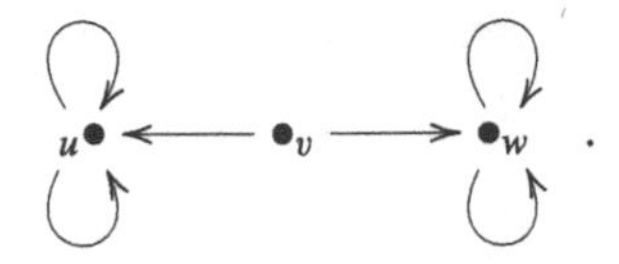

By Corollary 3.8.17 we see that $L_K(E)$ is purely infinite; note in particular that $v \in \{w_1, w_2\}$. On the other hand, $L_K(E)$ is non-simple because $\{u\}$ and $\{w\}$ are nontrivial hereditary saturated subsets of E.

We close the chapter by recording the following consequence of Theorem 3.8.16. Because Condition (2) of Corollary 3.8.17 easily gives that no vertex v of E can have $|CSP(v)| = 1$, Theorem 3.3.11 gives

Corollary 3.8.19 *Let E be an arbitrary graph and K any field. If $L_K(E)$ is purely infinite then $L_K(E)$ is an exchange ring.*

It is not known as of 2017 whether Corollary 3.8.19 can be extended to all purely infinite rings.

Chapter 4
General Ring-Theoretic Results

In the first three chapters we have explored a number of ideas and constructions which yield ring-theoretic information about Leavitt path algebras. In this chapter we continue this line of investigation. Specifically, in the first three sections we identify those graphs E for which the Leavitt path algebra $L_K(E)$ satisfies various standard ring-theoretic properties, including primeness, primitivity, one-sided chain conditions, semisimplicity, and self-injectivity. In the final section we explore the the stable rank of Leavitt path algebras.

4.1 Prime and Primitive Ideals in Leavitt Path Algebras of Row-Finite Graphs

The prime spectrum Spec(R) and the primitive spectrum Prim(R) of a ring R have played key roles in the history of ring theory, initially in the commutative setting, but importantly in the non-commutative setting as well. In this section we identify both the prime and primitive ideals of a Leavitt path algebra in terms of the graph E, where E is row-finite. In this setting, much of this work was completed in [41]. The prime ideal structure for Leavitt path algebras of arbitrary graphs has been given in [131], while the primitive Leavitt path algebras are described in [10]; further discussion of the latter algebras appears in Sect. 7.2 below.

We recall a few ring-theoretic definitions. A two-sided ideal P of a ring R is *prime* if $P \neq R$ and P has the property that for any two-sided ideals I, J of R, if $IJ \subseteq P$ then either $I \subseteq P$ or $J \subseteq P$. The ring R is called *prime* if $\{0\}$ is a prime ideal of R. It is easily shown that P is a prime ideal of R if and only if R/P is a prime ring. The set of all prime ideals of R is denoted by Spec(R). If R is a group-graded ring, a graded ideal P is *graded prime* if P satisfies the condition $IJ \subseteq P \Rightarrow I \subseteq P$ or $J \subseteq P$ for all graded two-sided ideals I, J of R. We denote the set of graded prime ideals of R by gr-Spec(R). It is shown in [119, Proposition II.1.4] that for a $\mathbb{Z}$-graded ring R, if

G. Abrams et al., *Leavitt Path Algebras*, Lecture Notes in Mathematics 2191,
DOI 10.1007/978-1-4471-7344-1_4

P is a graded ideal of R, then P is prime if and only if P is graded prime; we shall make use of this result throughout this section without explicit mention.

The prime ideals of the principal ideal domain $K[x,x^{-1}] \cong L_K(R_1)$ provide a model for the prime spectra of general Leavitt path algebras. The key property of R_1 in this setting is that it contains a unique cycle without exits. Specifically, $\mathrm{Spec}(K[x,x^{-1}])$ consists of the ideal $\{0\}$, together with ideals generated by the irreducible polynomials of $K[x,x^{-1}]$. These irreducible polynomials are in turn the polynomials of the form $x^n f(x)$, where $f(x)$ is an irreducible polynomial in the standard polynomial ring $K[x]$, and $n \in \mathbb{Z}$. (Note that x^n is a unit in $K[x,x^{-1}]$ for all $n \in \mathbb{Z}$.) In particular, there is exactly one graded prime ideal (namely, $\{0\}$) in $L_K(R_1)$. Moreover, all the remaining prime ideals of $L_K(R_1)$ are nongraded, (obviously) contain the graded ideal, and correspond to irreducible polynomials in $K[x,x^{-1}]$.

In [41], a correspondence is established between the prime spectrum $\mathrm{Spec}(L_K(E))$ of a Leavitt path algebra on the one hand, and a relatively simple set (built from the underlying graph together with $\mathrm{Spec}(K[x,x^{-1}])$) on the other. To construct this set, we recall a few basic definitions.

Definition 4.1.1 A subgraph F of a graph E is called *full* if for each $v, w \in F^0$,

$$\{f \in F^1 \mid s(f) = v,\ r(f) = w\} = \{e \in E^1 \mid s(e) = v,\ r(e) = w\}.$$

In other words, the subgraph F is full if whenever two vertices of E are in the subgraph, then all of the edges connecting those two vertices in E are also in F.

Recall that for vertices v, w in E^0, we write $v \geq w$ if there is a path $p \in \mathrm{Path}(E)$ for which $s(p) = v$ and $r(p) = w$.

Definition 4.1.2 Let E be an arbitrary graph. A nonempty full subgraph M of E is a *maximal tail* if it satisfies the following properties:

(MT1) If $v \in E^0$, $w \in M^0$, and $v \geq w$, then $v \in M^0$;
(MT2) if $v \in M^0$ and $s_E^{-1}(v) \neq \emptyset$, then there exists an $e \in E^1$ such that $s(e) = v$ and $r(e) \in M^0$; and
(MT3) if $v, w \in M^0$, then there exists a $y \in M^0$ such that $v \geq y$ and $w \geq y$.

Condition (MT3) is now more commonly called *downward directedness*.

In order to identify maximal tails, the result that follows will be very useful.

Lemma 4.1.3 *Let E be an arbitrary graph and let M be a full subgraph of E. Then M satisfies Conditions* (MT1) *and* (MT2) *if and only if $H = E^0 \setminus M^0 \in \mathscr{H}_E$.*

Proof Suppose first that M is a maximal tail. Consider $v \in H$ and $w \in E^0$ such that $v \geq w$. If $w \notin H$ then $w \in M^0$, and by Condition (MT1) we get $v \in M^0 = E^0 \setminus H$, a contradiction. This shows that H is hereditary. Now, let $v \in E^0$ with $0 < |s^{-1}(v)| < \infty$, and suppose that $r(s^{-1}(v)) \subseteq H$. If $v \notin H$ then by Condition (MT2), there exists an $e \in s^{-1}(v)$ such that $r(e) \notin H$, a contradiction. This proves that H is saturated.

Let us prove the converse. Take $v \in E^0$ and $w \in M^0$ such that $v \geq w$. If $v \notin M^0$ then, as H is hereditary, we get that $w \in H$. Consider now $v \in M^0$ with $0 < |s^{-1}(v)| < \infty$. If for every $e \in s^{-1}(v)$ we have that $r(e) \notin M^0$, then $r(s^{-1}(v)) \subseteq H$, and by saturation we obtain $v \in H$, a contradiction. □

In Example 4.1.7 below we present some specific computations regarding maximal tails. The following result gets us off the ground in our investigation of the prime ideals of $L_K(E)$. (A description of the quotient graph E/H, which plays a key role in this discussion, is given in Definition 2.4.11.)

Proposition 4.1.4 ([40, Proposition 5.6]) *Let E be a row-finite graph and K any field. Let H be a hereditary saturated subset of E^0. Then the (graded) ideal $I(H)$ of $L_K(E)$ is prime if and only if $M = E/H$ is a maximal tail in E, if and only if M is downward directed.*

In particular, $L_K(E)$ is a prime ring if and only if E is downward directed.

Proof By Lemma 4.1.3, Conditions (MT1) and (MT2) on the graph $M = E/H$ are equivalent to having $H \in \mathcal{H}_E$.

So we show that $I(H)$ is prime if and only if $M = E/H$ is downward directed. This is equivalent to showing that $I(H)$ is graded prime. So suppose M is downward directed, and suppose $I(H) \supseteq I_1 I_2$ for some graded ideals I_1, I_2 of $L_K(E)$. By Proposition 2.4.9 there exist $H_1, H_2 \in \mathcal{H}_E$ for which $I_1 = I(H_1)$ and $I_2 = I(H_2)$. If $H_1 \subseteq H$ then we are done. Otherwise, there exists a $v \in M^0 \cap H_1$. Now take any $w \in H_2$. If $w \notin H$ then by downward directedness there exists a $y \in M^0$ for which $v \geq y$ and $w \geq y$, which gives $y \in H_1 \cap H_2$, so that $y \in I(H_1 \cap H_2)$, which in turn by Corollary 2.5.11 gives $y \in I(H_1)I(H_2) \subseteq I(H)$. But this is impossible, since $y \in M^0 = E^0 \setminus H$. Thus $w \in H$, so that $H_2 \subseteq H$ as desired.

The converse is established in a similar manner.

The final statement is clear, as $\{0\} = I(\emptyset)$, and $E = E/\emptyset$. □

The analysis of prime ideals for Leavitt path algebras of non-row-finite graphs requires heavier machinery than that utilized in Proposition 4.1.4, owing to the existence of prime ideals arising from sets which include breaking vertices. See [131] for a complete description of this situation. However, the generalization of the final statement of Proposition 4.1.4 does in fact hold if $\{0\}$ is a prime ideal. Since the proof for the $\{0\}$ ideal is similar to that given above, we simply state that generalization.

Proposition 4.1.5 *Let E be an arbitrary graph and K any field. Then $L_K(E)$ is prime if and only if E is downward directed.*

For the remainder of this section, H will denote a hereditary saturated subset of E^0, and M will denote the quotient graph E/H. We will analyze below the prime ideals which arise in each of these three partitioning subsets of $\mathrm{Spec}(L_K(E))$:

- the graded prime ideals $I(H)$ for which M has Condition (L);
- the graded prime ideals $I(H)$ for which M does not have Condition (L); and
- the non-graded prime ideals.

Definition 4.1.6 Let E be an arbitrary graph.

(i) We let $\mathscr{M}(E)$ denote the set of maximal tails in E.
(ii) We let $\mathscr{M}_\gamma(E) \subseteq \mathscr{M}(E)$ denote the set of those maximal tails in E which satisfy Condition (L).
(iii) We let $\mathscr{M}_\tau(E)$ denote the complement $\mathscr{M}(E) \setminus \mathscr{M}_\gamma(E)$.

When the graph E is clear from context, we will sometimes simply write $\mathscr{M}_\gamma$ (resp., $\mathscr{M}_\tau$) for $\mathscr{M}_\gamma(E)$ (resp., $\mathscr{M}_\tau(E)$).

We note that by downward directedness, if $M \in \mathscr{M}_\tau$ (so that M contains some cycle having no exit), then there is a unique cycle c in M which has no exit. It is this property of the elements of $\mathscr{M}_\tau$ which will produce a behavior in the prime ideal structure of $L_K(E)$ which is analogous to the previously described behavior of the prime ideal structure of $L_K(\bullet \circlearrowright) \cong K[x, x^{-1}]$.

Example 4.1.7 Let E denote the graph pictured here.

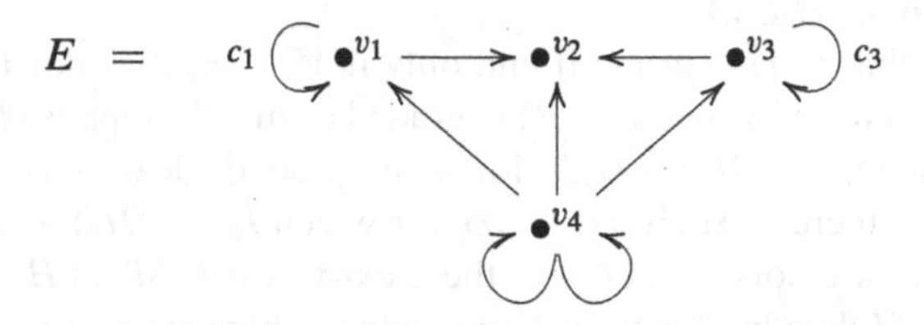

It is straightforward to see that there are four hereditary saturated subsets whose complements in E^0 are maximal tails: $H_1 = \emptyset$, $H_2 = \{v_1, v_2\}$, $H_3 = \{v_2, v_3\}$, and $H_4 = \{v_1, v_2, v_3\}$. We note that there are two additional hereditary saturated subsets of E: the set $H_5 = E^0$, and the set $H_6 = \{v_2\}$. Since E/E^0 is empty, E does not qualify as a maximal tail (by definition). Also, $E/\{v_2\}$ is not downward directed, since there is no vertex y in $E/\{v_2\}$ for which $v_1 \geq y$ and $v_3 \geq y$. So the two ideals $I(H_5) = L_K(E)$ and $I(H_6)$ are graded, but not prime.

Thus by Proposition 4.1.4 the four ideals $I(H_1), I(H_2), I(H_3)$, and $I(H_4)$ are precisely the graded prime ideals in $L_K(E)$. Furthermore, it is easy to see that the corresponding maximal tails have $M_1, M_4 \in \mathscr{M}_\gamma$, while $M_2, M_3 \in \mathscr{M}_\tau$.

Recast, Proposition 4.1.4 gives a description of the graded prime ideals in terms of various subsets of E^0, to wit, that there is a bijective correspondence

$$\text{gr-Spec}(L_K(E)) \longrightarrow \mathscr{M}(E) = \mathscr{M}_\gamma(E) \sqcup \mathscr{M}_\tau(E)$$

given by

$$P \mapsto E/P \cap E^0,$$

with inverse given by

$$M \mapsto I(E^0 \setminus M^0).$$

With this description of gr-Spec($L_K(E)$) in hand, we now analyze the set nongr-Spec($L_K(E)$) of non-graded prime ideals of $L_K(E)$.

Theorem 4.1.8 *Let E be a row-finite graph and K any field. Then there is a bijection*

$$\text{nongr-Spec}(L_K(E)) \longrightarrow \mathscr{M}_\tau(E) \times \text{nongr-Spec}(K[x,x^{-1}])$$

given by

$$P \mapsto (E/H, I(P_C)),$$

where $H = P \cap E^0$, $P = I(H \cup P_C)$, $P_C = \{p_c | c \in C\}$ *where c is a cycle having all exits inside H with* $c^0 \cap H = \emptyset$, *and* p_c *is an irreducible polynomial in* $K[x, x^{-1}]$.

The inverse of this bijection is the map

$$\mathscr{M}_\tau(E) \times \text{nongr-Spec}(K[x,x^{-1}]) \longrightarrow \text{nongr-Spec}(L_K(E))$$

given by

$$(M,\ I(p)) \mapsto I((E^0 \setminus M^0) \cup p_c(c)),$$

where c is the only cycle in M which has no exit in M.

Proof First take a prime ideal P in $L_K(E)$. By Proposition 2.8.11 the ideal P is the ideal generated by $H \cup P_C$, for H as presented in the statement. We claim that C contains only one cycle. Note that $C \neq \emptyset$ as P is a nongraded ideal. By Proposition 2.8.5(ii) we have that $P/I(H) = \oplus_{c\in C} I(p_c(c))$. This combined with the fact that

$$L_K(E/H)/(P/I(H)) \cong (L_K(E)/I(H))/(P/I(H)) \cong L_K(E)/P, \tag{4.1}$$

which gives that $P/I(H)$ is a prime ideal of $L_K(E/H)$, implies that C has only one element, call it c. Moreover, p_c must be irreducible, because $P/I(H)$ is a prime ideal in an algebra isomorphic to $M_{\Lambda_c}(K[x,x^{-1}])$ (see the proof of parts (ii) and (iii) of Proposition 2.8.5).

Now consider $M \in \mathscr{M}_\tau(E)$ and a cycle c as described in the statement, and let p be an irreducible polynomial generating the prime ideal $I(p)$. We use (4.1) to conclude that $I((E^0 \setminus M^0) \cup p(c))$ is a prime ideal, which, clearly, is not graded.

Finally, we leave the reader to prove that the two maps are inverses of each other. □

Example 4.1.9 We return to the graph E presented in Example 4.1.7. We have now built the machinery to explicitly describe $\mathrm{Spec}(L_K(E))$, as pictured here.

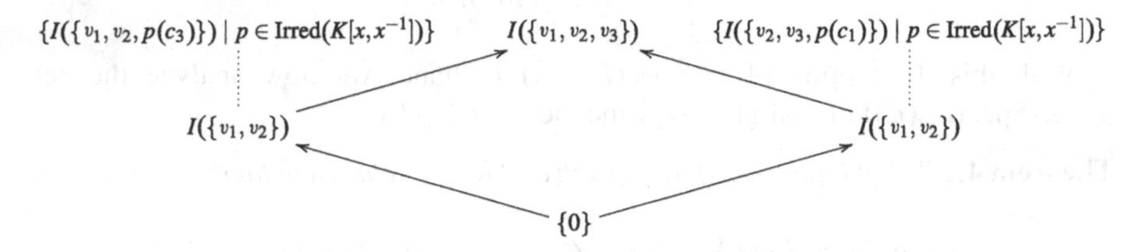

An important subset of the set of prime ideals of a ring R is the set $\mathrm{Prim}(R)$ of primitive ideals. As a reminder, a two-sided ideal P in a ring R is left *primitive* if there exists a simple left R-module S for which $P = \mathrm{Ann}_R(S)$. The ring R is *left primitive* if $\{0\}$ is a left primitive ideal of R. It is easy to show that the two-sided ideal P is left primitive in R if and only if R/P is a left primitive ring. Although left primitivity and its obvious analogous notion of right primitivity do not coincide in general, the two notions do coincide for Leavitt path algebras (since $L_K(E) \cong L_K(E)^{op}$ by Corollary 2.0.9, or simply by using that $L_K(E)$ has an involution), so we will simply talk of *primitive* Leavitt path algebras. It is easy to show that for any ring R, any primitive ideal is necessarily prime. Similarly, it is straightforward to establish that the only commutative primitive rings are fields (so that, in particular, $K[x, x^{-1}]$ is not primitive), and that, if R is primitive, then eRe is primitive for any nonzero idempotent $e \in R$.

We begin by identifying those row-finite graphs E for which $L_K(E)$ is a primitive ring, that is, for which $\{0\}$ is a primitive ideal of $L_K(E)$.

Theorem 4.1.10 *Let E be a row-finite graph and K any field. Then $R = L_K(E)$ is primitive if and only if*

(i) *E is downward directed, and*
(ii) *E satisfies Condition* (L).

Proof First, suppose E satisfies the two conditions. By Proposition 4.1.4, downward directedness yields that $L_K(E)$ is prime. Now invoking [110, Lemmas 2.1 and 2.2] and the primeness of $L_K(E)$, we embed $L_K(E)$ as a two-sided ideal in a unital prime K-algebra (which we denote by $L_K(E)_1$) in such a way that the primitivity of $L_K(E)$ follows by establishing the primitivity of $L_K(E)_1$. By Lam [107, Lemma 11.28], a unital ring A is left primitive if and only if there is a left ideal $M \neq A$ of A such that for every nonzero two-sided ideal I of A, $M + I = A$. Using this, we now establish the primitivity of $L_K(E)_1$.

To that end, let v be any vertex in E, and let $T(v) = \{u \in E^0 \mid v \geq u\}$ as usual. Since E is row-finite, the set $T(v)$ is at most countable. So we may label the elements of $T(v)$ as $\{v_1, v_2, \ldots\}$. We define a sequence $\lambda_1, \lambda_2, \ldots$ of paths in E having these two properties for each $i \in \mathbb{N}$: λ_i is an initial subpath of λ_j whenever $i \leq j$, and $v_i \geq r(\lambda_i)$. To do so, define $\lambda_1 = v_1$. Now suppose $\lambda_1, \ldots, \lambda_n$ have been defined with the indicated properties for some $n \in \mathbb{N}$. By downward directedness, there is a vertex u_{n+1} in E for which $r(\lambda_n) \geq u_{n+1}$ and $v_{n+1} \geq u_{n+1}$. Let p_{n+1} be a path from $r(\lambda_n)$ to u_{n+1}, and define $\lambda_{n+1} = \lambda_n p_{n+1}$. Then the inductively defined set $\{\lambda_i \mid i \in \mathbb{N}\}$ is clearly seen to have the two desired properties.

We easily get that $\lambda_i\lambda_i^*\lambda_t\lambda_t^* = \lambda_t\lambda_t^*$ for each pair of positive integers $t \geq i$, since then λ_i is a subpath of λ_t. Now define the left $L_K(E)_1$-ideal M by setting

$$M = \sum_{i=1}^{\infty} L_K(E)_1(1 - \lambda_i\lambda_i^*).$$

We first claim that $M \neq L_K(E)_1$. On the contrary, suppose $1 \in M$. Then there exist $n \in \mathbb{N}$ and $r_1, \ldots, r_n \in L_K(E)_1$ for which $1 = \sum_{i=1}^n r_i(1 - \lambda_i\lambda_i^*)$. Multiplying this equation on the right by $\lambda_n\lambda_n^*$, and using the previous observation, yield $\lambda_n\lambda_n^* = 0$ and so $\lambda_n = \lambda_n\lambda_n^*\lambda_n = 0$, a contradiction. Thus M is indeed a proper left ideal of $L_K(E)_1$.

We now show that $M + I = L_K(E)_1$ for all nonzero two-sided ideals I of $L_K(E)_1$. Since $L_K(E)_1$ is prime and $L_K(E)$ embeds in $L_K(E)_1$ as a two-sided ideal, we have $I \cap L_K(E)$ is a nonzero two-sided ideal of $L_K(E)$. So Condition (L) on E, together with Proposition 2.9.13, implies that I contains some vertex, call it w. By downward directedness there exists a $u \in E^0$ for which $v \geq u$ and $w \geq u$. But $v \geq u$ gives by definition that $u = v_n$ for some $n \in \mathbb{N}$, so that $w \geq v_n$. By the construction of the indicated sequence of paths we have $v_n \geq r(\lambda_n)$, so that there is a path q in E for which $s(q) = w$ and $r(q) = r(\lambda_n)$. Since $w \in I$ this gives $r(\lambda_n) \in I$, so that $\lambda_n\lambda_n^* = \lambda_n \cdot r(\lambda_n) \cdot \lambda_n^* \in I$. But then $1 = (1 - \lambda_n\lambda_n^*) + \lambda_n\lambda_n^* \in M + I$, so that $M + I = L_K(E)_1$ as desired. Thus the left ideal M of $L_K(E)_1$ possesses the two required properties, which establishes the primitivity of $L_K(E)_1$, and thus the primitivity of $L_K(E)$.

Conversely, suppose $R = L_K(E)$ is primitive. Since R is then in particular a prime ring, E is downward directed by Proposition 4.1.4. We argue by contradiction that E has Condition (L) as well, for, if not, there is a cycle c based at a vertex v in E having no exits. But then by Lemma 2.2.7, the corner ring $vRv \cong K[x, x^{-1}]$ is not primitive. Since a nonzero corner of a primitive ring must again be primitive, we reach the desired contradiction, and the result follows. □

With the previous result in hand, we are now in position to identify the primitive ideals of a row-finite Leavitt path algebra $L_K(E)$.

Theorem 4.1.11 *Let E be a row-finite graph and K any field.*

(i) *Let P be a graded prime ideal of $L_K(E)$, and let $M = E/P \cap E^0$. Then P is primitive if and only if M satisfies Condition (L); i.e., if and only if $M \in \mathscr{M}_\gamma(E)$.*
(ii) *Every non-graded prime ideal of $L_K(E)$ is primitive.*

Proof (i) Let $H = P \cap E^0$. Then $M = E/H$ is downward directed, as P is prime. Since P is graded and E is row-finite we have $P = I(H)$ by Theorem 2.5.9. But then $L_K(E)/P = L_K(E)/I(H) \cong L_K(E/H) = L_K(M)$, with the isomorphism following from Corollary 2.4.13(i). So $L_K(E)/P \cong L_K(M)$, where M satisfies

Condition (L) and is downward directed. Thus $L_K(M)$ (and hence P) is primitive by Theorem 4.1.10.

On the other hand, if M does not have Condition (L) then let c be the (necessarily unique) cycle without exits in M, and suppose c is based at the vertex v. By Lemma 2.2.7 we obtain that $K[x, x^{-1}] \cong vL_K(E)v$, which is not primitive. As nonzero corners of primitive rings are primitive, we then get the result.

(ii) Let P be a prime non-graded ideal of $L_K(E)$. By Theorem 4.1.8, we have

$$P = I((P \cap E^0) \cup p_c(c)),$$

for p_c as explained therein. Take $w = s(c)$ and let u denote the nonzero idempotent $w + P$ in the prime ring $L_K(E)/P$, let $\varphi : K[x, x^{-1}] \to wL_K(E)w$ denote the isomorphism described in Lemma 2.2.7 (so that $\varphi(x) = c$), and let

$$\overline{\varphi} : K[x, x^{-1}] \to (wL_K(E)w + P)/P = u(L_K(E)/P)u$$

denote the quotient map. But the description of P yields that $\text{Ker}(\overline{\varphi}) \supseteq I(p_c)$ for the irreducible polynomial p_c, and since $\overline{\varphi}$ is not the zero map (as $w \notin \text{Ker}(\overline{\varphi})$), the maximality of $I(p_c)$ gives $\text{Ker}(\overline{\varphi}) = I(p_c)$. So the nonzero corner $u(L_K(E)/P)u$ of the prime ring $L_K(E)/P$ is isomorphic to $K[x, x^{-1}]/I(p_c)$, and hence is a field, and so in particular is primitive. We now apply [110, Theorem 1] to conclude that P is a primitive ideal of $L_K(E)$. □

We conclude this section by returning again to the graph E presented in Example 4.1.7. By Theorem 4.1.11, the primitive ideals of $L_K(E)$ consist of

$$\{0\}, \ I(\{v_1, v_2, v_3\}),$$

$$\{I(\{v_1, v_2, f(c_3)\}) \mid f \in \text{Irred}(K[x, x^{-1}])\}, \text{and } \{I(\{v_2, v_3, f(c_1)\}) \mid f \in \text{Irred}(K[x, x^{-1}])\},$$

while the prime ideals $I(\{v_1, v_2\})$ and $I(\{v_2, v_3\})$ are nonprimitive. The graded primitive ideals are $\{0\}$ and $I(\{v_1, v_2, v_3\})$.

4.2 Chain Conditions on One-Sided Ideals

In this section we consider the one-sided chain conditions (artinian and noetherian) in the context of Leavitt path algebras. As one consequence of this investigation we will obtain a characterization of the semisimple Leavitt path algebras. Much of the discussion in this section follows the presentation made in [7].

In the context of unital rings, the one-sided chain conditions are unambiguously described. Specifically, a unital ring R is left artinian (resp., left noetherian) if, for every chain of left ideals $I_1 \supseteq I_2 \supseteq \dots$ (resp., $I_1 \subseteq I_2 \subseteq \dots$) of R, there exists an integer n for which $I_n = I_t$ for all $t \geq n$. It is well known that, for unital rings, R is left artinian (resp., noetherian) if and only if every finitely generated left R-module is artinian (resp., noetherian), if and only if every corner eRe of R is left artinian (resp., noetherian).

There are natural notions of the artinian and noetherian conditions for rings with enough idempotents, in particular, notions which apply to Leavitt path algebras $L_K(E)$ for arbitrary graphs E and fields K. Here we choose to cast all definitions and results for left modules; by Corollary 2.0.9, appropriately symmetric results hold for right modules as well. We recall (Definition 1.2.10) that a left *R-module M* over a ring with enough idempotents R is a module in the usual sense, but with the ordinary unitary condition replaced by the condition that $RM = M$.

If R is a non-unital ring with a (necessarily infinite) set of enough idempotents E, then the decomposition $R = \oplus_{e\in E} Re$ shows that R can never be left artinian (resp., noetherian) in the usual sense. Thus the standard definition of a left artinian ring is not the germane one in this context. However, the following definition gives a natural recasting of this notion which appropriately extends the chain conditions from the unital case.

Definition 4.2.1 Let R be a ring with enough idempotents. We say R is *categorically left artinian* (resp., *categorically left noetherian*) if every finitely generated left R-module is artinian (resp., noetherian).

Using the fact that the left regular module R is a generator for the category $R - Mod$ for any ring R with enough idempotents E, it is easy to verify that R is categorically left artinian (resp., noetherian) if and only if each Re is a left artinian (resp., noetherian) R-module for each $e \in E$. In particular, if R is a unital ring, then R is left artinian (resp., noetherian) if and only if R is categorically left artinian (resp., noetherian).

Let Λ be any set. For any $i \in \Lambda$ and unital ring S let $e = e_{ii}$ denote the standard matrix idempotent in the matrix ring $R = \mathrm{M}_\Lambda(S)$ (Notation 2.6.3). As any field K is (left) artinian, and as the Laurent polynomial algebra $K[x, x^{-1}]$ is (left) noetherian for any field K, we get

Lemma 4.2.2 *Let K be any field.*

(i) *Any ring of the form* $\bigoplus_{i\in\Upsilon} \mathrm{M}_{X_i}(K)$, *where* Υ *and* X_i *are arbitrary sets, is categorically left artinian.*

(ii) *Any ring of the form* $\left(\bigoplus_{i\in\Upsilon_1} \mathrm{M}_{X_i}(K)\right) \oplus \left(\bigoplus_{j\in\Upsilon_2} \mathrm{M}_{Y_j}(K[x, x^{-1}])\right)$, *where* Υ_1, Υ_2, X_i, *and* Y_j *are arbitrary sets, is categorically left noetherian.*

A second germane notion in the context of extending chain conditions to rings with enough idempotents is the following.

Definition 4.2.3 The ring R is called *locally left artinian* (resp., *locally left noetherian*) if for any finite subset X of R there exists an $e = e^2 \in R$ for which $X \subseteq eRe$, and eRe is left artinian (resp., left noetherian).

By the definition of a set of enough idempotents, it is easy to see that a ring R is locally left artinian (resp., noetherian) precisely when R has a set of enough idempotents E for which eRe is left artinian (resp., noetherian) for each $e \in E$.

Clearly, if R is unital then R is locally left artinian if and only if R is left artinian; it was noted above that in this situation R is equivalently categorically left artinian as well. However, in the non-unital setting the categorically artinian and locally artinian properties need not be the same. For instance, let $T = \mathrm{T}_{\mathbb{N}}(K) \subseteq \mathrm{M}_{\mathbb{N}}(K)$ denote the K-subalgebra of $\mathrm{M}_{\mathbb{N}}(K)$ consisting of lower triangular matrices. Clearly T contains a set of enough idempotents (the same set as in $\mathrm{M}_{\mathbb{N}}(K)$, the matrix units $\{e_{ii} \mid i \in \mathbb{N}\}$). Then T is locally artinian, since for each matrix idempotent $f = \sum_{i=1}^{m} e_{ii}$ the algebra fTf is finite-dimensional. However, the finitely generated left T-module Te_{11} is not left artinian, since it is easy to check that $Te_{11} \supsetneqq Te_{21} \supsetneqq Te_{31} \supsetneqq \dots$. We do, however, get the converse.

Lemma 4.2.4 *Let R be a ring with enough idempotents. If R is categorically left artinian (resp., noetherian), then R is locally left artinian (resp., noetherian).*

Proof Let E be a set of enough idempotents for R. We prove the artinian case, the noetherian case being virtually identical. It suffices to show that eRe is left artinian for every $e \in E$. By hypothesis the finitely generated left ideal Re is artinian. Now consider a decreasing sequence of left eRe-ideals $I_1 \supseteq I_2 \supseteq \dots$. Then $RI_1 \supseteq RI_2 \supseteq \dots$ is a decreasing sequence of R-submodules of Re, and hence stabilizes, so that $RI_k = RI_{k+1} = \dots$ for some integer k, which in turn yields $eRI_k = eRI_{k+1} = \dots$. But for each positive integer j we have $eRI_j = I_j$ (because $I_j \subseteq eRe$ gives $eI_j = I_j$, whence $eRI_j = eReI_j = I_j$), so that we get $I_k = I_{k+1} = \dots$, as desired. □

Definitions 4.2.5 Let E be any graph. Recall (Definitions 2.9.4) that by an *infinite path* in E we mean a sequence $\gamma = e_1, e_2, \dots$ of edges of E for which $r(e_n) = s(e_{n+1})$ for all $n \in \mathbb{N}$. In this situation we typically write $\gamma = e_1 e_2 \cdots$, or $\gamma = (e_n)_{n=1}^{\infty}$.

(i) An infinite path $\gamma = (e_n)_{n=1}^{\infty}$ is called an *infinite sink in E* if γ has neither bifurcations nor cycles; that is, if $\gamma^0 \subseteq P_l(E)$, the set of line points of E.
(ii) An infinite path $(e_n)_{n=1}^{\infty}$ *ends in a sink* if there exists an $m \geq 1$ such that the infinite path $(e_n)_{n \geq m}$ is an infinite sink in E.
(iii) An infinite path $(e_n)_{n=1}^{\infty}$ *ends in a cycle* if there exists an $m \geq 1$ and a cycle c in E such that the infinite path $(e_n)_{n=m}^{\infty}$ equals the infinite path $ccc\cdots$.

If E contains an infinite emitter v, emitting edges $\{e_i \mid i \in I\}$, then $L_K(E)$ cannot be categorically left artinian, nor categorically left noetherian. This is clear, as the finitely generated left ideal $L_K(E)v$ contains the infinite collection of independent submodules $L_K(E)e_ie_i^*$. (The obvious analogous conditions hold on the right as well.) Thus any Leavitt path algebra satisfying a one-sided categorical chain condition must necessarily be row-finite. This observation allows us to anticipate at least one of the graph-theoretic conditions given in the theorem below.

As we will see, paths with exits (including cycles with exits) will play a significant role in this discussion. We establish the following result, whose proof will provide the template for a number of related results in the sequel.

Lemma 4.2.6 *Let E be an arbitrary graph and K any field. Suppose c is a cycle in E based at v, and suppose f is an exit for c with $s(f) = v$. Then*

$$L_K(E)cc^* \supsetneqq L_K(E)c^2(c^*)^2 \supsetneqq L_K(E)c^3(c^*)^3 \supsetneqq \cdots$$

is a non-stabilizing chain of left ideals of $L_K(E)$.

Proof The inclusions follow from the observation that $c^{i+1}(c^*)^{i+1} = c^{i+1}(c^*)^{i+1} \cdot c^i(c^*)^i$ for all $i \in \mathbb{N}$.

Note that, since f is an exit for c, we have $c^*f = 0$ in $L_K(E)$. That the inclusions are proper is then established as follows. Assume otherwise; then $c^n(c^*)^n = rc^{n+1}(c^*)^{n+1}$ for some $r \in L_K(E)$ and $n \in \mathbb{N}$. Multiplying this equation on the right by c^nf yields $c^nf = rc^{n+1}(c^*)f = 0$. But $c^nf \neq 0$ in $L_K(E)$ by Corollary 1.5.13. □

We now have all the necessary ingredients in hand to prove the following one-sided chain condition result, in which we characterize the left artinian Leavitt path algebras by describing them in categorical, ring-theoretic, graph-theoretic, and explicit terms. In addition, we give a characterization of these algebras which utilizes properties of their finitely generated projective modules.

Theorem 4.2.7 *Let E be an arbitrary graph and K any field. The following are equivalent.*

(1) *$L_K(E)$ is semisimple; that is, every left $L_K(E)$-module is isomorphic to a direct sum of simple left $L_K(E)$-modules.*
(2) *$L_K(E)$ is categorically left artinian.*
(3) *$L_K(E)$ is locally left artinian.*
(4) *E is acyclic, row-finite, and every infinite path in E ends in a sink.*
(5) *$L_K(E) = I(P_l(E))$; that is, the ideal generated by the set of line points of E is all of $L_K(E)$.*
(6) *$L_K(E) \cong \bigoplus_{i\in\Upsilon} \mathrm{M}_{X_i}(K)$ for some (possibly infinite) sets Υ and $\{X_i \mid i \in \Upsilon\}$.*
(7) *$L_K(E)$ is von Neumann regular and $\mathcal{V}(L_K(E)) \cong (\mathbb{Z}^+)^{(\Upsilon)}$ for some set Υ; that is, $\mathcal{V}(L_K(E))$ is a direct sum of* card(Υ) *copies of the monoid $\mathbb{Z}^+$.*

Proof (1) ⇒ (2) is clear (since every finitely generated left $L_K(E)$-module is a direct sum of simples).

(2) ⇒ (3) follows from Lemmas 4.2.4 and 1.2.12(v).

(3) ⇒ (4). We prove all three conditions by contradiction. Suppose first that E contains a cycle; let c be such, based at the vertex v. There are two cases. If c has no exit, then by Lemma 2.2.7 $vL_K(E)v \cong K[x, x^{-1}]$, which is not left artinian, violating the hypothesis. On the other hand, if c has an exit, then (by an argument identical to

that used in the proof of Lemma 4.2.6) the following is a non-stabilizing sequence of left ideals in $vL_K(E)v$:

$$vL_K(E)vcc^* \supsetneqq vL_K(E)vc^2(c^*)^2 \supsetneqq vL_K(E)vc^3(c^*)^3 \supsetneqq \dots ,$$

again violating the hypothesis that $vL_K(E)v$ is left artinian.

Next, suppose that E contains an infinite emitter; let v be such, and pick some countably infinite subset $\{e_i \mid i \in \mathbb{N}\}$ of $s^{-1}(v)$. Then the following is a non-stabilizing sequence of left ideals in $vL_K(E)v$:

$$\oplus_{i=1}^{\infty} vL_K(E)ve_ie_i^*v \supsetneqq \oplus_{i=2}^{\infty} vL_K(E)ve_ie_i^*v \supsetneqq \oplus_{i=3}^{\infty} vL_K(E)ve_ie_i^*v \supsetneqq \dots ,$$

violating the hypothesis that $vL_K(E)v$ is left artinian.

Finally, suppose that there exists an infinite path γ in E which does not end in a sink. Let $v = s(\gamma)$. Since E is row-finite and acyclic, γ^0 must contain infinitely many bifurcation vertices. We decompose γ as an infinite sequence of paths $\gamma = \gamma_1\gamma_2\gamma_3\cdots$ in such a way that there exists a bifurcation at $r(\gamma_i)$ for every i. But then

$$vL_K(E)v\gamma_1\gamma_1^*v \supsetneqq vL_K(E)v\gamma_1\gamma_2\gamma_2^*\gamma_1^*v \supsetneqq vL_K(E)v\gamma_1\gamma_2\gamma_3\gamma_3^*\gamma_2^*\gamma_1^*v \supsetneqq \cdots$$

is a non-stabilizing chain of left ideals of $vL_K(E)v$, as can be established easily using the same ideas as in the proof of Lemma 4.2.6.

(4) $\Rightarrow$ (5). By Theorem 2.6.14 it is enough to show that $E^0 = \overline{P_l(E)}$, the saturated closure of the set of line points of E. Suppose on the contrary that there exists a $v_1 \in E^0$ with $v_1 \notin \overline{P_l(E)}$. Then v_1 is not a line point, and as such cannot be a sink, so that $s^{-1}(v_1) \neq \emptyset$. Now, using the hypothesis that E is row-finite together with the saturated condition on $\overline{P_l(E)}$, $v_1 \notin \overline{P_l(E)}$ yields that $r(s^{-1}(v_1)) \not\subseteq \overline{P_l(E)}$, so that there exists an $e_1 \in E^1$ with $s(e_1) = v_1$ and $r(e_1) = v_2 \notin \overline{P_l(E)}$. We repeat this process, starting now with v_2, and obtain some $e_2 \in E^1$ for which $s(e_2) = v_2$ and $r(e_2) = v_3 \notin \overline{P_l(E)}$. Since E is acyclic by hypothesis, the vertices $\{v_1, v_2, v_3, \dots\}$ are distinct. In other words, using this process we can build an infinite path $\gamma = e_1e_2e_3\cdots$ for which all the vertices appearing in the path are distinct, and there is a bifurcation at each vertex in the path. But then γ is an infinite path which does not end in a sink, contrary to hypothesis.

(5) $\Rightarrow$ (6) is immediate from Theorem 2.6.14.

(6) $\Rightarrow$ (1) is well known.

Thus we have established the equivalence of statements (1) through (6). The implication (6) $\Rightarrow$ (7) is well known. (Indeed, the sets denoted by Υ which appear in statements (6) and (7) are equal.) So to complete the proof of the theorem it suffices to show that (7) $\Rightarrow$ (4).

(7) $\Rightarrow$ (4). Assuming (7), have that the von Neumann regularity of $L_K(E)$ yields that E is acyclic by Theorem 3.4.1.

To establish the other two properties of E, we start by making this observation: in the monoid $M = (\mathbb{Z}^+)^{(\Upsilon)}$, each nonzero element x has the property that there is a bound on the size of the set $N_x = \{n \in \mathbb{N} \mid x \text{ can be written as a sum of } n \text{ nonzero elements of } M\}$. Using this, we now show by contradiction that the other two properties hold. If E is not row-finite then there exist a vertex v and some countably infinite subset $\{e_i \mid i \in \mathbb{N}\}$ of $s^{-1}(v)$. But then in $\mathcal{V}(L_K(E))$ we have

$$[v] = [v - e_1e_1^*] + [e_1e_1^*] = [v - e_1e_1^*] + [e_1e_1^* - e_2e_2^*] + [e_2e_2^*] = \dots.$$

Since each expression is nonzero in $\mathcal{V}(L_K(E))$ we have violated the indicated property.

On the other hand, suppose E has an infinite path γ which does not end in a sink, and write $v = s(\gamma)$. We proceed as in the proof of (3) $\Rightarrow$ (4); using that E has been shown to be acyclic, we may write γ as an infinite sequence of paths $\gamma = \gamma_1\gamma_2\gamma_3\cdots$ in such a way that we have a bifurcation at $r(\gamma_i)$ for every i. For each $n \in \mathbb{N}$, define $g_n = \gamma_1\gamma_2\cdots\gamma_n\gamma_n^*\cdots\gamma_2^*\gamma_1^*v \in L_K(E)$. Using the previously described properties of the set $\{g_n \mid n \in \mathbb{N}\}$, we see that $g_n - g_{n+1}$ is an idempotent for each $n \in \mathbb{N}$, and that we get the following equations in $\mathcal{V}(L_K(E))$ for each $m \in \mathbb{N}$:

$$[g_1]=[g_2]+[g_1-g_2]=[g_3]+[g_2-g_3]+[g_1-g_2]=\dots=[g_m]+[g_{m-1}-g_m]+\dots+[g_2-g_3]+[g_1-g_2].$$

This violates the indicated property of $\mathcal{V}(L_K(E))$, and establishes the Theorem. □

Using properties of $\mathcal{V}(L_K(E))$, we show that statement (7) of Theorem 4.2.7 may be replaced by a seemingly much weaker statement. An element x in an abelian monoid $(M, +)$ is an *atom* if $x \neq 0$, and if $x = m + m'$ in M then $m = 0$ or $m' = 0$. M is called *atomic* if there exists a subset A of M for which A consists of atoms of M, and every element of M is a (finite) sum of elements taken from A.

Recall from Sect. 3.6 that an abelian monoid $(M, +)$ is called *conical* if for any $x, y \in M$, $x + y = 0$ if and only if $x = y = 0$. The definition of a *refinement* monoid is given in Definitions 3.6.1. (We note that in the current section we will use 0 to denote the neutral element of $(M, +)$; in the previous discussion we had used z for this element.)

Lemma 4.2.8 *Let $(M, +)$ be an abelian, atomic, conical, refinement monoid. Then each nonzero element $m \in M$ has the property that there is a bound on the size of the set $N_m = \{n \in \mathbb{N} \mid m \text{ can be written as the sum of } n \text{ nonzero elements of } M\}$. In this case, $|N_m|$ is the number of terms which appear in the representation of m as a sum of atoms of M.*

Proof Suppose a is an atom in M, and suppose $a = \sum_{i=1}^{t} z_i$ is the sum of nonzero elements of M. Then using the conical property of M, we necessarily get $a = z_j$ for some $1 \leq j \leq t$, and $z_i = 0$ for all $i \neq j$. Now let $m \neq 0$ in M, and write

$m = \sum_{i=1}^{N} a_i$ with each a_i an atom. Suppose also that $m = \sum_{i=1}^{t} m_i$ in M. Since M is a refinement monoid, we have a refinement matrix of the form:

	m_1	$\cdots$	m_t
a_1	$z_{1,1}$	$\cdots$	$z_{1,t}$
$\vdots$	$\vdots$	$\ddots$	$\vdots$
a_N	$z_{N,1}$	$\cdots$	$z_{N,t}$

By the previous observation, each row contains exactly one nonzero entry. Thus there are exactly N nonzero entries in the table. We conclude that at most N of the expressions $\{m_j \mid 1 \leq j \leq t\}$ can be nonzero, thus establishing the result. □

Corollary 4.2.9 *Let E be an arbitrary graph and K any field. Then $L_K(E)$ is semisimple if and only if $L_K(E)$ is von Neumann regular and $\mathcal{V}(L_K(E))$ is atomic.*

Proof Suppose $L_K(E)$ is semisimple. Then $L_K(E)$ is clearly von Neumann regular. In addition, if $L_K(E) = \oplus_{i\in I} T_i$ is a decomposition of $L_K(E)$ into a direct sum of simple left ideals, then it is well known that $\mathcal{V}(L_K(E)) \cong (\mathbb{Z}^+)^{(\Upsilon)}$ for some Υ, which is clearly atomic.

Conversely, by Theorem 4.2.7, it suffices to show that if $L_K(E)$ is von Neumann regular and $\mathcal{V}(L_K(E))$ is atomic, then E is row-finite, acyclic, and every infinite path in E ends in a sink. The acyclic property of E follows from the hypothesis that $L_K(E)$ is von Neumann regular, by Theorem 3.4.1. For any ring R we have that the monoid $\mathcal{V}(R)$ is conical. By Theorem 3.6.21 we have that the monoid $\mathcal{V}(L_K(E))$ is a refinement monoid. Thus, together with the atomic hypothesis, we have that $\mathcal{V}(L_K(E))$ satisfies the hypotheses of Lemma 4.2.8. Using this, we now argue exactly as in the proof of (7) $\Rightarrow$ (4) of Theorem 4.2.7 to conclude both that E is row-finite, and that every infinite path in E ends in a sink. □

Remark 4.2.10 We note that K-algebras of the form $\bigoplus_{i\in\Upsilon} \mathrm{M}_{X_i}(K)$ (where Υ, X_i are sets of arbitrary size) which appear in Theorem 4.2.7 do in fact arise as Leavitt path algebras, see Corollary 2.6.6.

The second part of this section is devoted to the characterization of categorically left noetherian Leavitt path algebras, equivalently, of locally left noetherian Leavitt path algebras, in terms of the underlying graph. Moreover, we will describe them up to K-algebra isomorphism. We note that if v is an infinite emitter, with $\{e_i \mid i \in \mathbb{N}\}$ an infinite subset of $s^{-1}(v)$, then

$$L_K(E)e_1e_1^* \subsetneqq L_K(E)e_1e_1^* \oplus L_K(E)e_2e_2^* \subsetneqq L_K(E)e_1e_1^* \oplus L_K(E)e_2e_2^* \oplus L_K(E)e_3e_3^* \subsetneqq \ldots$$

is a strictly increasing chain of submodules of the cyclic left ideal $L_K(E)v$; thus any categorically noetherian (or locally noetherian) Leavitt path algebra must be row-finite. The other conditions that the graph must satisfy are that the cycles have no exits, and that every infinite path ends in a sink or in a cycle.

Definition 4.2.11 We say that a graph E satisfies *Condition* (NE) if no cycle in E has an exit.

Theorem 4.2.12 *Let E be an arbitrary graph and K any field. The following are equivalent.*

(1) *$L_K(E)$ is categorically left noetherian.*
(2) *$L_K(E)$ is locally left noetherian.*
(3) *E is row-finite, satisfies Condition* (NE), *and every infinite path in E ends either in a sink or in a cycle.*
(4) *$L_K(E) = I(P_l(E) \cup P_c(E))$, the ideal generated by the line points together with the vertices which lie on cycles without exits.*
(5) *$L_K(E) \cong \bigoplus_{i\in\Upsilon_1} \mathrm{M}_{X_i}(K) \oplus \bigoplus_{j\in\Upsilon_2} \mathrm{M}_{Y_j}(K[x,x^{-1}])$, where Υ_1 and X_i are the sets Γ and Λ_{v_i} (respectively) described in Theorem 2.6.14, and Υ_2 and Y_j are the sets Υ and Λ_{v_i} (respectively) described in Theorem 2.7.3.*

Proof (1) $\Rightarrow$ (2) follows by Lemma 4.2.4.

(2) $\Rightarrow$(3). Assume that c is a cycle in E based at a vertex v, and that c has an exit at v. It is not difficult to check (again using the idea in the proof of Lemma 4.2.6) that

$$vL_K(E)v(v - cc^*) \subsetneqq vL_K(E)v(v - c^2(c^*)^2) \subsetneqq \dots$$

is an infinite ascending chain of left ideals of $vL_K(E)v$. But this contradicts the locally noetherian hypothesis, and thus shows that E satisfies Condition (NE). Suppose now that γ is an infinite path which does not end either in a sink or in a cycle. In this situation γ cannot contain any closed path, as follows. Assume to the contrary that $\gamma = \gamma_1 p \gamma_2$, with p being a closed path; then, as E has been shown to satisfy (NE), p must in fact be a cycle and $\gamma_2 = ppp\cdots$, so that γ does end in a cycle, contrary to hypothesis. Now, since γ does not end in a sink either (and does not contain cycles), γ^0 contains infinitely many bifurcation vertices, so that we can write $\gamma = \gamma_1\gamma_2\gamma_3\cdots$ for γ_i paths such that $r(\gamma_i)$ is a bifurcation vertex for all i. Then by again using an argument analogous to that given in Lemma 4.2.6, we have the non-stabilizing chain of left ideals of $vL_K(E)v$ given by

$$vL_K(E)v(v - \gamma_1\gamma_1^*) \subsetneqq vL_K(E)v(v - \gamma_1\gamma_2\gamma_2^*\gamma_1^*) \subsetneqq \cdots$$

(3) $\Rightarrow$ (4). We will use Lemma 2.4.1, which describes the form of elements in the ideal generated by a hereditary subset of vertices. Define $H := P_l(E) \cup P_c(E)$. We want to show that $L_K(E) = I(H)$. Assume that this is not the case, and consider an element $x = \sum_{i=1}^m k_i\gamma_i\lambda_i^* \in L_K(E) \setminus I(H)$. Let $j \in \{1,\dots,m\}$ be such that $\gamma_j\lambda_j^* \notin I(H)$. Define $v_1 := r(\gamma_j)$. Then $v_1 \notin H$. In particular, v_1 is neither a sink, nor is in a cycle without exits. But as E satisfies Condition (NE), this is equivalent to saying that v_1 is neither a sink, nor is in any cycle. If $s^{-1}(v_1) \subseteq I(H)$, then by the row-finiteness of E we have $v_1 = \sum_{e\in s^{-1}(v)} ee^* \in I(H)$, implying $\gamma_j\lambda_j = \gamma_j v_1 \lambda_j \in I(H)$,

a contradiction. Therefore, there exists an $e_1 \in s^{-1}(v)$ such that $s(e_1) = v_1$ and $v_2 := r(e_2) \notin I(H)$. Again we get $v_2 \notin H$, which implies as before that v_2 is neither a sink nor is in a cycle. Repeating this process we find an infinite path $e_1 e_2 \cdots$ which does not end either in a sink or in a cycle, contrary to hypothesis. This proves (4).

(4) $\Rightarrow$ (5). Since $L_K(E) = I(P_l(E) \cup P_c(E))$, apply Proposition 2.4.7 to get $I(P_l(E) \cup P_c(E)) = I(P_l(E)) \oplus I(P_c(E))$. Then Theorems 2.6.14 and 2.7.3 yield (5).

(5) $\Rightarrow$ (1) is immediate from Lemma 4.2.2. □

We continue this section by noting separately the description of left artinian (resp., left noetherian) Leavitt path algebras for finite graphs; these follow easily from Theorems 4.2.7 and 4.2.12. Much of the artinian result has already been presented in the Finite Dimension Theorem 2.6.17.

Corollary 4.2.13 *Let E be an arbitrary graph and K any field. The following are equivalent.*

(1) $L_K(E)$ *is unital and semisimple.*
(2) $L_K(E)$ *is left artinian.*
(3) $L_K(E)$ *is finite-dimensional.*
(4) *E is finite and acyclic.*
(5) $L_K(E) \cong \bigoplus_{i \in \Upsilon} \mathrm{M}_{n_i}(K)$, *where Υ is a finite set and $n_i \in \mathbb{N}$. (Specifically, $|\Upsilon|$ is the number of sinks in E, and for each $i \in \Upsilon$, n_i is the number of paths in E which end in the sink corresponding to i.)*

Corollary 4.2.14 *Let E be an arbitrary graph and K any field. The following are equivalent.*

(1) $L_K(E)$ *is left noetherian.*
(2) *E is finite and satisfies Condition* (NE).
(3) $L_K(E) \cong \bigoplus_{i \in \Upsilon_1} \mathrm{M}_{n_i}(K) \oplus \bigoplus_{j \in \Upsilon_2} \mathrm{M}_{m_j}(K[x, x^{-1}])$, *where Υ_1, Υ_2, X_i, and Y_j are finite sets. (Specifically, $|\Upsilon_1|$ is the number of sinks in E, $|\Upsilon_2|$ is the number of (necessarily disjoint) cycles in E, for each $i \in \Upsilon_1$ n_i is the number of paths which end in the sink corresponding to i, and for each $j \in \Upsilon_2$ m_j is the number of paths which end in the cycle corresponding to j.)*

Condition (3) in Corollary 4.2.13 has an appropriate analog which may be added to Corollary 4.2.14, a discussion of which takes up much of the remainder of this section.

Definition 4.2.15 Let K be a field. We say that a $\mathbb{Z}$-graded K-algebra $A = \bigoplus_{n \in \mathbb{Z}} A_n$ is *locally finite* if $\dim_K(A_n)$ is finite for all $n \in \mathbb{Z}$.

Of course, any finite-dimensional graded K-algebra is locally finite; clearly so are the algebras $K[x]$ and $K[x, x^{-1}]$.

Lemma 4.2.16 *Suppose E is a finite graph which satisfies Condition (NE). Then the saturated closure Λ of the set $P_l(E) \sqcup P_c(E)$ is all of E^0.*

Proof Recall that $P_l(E)$ is the set of line points in E, and that $P_c(E)$ is the set of vertices which lie on cycles without exits. Suppose to the contrary that there is some vertex $v \in E^0$ which is not in Λ. Then v cannot be a sink (because every sink is a line point). So $s^{-1}(v)$ is nonempty. By the saturation of Λ, there then necessarily exists some $e_1 \in s^{-1}(v)$ for which $r(e_1) \notin \Lambda$. Now repeat the argument to produce a sequence of edges $e_1, e_2, \ldots$ in E for which $r(e_i) \notin \Lambda$ for all $i \geq 1$. There are two possibilities. If $r(e_i) = r(e_j)$ for some $i \neq j$, then there is a closed path $e_{i+1} \cdots e_j$; but E has Condition (NE), so that each of $r(e_i)$ and $r(e_j)$ must lie on some cycle without exits, and thus are in $P_c(E) \subseteq \Lambda$, a contradiction. On the other hand, if $r(e_i) \neq r(e_j)$ for all i, j then E would have infinitely many distinct vertices, contrary to hypothesis. □

We now add the aforementioned fourth equivalent condition to Corollary 4.2.14.

Theorem 4.2.17 ([8, Theorems 3.8 and 3.10]) *Let E be an arbitrary graph and K any field. The following are equivalent.*

(1) *$L_K(E)$ is locally finite.*
(2) *$L_K(E)$ is left noetherian.*
(3) *E is finite and satisfies Condition* (NE).
(4) $L_K(E) \cong \left(\bigoplus_{i \in \Upsilon_1} \mathrm{M}_{n_i}(K)\right) \oplus \left(\bigoplus_{j \in \Upsilon_2} \mathrm{M}_{m_j}(K[x, x^{-1}])\right)$, *where Υ_1, Υ_2, X_i, and Y_j are finite sets.*

Proof We establish (1) ⇔ (2). To prove (2) ⇒ (1), we have by Lemma 4.2.16 that the saturated closure of $P_l(E) \sqcup P_c(E)$ is all of E^0. Then Corollary 2.7.5(i) gives the result.

We now establish (1) ⇒ (2). If $L_K(E)$ is locally finite, then E^0 must be finite because otherwise E^0 would be a linearly independent set of elements in $L_K(E)_0$. Moreover, E must be row-finite because if $v \in E^0$ were an infinite emitter, then the set $\{ee^* \mid e \in s^{-1}(v)\}$ would be a linearly independent set of elements in $L_K(E)_0$. Finally, we show that E satisfies Condition (NE). Assume that there exists a cycle $c = e_1 \cdots e_m$ based at a vertex v which has an exit, say f, at v. We claim that $\{c^n(c^*)^n \mid n \in \mathbb{N}\}$ is a linearly independent set of elements in $L_K(E)_0$. Indeed, let $k_1, \ldots, k_m \in K$ be such that $\sum_{i=1}^m k_i c^i (c^*)^i = 0$. Multiply on the right-hand side by cf to get $k_1 cf + \sum_{i=2}^m k_i c^i (c^*)^{i-1} f = 0$. But then $k_1 cf = 0$, as $c^* f = 0$ in $L_K(E)$. This then gives $k_1 = 0$ by Corollary 1.5.13. Reasoning in a similar way we get $k_i = 0$ for every i, establishing the result. □

4.3 Self-Injectivity

In this section we establish the perhaps-surprising result that the self-injective Leavitt path algebras are precisely the semisimple ones.

Definitions 4.3.1 A left R-module A is called *injective* if for every pair of left R-modules M, N, every R-homomorphism $\eta : M \to A$, and every R-monomorphism

$f : M \to N$, there exists an R-homomorphism $h : N \to A$ such that the following diagram is commutative.

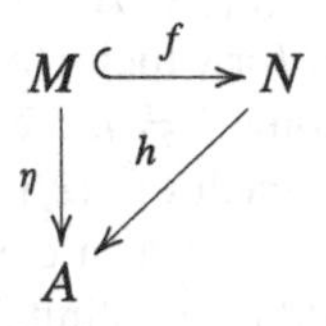

A ring R is said to be *left* (respectively *right*) *self-injective* if ${}_RR$ (respectively R_R) is an injective left (respectively right) R-module.

Because $L_K(E)$ is isomorphic to its opposite algebra (Corollary 2.0.9), the notions of left self-injectivity and right self-injectivity coincide in the context of Leavitt path algebras. (See also the introductory remarks in [16].) Accordingly, we will use only the phrase "self-injective" in this discussion; we will continue as in previous sections to present results in terms of left modules.

Remark 4.3.2 Definitions 4.3.1 of course agree with the usual notion of an injective object in any abelian category. We note that although a module over a non-unital ring R can be viewed as a module over its unitization R^1, injectivity is not necessarily preserved in this process. As a quick example, let K be a field and let $R = \oplus_{i=1}^{\infty} R_i$, with each $R_i = K$. Clearly R is a non-unital ring with enough idempotents. Since ${}_RR$ is a direct sum of simple left R-modules, every left R-module M (which, by definition, satisfies $RM = M$), and in particular R itself, is injective as a left R-module. But ${}_{R^1}R$ is not injective, since otherwise the embedding ${}_{R^1}R \to {}_{R^1}R^1$ would split, which would yield the contradiction that the non-finitely-generated module ${}_{R^1}R$ is a direct summand of the finitely generated module ${}_{R^1}R^1$.

(This Remark shows in particular that a comment made in [79, p. 67] is not valid.)

Following a proof similar to one used in the context of unital rings (e.g., [137, Theorem 3.30]), one can show that the Baer Criterion for injectivity is valid for rings with local units. For completeness, we include a proof of that result. We remind the reader that we write homomorphisms of left R-modules on the right (e.g., $(m)f$).

Proposition 4.3.3 (The Baer Criterion for Rings with Local Units) *Let R be a ring with local units. The left R-module A is injective if and only if for any left ideal I of R and any R-homomorphism $\eta : I \to A$ there is an R-homomorphism $h : R \to A$ such that $h_{|_I} = \eta$. (In other words, to verify the injectivity of A, it suffices to show that the appropriate extension property is satisfied with respect to the embedding monomorphism $I \subseteq R$ for each left ideal I of R.)*

Proof We need only prove the "if" part. We start by noting that if N is a left R-module, then by definition we have $RN = N$, so if $n \in N$ then $n = \sum_{i=1}^{t} r_i n_i \in RN$. But since R has local units, we in fact have $n \in Rn$, which is easily seen by choosing $e \in R$ for which $er_i = r_i$ for $1 \le i \le t$.

Suppose $f : M \to N$ is an R-monomorphism from a submodule M of an R-module N in R-Mod, and let $\eta : M \to A$ be an R-homomorphism. Consider the family

$$\mathscr{F} = \{(M_i, h_i) \mid (M)f \subseteq M_i \subseteq N, \text{ and } h_i \in \mathrm{Hom}_R(M_i, A) \text{ with } (m)fh_i = (m)\eta \text{ for all } m \in M\}.$$

Since f is a monomorphism, f^{-1} is well-defined on $(M)f$; thus $((M)f, f^{-1}\eta) \in \mathscr{F}$. By defining a partial order on $\mathscr{F}$ by setting $(M_i, h_i) \leq (M_j, h_j)$ if $M_i \subseteq M_j$ and $h_{j|_{M_i}} = h_i$, we appeal to Zorn's Lemma to obtain a maximal element (M^*, h^*) in $\mathscr{F}$. We claim that $M^* = N$. Suppose, by way of contradiction, that there is an element $n \in N$ such that $n \notin M^*$. Let $I = \{r \in R \mid rn \in M^*\}$; then I is easily seen to be a left ideal of R. Consider the homomorphism $\varphi : I \to A$ given by $(i)\varphi = (in)h^*$. By hypothesis, φ has an extension $\overline{\varphi} : R \to A$. As noted at the outset of the proof we have $n \in Rn$, and so $M^* + Rn \supsetneqq M^*$. Now define the map $h : M^* + Rn \to A$ by setting $(m^* + rn)h = (m^*)h^* + (r)\overline{\varphi}$. It is a straightforward computation to show that h is well-defined; once established, h is evidently an R-homomorphism. But clearly $h_{|_{M^*}} = h^*$, so that $(M^* + Rn, h)$ violates the maximality of (M^*, h^*) in $\mathscr{F}$. Hence $M^* = N$, and thus A is injective. □

We will use the Baer Criterion now to establish that corners of a class of left/right self-injective rings are left/right self-injective rings.

Lemma 4.3.4 *Let R be a ring with local units which is semiprime and left self-injective. Then for every nonzero idempotent $\epsilon \in R$, the corner $\epsilon R\epsilon$ is a (unital) left self-injective ring.*

Proof We will see that the Baer Criterion is satisfied by ${}_{\epsilon R\epsilon}\epsilon R\epsilon$. Let T be a left ideal of $\epsilon R\epsilon$ and assume that $\eta : T \to \epsilon R\epsilon$ is a homomorphism of left $\epsilon R\epsilon$-modules. Let RT denote the left ideal of R generated by T.

Consider the map: $\overline{\eta} : RT \to R$ defined by $\sum_i r_i y_i \mapsto \sum_i r_i (y_i)\eta$. We show that $\overline{\eta}$ is well-defined. Indeed, assume $\sum_i r_i y_i = \sum_j s_j z_j$, for $r_i, s_j \in R$ and $y_i, z_j \in T$. In particular, since $T \subseteq \epsilon R\epsilon$ we have $(\sum_i r_i y_i)\epsilon = \sum_i r_i y_i$ and $(\sum_i s_j z_j)\epsilon = \sum_i s_j z_j$. Now for every $a \in R$ we have $\epsilon a \sum_i r_i y_i = \epsilon a \sum_j s_j z_j$, that is, $\sum_i \epsilon a r_i y_i = \sum_j \epsilon a s_j z_j$, which are elements in T because $\epsilon a r_i y_i = \epsilon a r_i (\epsilon y_i) \in \epsilon R\epsilon T \subseteq T$, and similarly $\epsilon a s_j z_j \in T$. Apply η to get $(\sum_i \epsilon a r_i y_i)\eta = (\sum_j \epsilon a s_j z_j)\eta$. Now, use that η is a homomorphism of left $\epsilon R\epsilon$-modules to obtain $\epsilon a \sum_i r_i \epsilon (y_i)\eta = \epsilon a \sum_j s_j \epsilon (z_j)\eta$, that is, $\epsilon a \sum_i r_i (y_i)\eta = \epsilon a \sum_j s_j (z_j)\eta$. Equivalently, $\epsilon a \left(\sum_i r_i (y_i)\eta - \sum_j s_j (z_j)\eta\right) = 0$.

Denote $\sum_i r_i (y_i)\eta - \sum_j s_j (z_j)\eta$ by b; we claim that $b = 0$. We have shown that $b\epsilon = b$, and $\epsilon R b = 0$. Now consider the two-sided ideal RbR of R, and note that $(RbR)^2 \subseteq RbRbR \subseteq Rb\epsilon RbR = \{0\}$. So the semiprimeness of R yields that $RbR = \{0\}$, and so $b = 0$ as R has local units. That is, $b = \sum_i r_i (y_i)\eta - \sum_j s_j (z_j)\eta = 0$, which gives that $\overline{\eta}$ is well-defined.

Since R is self-injective, by the Baer Criterion for rings with local units 4.3.3 there exists a homomorphism of left R-modules $\overline{h} : R \to R$ extending $\overline{\eta}$. Define $h : I \to \epsilon R\epsilon$ by setting $(y)h = (y)\overline{h}$. Then h is a homomorphism of left $\epsilon R\epsilon$-modules which extends η. This shows that the corner $\epsilon R\epsilon$ is a left self-injective (unital) ring. □

Proposition 4.3.5 *Let E be an arbitrary graph and K any field. If $L_K(E)$ is self-injective then $L_K(E)$ is von Neumann regular. In particular, E is necessarily acyclic.*

Proof We show that for every nonzero idempotent $\epsilon \in L_K(E)$, the corner $\epsilon L_K(E)\epsilon$ is a von Neumann regular ring. Let ϵ be such an element. By Lemma 4.3.4 the ring $\epsilon L_K(E)\epsilon$ is left self-injective. By Lam [108, Corollary 13.2(2)] the ring $\epsilon L_K(E)\epsilon / J(\epsilon L_K(E)\epsilon)$ is von Neumann regular, where $J(\epsilon L_K(E)\epsilon)$ is the Jacobson radical of $\epsilon L_K(E)\epsilon$. By Jacobson [99, Proposition 1] (which holds for not-necessarily-unital rings), $J(\epsilon L_K(E)\epsilon) = \epsilon J(L_K(E))\epsilon$. Since by Proposition 2.3.2 we have $J(L_K(E)) = \{0\}$, and thereby $J(\epsilon L_K(E)\epsilon) = \{0\}$, we thus conclude that $\epsilon L_K(E)\epsilon$ is a von Neumann regular ring.

Now since $L_K(E)$ is a ring with local units, for every $a \in L_K(E)$ there is an idempotent $\epsilon \in L_K(E)$ such that $a = \epsilon a \epsilon$. Since $a \in \epsilon L_K(E)\epsilon$, by the previous paragraph there exists a $b \in \epsilon L_K(E)\epsilon$ such that $aba = a$. Hence $L_K(E)$ is von Neumann regular. By Theorem 3.4.1, this yields that the graph E is acyclic. □

With Proposition 4.3.5 in hand, in order to establish that the self-injectivity of $L_K(E)$ implies semisimplicity, we need only show (by the implication (4) $\Rightarrow$ (1) of Theorem 4.2.7) that E is row-finite, and that every infinite path in E ends in a sink. There are two possible approaches one may utilize in establishing both of these statements: a "first principles" approach, and a "counting dimensions" approach. For completeness of exposition, we use one approach to establish the first condition, and the other approach to establish the second.

We use the first principles approach to establish the following.

Proposition 4.3.6 *Let E be an arbitrary graph and K any field. If $L_K(E)$ is self-injective then E is row-finite.*

Proof Since $L_K(E)$ is self-injective then so too is the corner $vL_K(E)v$ for any $v \in E^0$, by Lemma 4.3.4. Suppose otherwise that $v \in E^0$ is an infinite emitter, and let $\{e_n \mid n \in \mathbb{N}\}$ be an infinite subset of $s^{-1}(v)$. Then $\{e_n e_n^* \mid n \in \mathbb{N}\}$ is an infinite orthogonal set of idempotents in $vL_K(E)v$. Consider the left ideal $I = \oplus_{n\in\mathbb{N}} vL_K(E)ve_n e_n^*$ of $vL_K(E)v$. Define $\varphi : I \to vL_K(E)v$ to be the identity map on even-indexed summands, and zero on the odds; that is, φ is defined by setting $(e_n e_n^*)\varphi = e_n e_n^*$ if n is even, and $(e_n e_n^*)\varphi = 0$ if n is odd, and extending to all of I.

Since $vL_K(E)v$ is left self-injective, there exists an extension $\overline{\varphi} : vL_K(E)v \to vL_K(E)v$ of φ to all of $vL_K(E)v$. Since $vL_K(E)v$ is unital, there exists an $x \in vL_K(E)v$ for which $(i)\varphi = i \cdot x$ for all $i \in I$. In particular,

(i) $e_n e_n^* \cdot x = e_n e_n^*$ when n is even, and
(ii) $e_n e_n^* \cdot x = 0$ when n is odd.

We argue that this is impossible. For let $x = \sum_{j=1}^{t} k_j \alpha_j \beta_j^* \in vL_K(E)v$, where α_j and β_j are paths in E with $s(\alpha_j) = s(\beta_j) = v$, and $r(\alpha_j) = r(\beta_j)$. Let S' denote the subset of $\{1, 2, \ldots, t\}$ consisting of those j for which $\ell(\alpha_j) \geq 1$, and let $S = \{1, 2, \ldots, t\} \setminus S'$. So S is the set of those $j \in \{1, 2, \ldots, t\}$ for which $\alpha_j = v$. Note that for $j \in S'$ we have $e_n e_n^* \alpha_j = 0$ for all $n \geq M_j$ (for some $M_j \in \mathbb{N}$). Let M be the maximum of $\{M_j \mid j \in S'\}$. Write $x = y + y'$, where

$$y = \sum_{j \in S} k_j \alpha_j \beta_j^* = \sum_{j \in S} k_j v \beta_j^* = \sum_{j \in S} k_j \beta_j^* \in vKE^*,$$

and $y' = \sum_{j \in S'} k_j \alpha_j \beta_j^*$. Then for all $n \geq M$ we have $e_n e_n^* \cdot y = e_n e_n^* \cdot x$, i.e.,

(i) $e_n e_n^* \cdot y = e_n e_n^*$ when $n \geq M$ is even, and
(ii) $e_n e_n^* \cdot y = 0$ when $n \geq M$ is odd.

Let $P \geq M$ be a fixed odd integer. Then by (ii) we have $e_P e_P^* \cdot y = 0$, which by left multiplication by e_P^* gives $e_P^* \cdot y = 0$. But since $r(e_P^*) = v$, this product together with $y \in vLE^*$ yields $y = 0$ by Lemma 2.7.8 (applied to the path algebra KE^*), a contradiction to (i). □

Remark 4.3.7 The counting dimensions approach to the proof of Proposition 4.3.6 is a rather deep analysis of the K-dimensions of various sets of homomorphisms. Specifically, one shows that the existence of an infinite emitter v in E leads to a submodule $S = \oplus_{n \in \mathbb{N}} L_K(E) e_n e_n^*$ of $L_K(E)v$; the injectivity of $L_K(E)v$ then gives an epimorphism of K-vector spaces $\phi^* : \mathrm{Hom}_{L_K(E)}(L_K(E)v, L_K(E)v) \to \mathrm{Hom}_{L_K(E)}(S, L_K(E))$. In addition, $\mathrm{Hom}_{L_K(E)}(S, S)$ embeds in $\mathrm{Hom}_{L_K(E)}(S, L_K(E))$.

On the other hand, by keeping track of various homomorphisms in the indicated sets, one shows that there exists an infinite cardinal σ for which the K-dimension of $\mathrm{Hom}_{L_K(E)}(S, S)$ is at least 2^σ, while the K-dimension of $\mathrm{Hom}_{L_K(E)}(L_K(E)v, L_K(E)v) \cong vL_K(E)v$ is at most σ. This contradicts the existence of the epimorphism ϕ^*.

Definitions 4.3.8 Let R be a (not-necessarily-unital) ring. For a left ideal I of R, the *uniform dimension* of I, denoted u-dim(I), is defined to be the maximum of the set

$$\{|\Lambda| \mid \Lambda \text{ is a set for which there exists a family of left ideals } \{I_i\}_{i \in \Lambda} \text{ of } R \text{ such that } \oplus_{i \in \Lambda} I_i \subseteq I\}.$$

For an element $a \in R$, the *left uniform dimension of* a, denoted by u-dim$_l(a)$, is the uniform dimension of the left ideal Ra.

The key observation in utilizing the counting dimensions approach to establish Proposition 4.3.10 is the following.

Proposition 4.3.9 *Let E be an arbitrary graph and K any field. If the Leavitt path algebra $L_K(E)$ is self-injective, then every element of $L_K(E)$ has finite left uniform dimension. In particular, for every $v \in E^0$, the left ideal $L_K(E)v$ cannot contain an infinite set of nonzero orthogonal idempotents.*

Proof By Proposition 4.3.6, the graph E is row-finite. Let $a \in L_K(E)$. Since $L_K(E)$ has local units, there exists an idempotent $\epsilon \in L_K(E)$ such that $a = a\epsilon$, hence $L_K(E)a \subseteq L_K(E)\epsilon$. We want to prove that u-dim$_l(\epsilon) < \infty$, from which the statement will follow.

Write $\epsilon = \sum k_j\alpha_j\beta_j^*$, where α_j and β_j are paths in E and $k_j \in K^\times$. Let $v_1, \ldots, v_m$ be the vertices that appear as $s(\alpha_j)$ or $s(\beta_j)$ of the finitely many paths α_j and β_j. Then every element of $\epsilon L_K(E)\epsilon$ is of the form $\sum_{i=1}^{t} k_i'\lambda_i\mu_i^*$ where $k_i' \in K^\times$, $\lambda_i, \mu_i \in \text{Path}(E)$, and $s(\lambda_i) = s(\mu_i) \in \{v_1, \ldots, v_m\}$. Since E is row-finite, the cardinality of paths of a fixed length n beginning with any of the vertices $v_1, \ldots, v_m$ is finite, and hence the cardinality of the set of all paths of finite length beginning at any of the vertices $v_1, \ldots, v_m$ is at most countable. Since expressions of the form $\alpha\beta^*$ where α and β start at one of these vertices forms a generating set for $\epsilon L_K(E)\epsilon$ as a K-vector space, we then conclude that the K-dimension of $\epsilon L_K(E)\epsilon$ is at most countable.

Suppose on the contrary $L_K(E)\epsilon$ contains an infinite family of left $L_K(E)$-ideals $\{A_k \mid k \in \Lambda\}$ of the indicated type. So $\bigoplus_{k\in\Lambda} A_k$ is a left ideal of $L_K(E)$ contained in $L_K(E)\epsilon$. We see that there are uncountably many K-linearly independent homomorphisms in $\text{Hom}_{L_K(E)}(\bigoplus_{k\in\Lambda} A_k, \bigoplus_{k\in\Lambda} A_k)$, since for each of the (uncountably many) subsets T of Λ, let $\varphi_T \in \text{Hom}_{L_K(E)}(\bigoplus_{k\in\Lambda} A_k, \bigoplus_{k\in\Lambda} A_k)$ be the function which is the identity on A_k if $k \in T$, and is 0 otherwise. Since the direct summand $L_K(E)\epsilon$ of $L_K(E)$ is an injective $L_K(E)$-module, the inclusion map $\iota : \bigoplus_{k\in\Lambda} A_k \to L_K(E)\epsilon$ yields an epimorphism of K-vector spaces $\iota^* : \text{Hom}_{L_K(E)}(L_K(E)\epsilon, L_K(E)\epsilon) \to \text{Hom}_{L_K(E)}(\bigoplus_{k\in\Lambda} A_k, \bigoplus_{k\in\Lambda} A_k)$. But this is not possible, since $\text{Hom}_{L_K(E)}(L_K(E)\epsilon, L_K(E)\epsilon) \cong \epsilon L_K(E)\epsilon$ has countable K-dimension by the previous paragraph. Hence $L_K(E)\epsilon$, and so too $L_K(E)a$, must have finite uniform dimension. □

Proposition 4.3.10 *Let E be an arbitrary graph and K any field. If $L_K(E)$ is self-injective, then every infinite path in E ends in a line point.*

Proof Suppose that γ is an infinite path in E. Since by Proposition 4.3.5 we have that E is acyclic, if γ is an infinite path in E which does not end in a sink then necessarily γ can be written as $\gamma_0\gamma_1\gamma_2\gamma_3\cdots$, where γ_i is a path of length at least 1 and $r(\gamma_i)$ is a bifurcation vertex for each $i \in \mathbb{N}$. Let v_i denote $s(\gamma_i)$ for $i \in \mathbb{Z}^+$; let v denote v_0.

For each $n \in \mathbb{N}$ let f_n denote an edge in E for which $s(f_n) = v_n$, but f_n is not the first edge of γ_n. (Such exists by the bifurcation property.) For each $n \in \mathbb{Z}^+$ define $\Gamma_n = \gamma_0\gamma_1\cdots\gamma_n$. It is then easy to show that the set $\{\Gamma_n f_n f_n^* \Gamma_n^* \mid n \in \mathbb{N}\}$ is an orthogonal set of nonzero idempotents in $L_K(E)v$. (The orthogonality follows from the bifurcation property.) But this violates Proposition 4.3.9. Therefore no such γ exists, and the result follows. □

Remark 4.3.11 The first principles approach to establishing Proposition 4.3.10 proceeds in much the same way as the proof of Proposition 4.3.6: specifically, one uses the set $\{\Gamma_n f_n f_n^* \Gamma_n^* \mid n \in \mathbb{N}\}$ in a manner similar to the way the set $\{e_n e_n^* \mid n \in \mathbb{N}\}$ was used, and then subsequently shows that an element which induces the indicated homomorphism via right multiplication cannot exist. Completing the first principles proof in this case requires some additional work, but in the end contains essentially the same ideas as in its counterpart.

We now have all the necessary tools in hand to prove the main result of this section.

Theorem 4.3.12 *Let E be an arbitrary graph and K any field. Then $L_K(E)$ is semisimple if and only if $L_K(E)$ is self-injective.*

Proof It is well known that for a ring R, if ${}_RM$ is a semisimple left R-module then every R-submodule of M is a direct summand of M. (The standard Zorn's Lemma argument used for modules over unital rings holds verbatim in the more general setting of modules over rings with local units.) So the Baer Criterion 4.3.3 is automatically satisfied for semisimple rings with local units, which yields one implication.

Conversely, assume that $L_K(E)$ is self-injective. Then E is acyclic (Proposition 4.3.5), E is row-finite (Proposition 4.3.6), and every infinite path in E ends in a sink (Proposition 4.3.10). Now implication (4) $\Rightarrow$ (1) of Theorem 4.2.7 gives the result. □

4.4 The Stable Rank

The notion of the stable rank of a ring was introduced by Bass [43] in order to study stabilization problems in algebraic K-theory. Later Vaserstein [150] showed several important properties of the stable rank, and related it to dimension theory through the determination of the stable rank of rings of continuous functions. Stable rank also has important connections with cancellation conditions on modules [153].

In this section we will prove that the only possible values of the stable rank for a Leavitt path algebra are 1, 2 or ∞, and that it is possible to determine this value by looking at the graph. Indeed, it is known (and we will re-establish) that these three values appear as the stable ranks of the three primary colors of Leavitt path algebras: the stable rank of K is 1, the stable rank of $K[x, x^{-1}]$ is 2, and the stable rank of $L_K(1, n)$ is ∞. Later, in Chap. 5, we will see that 1, 2 and ∞ are also the only possible values of the stable rank of a graph C^*-algebra, and they too can be read from the underlying graph. However, for a given graph E, the stable ranks of $L_{\mathbb{C}}(E)$ and $C^*(E)$ may differ. Historically, stable rank was one of the first properties that was shown to differ in the contexts of Leavitt path algebras and of graph C^*-algebras.

We will focus on verifying these results about the stable rank of Leavitt path algebras in the situation where the graph E is row-finite, but without restriction on the cardinality of E^0. Along the way, we will include some general results about the stable rank of arbitrary rings with local units, including Lemmas 4.4.6 and 4.4.9, and Corollary 4.4.17. Most of the results contained in this section, including the results about stable rank for arbitrary rings with local units, appear in [32].

The following definitions can be found in [150].

Definitions 4.4.1 Let R be a ring and suppose that S is a unital ring containing R as an ideal. A column vector $\mathbf{b} = (b_i)_{i=1}^n$ in S^n is called *R-unimodular* if $b_1 - 1 \in R$, $b_i \in R$ $(2 \le i \le n)$, and there exists a row vector $\mathbf{a} = (a_i)_{i=1}^n$ in S^n with $a_1 - 1 \in R$, $a_i \in R$ $(2 \le i \le n)$ such that $\sum_{i=1}^n a_i b_i = 1$. The *stable rank* of R (denoted by $\mathrm{sr}(R)$)

is the least natural number m for which, for any R-unimodular vector $(b_i)_{i=1}^{m+1}$, there exist $v_i \in R$ $(1 \leq i \leq m)$ such that the vector $(b_i + v_i b_{m+1})_{i=1}^{m}$ is R-unimodular. If such a natural number m does not exist we say that the stable rank of R is infinite, and write $\mathrm{sr}(R) = \infty$.

It can be shown that the definition of the stable rank of R does not depend on the choice of the unital overring S. We will use the following elementary lemma, due to Vaserstein.

Lemma 4.4.2 ([149, Lemma 2.0]) *Let $b_1 - 1 \in R$ and $b_i \in R$ for $2 \leq i \leq n$, where R is a ring and S is a unital ring containing R as a two-sided ideal. The following are equivalent.*

(1) The vector $\mathbf{b} = (b_i)_{i=1}^{n}$ is R-unimodular.
(2) $\sum_{i=1}^{n} Sb_i = S$.
(3) $\sum_{i=1}^{n} Rb_i = R$.

Proof (1) $\Longrightarrow$ (2) $\Longrightarrow$ (3) are clear. To show that (3) implies (1), take elements $u_i \in R$ such that $\sum_{i=1}^{n} u_i b_i = -b_1 + 1$. Then we have

$$(1 + u_1)b_1 + \sum_{i=2}^{n} u_i b_i = 1,$$

so that $\mathbf{b}$ is R-unimodular. □

Some properties which will be very useful for us, and whose proofs (except for one) can be found in [150], are the following.

Theorem 4.4.3 *Let R be a ring.*

(i) *For any set of rings $\{R_i \mid i \in \Lambda\}$, if $R = \prod_{i \in \Lambda} R_i$ then* $\mathrm{sr}(R) = \max_{i \in \Lambda}\{\mathrm{sr}(R_i)\}$.
(ii) *For every $m \in \mathbb{N}$,* $\mathrm{sr}(\mathrm{M}_m(R)) = \lceil (\mathrm{sr}(R) - 1)/m \rceil + 1$, *where $\lceil a \rceil$ denotes the smallest integer $\geq a$. This includes the statement that if* $\mathrm{sr}(R) = \infty$, *then* $\mathrm{sr}(\mathrm{M}_m(R)) = \infty$ *for all $m \in \mathbb{N}$.*
(iii) *For any ideal I of R,*

$$\max\{\mathrm{sr}(I), \mathrm{sr}(R/I)\} \leq \mathrm{sr}(R) \leq \max\{\mathrm{sr}(I), \mathrm{sr}(R/I) + 1\}.$$

(iv) *Let $\{R_i, \varphi_{ij}\}_{i,j \in I}$ be a directed system in the category of not-necessarily-unital rings. Then*

$$\mathrm{sr}(\varinjlim_{i \in I} R_i) \leq \liminf_{i \in I} (\mathrm{sr}(R_i)).$$

Proof (i), (ii), and (iii) are shown in [150], and (iv) follows from the definitions. □

Examples 4.4.4

(i) If K is a field, then its stable rank is 1. Moreover, by Theorem 4.4.3(ii), the stable rank of $\mathrm{M}_n(K)$ is 1 for every $n \in \mathbb{N}$.
(ii) If R is a purely infinite simple unital ring, then its stable rank is ∞ (see [35, Proposition 3.10]).

Lemma 4.4.5 *Let E be an acyclic graph and K any field. Then the stable rank of $L_K(E)$ is* 1.

Proof Suppose first that the graph E is finite. Then, by the Finite Dimension Theorem 2.6.17, $L_K(E)$ is isomorphic to $\oplus_{i=1}^m \mathrm{M}_{n_i}(K)$, where $m, n_i \in \mathbb{N}$. Whence, by Theorem 4.4.3(i) and Examples 4.4.4, the stable rank of $L_K(E)$ is 1. Now suppose E is infinite. By Theorem 3.4.1, the algebra $L_K(E)$ is locally K-matricial, that is, $L_K(E) = \varinjlim_{i \in I} L_K(F_i)$, where each F_i is a finite and acyclic graph. By Theorem 4.4.3(iv), we have

$$\mathrm{sr}(L_K(E)) \leq \liminf_{i \in I} (\mathrm{sr}(L_K(F_i))).$$

Now use the first step of the proof and the displayed inequality to yield the desired result. □

We recall here the definitions of the relations $\leq$, $\sim$ and $\precsim$ for idempotents of a ring, which were introduced in Chap. 3. The partial order $\leq$ on idempotents is defined by declaring $e \leq f$ if and only if $e = ef = fe$. The equivalence relation $\sim$ is defined by $e \sim f$ if and only if there are elements $x, y \in R$ (which indeed can be chosen so that $x \in eRf$ and $y \in fRe$) such that $e = xy$ and $f = yx$. The pre-order $\precsim$ is defined by $e \precsim f$ if and only if there are elements $x \in eRf$ and $y \in fRe$ such that $e = xy$. Note that the latter condition implies that yx is an idempotent such that $yx \leq f$.

A set of local units E for a ring R is called an *ascending local unit* if there is an upward directed set Λ for which $E = \{p_\alpha \mid \alpha \in \Lambda\}$, such that $p_\alpha \leq p_\beta$ whenever $\alpha \leq \beta$ in Λ. Any ring with local units contains an ascending local unit: simply take as Λ the set of all the idempotents of R, and define the order induced from the order of idempotents (i.e., $e \leq f$ if $ef = fe = e$). Then define $p_\alpha = \alpha$ for $\alpha \in \Lambda$.

Lemma 4.4.6 *Let R be a ring with ascending local unit $\{p_\alpha\}_{\alpha \in \mathscr{F}}$. If for every $\alpha \in \mathscr{F}$ there exists a $\beta > \alpha$ such that $p_\alpha \precsim p_\beta - p_\alpha$, then* $\mathrm{sr}(R) \leq 2$.

Proof Fix a unital ring S which contains R as a two-sided ideal. Let $a_1, a_2, a_3, b_1, b_2, b_3 \in S$ such that $a_1 - 1, a_2, a_3, b_1 - 1, b_2, b_3 \in R$, while $a_1b_1 + a_2b_2 + a_3b_3 = 1$. By hypothesis, there exists an $\alpha \in \mathscr{F}$ such that $a_1 - 1, a_2, a_3, b_1 - 1, b_2, b_3 \in p_\alpha R p_\alpha$. Let $\beta > \alpha$ such that $p_\alpha \precsim p_\beta - p_\alpha$. Then there exists a $q \sim p_\alpha$ with $q \leq p_\beta - p_\alpha$. In particular, $qp_\alpha = p_\alpha q = 0$. Now, there exist $u \in p_\alpha R q$, $v \in qRp_\alpha$ such that $uv = p_\alpha$, $vu = q$, $u = p_\alpha u = uq$ and $v = qv = vp_\alpha$.

Fix $v_1 = 0$, $v_2 = u$, $c_1 = b_1$, and $c_2 = b_2 + vb_3$. Notice that $(a_1 + a_3v_1) - 1$, $c_1 - 1, (a_2 + a_3v_2), c_2 \in R$. Also, $a_3uvb_3 = a_3p_nb_3 = a_3b_3$, $a_3ub_2 = a_3uq_np_nb_2 = 0$, and $a_2vb_3 = a_2p_nq_nvb_3 = 0$. Hence,

$$(a_1 + a_3v_1)c_1 + (a_2 + a_3v_2)c_2 = a_1b_1 + a_2b_2 + a_3b_3 = 1.$$

Thus, any unimodular 3-row is reducible, whence the result holds. □

Definition 4.4.7 Let E be a graph. For every $v \in E^0$, we define

$$M(v) = \{w \in E^0 \mid w \geq v\}.$$

We say that $v \in E^0$ is *left infinite* if $\mathrm{card}(M(v)) = \infty$.

Proposition 4.4.8 *Let E be a row-finite graph and K any field. Suppose that $X \subseteq E^0$ is a set of vertices with $|CSP(v)| \geq 2$ for all $v \in X$, and that $E = \overline{X}$. If each $v \in X$ is left infinite, then* $\mathrm{sr}(L_K(E)) = 2$.

Proof We are going to check the condition in Lemma 4.4.6. Let $\mathscr{F}$ be the directed family of all the finite subsets of E^0. For $A \in \mathscr{F}$, set $p_A = \sum_{v \in A} v \in L_K(E)$. Then $\{p_A\}_{A \in \mathscr{F}}$ is an ascending local unit for $L_K(E)$.

Observe that all vertices in X are properly infinite, by Lemma 3.8.11. If A is a finite subset of $T(X)$, then for each $v \in A$ there is a $w \in X$ such that $v \precsim w$. Using that the vertices in X are properly infinite, we see that there are *distinct* $w_1, \ldots, w_m \in X$ such that $p_A \precsim \sum_{i=1}^m w_i$. Now if $v \in S(T(X))$ (the saturated closure of $T(X)$, see Definition 2.0.6), there is a finite number of vertices $v_1, \ldots, v_r$ in $T(X)$ such that

$$v \precsim k_1 \cdot v_1 \oplus \cdots \oplus k_r \cdot v_r$$

for some positive integers $k_1, \ldots, k_r$. As before we deduce the existence of a finite number of distinct vertices $z_1, \ldots, z_s$ in X such that $v \precsim \sum_{i=1}^s z_i$. By induction, one shows a similar result for any vertex v in $S^n(T(X))$, for all n, and therefore for any vertex of E. Using again that the vertices in X are properly infinite, we conclude that given $A \in \mathscr{F}$, there exists a finite subset B of X such that $p_A \precsim p_B$.

It therefore suffices to check that given finite subsets A and B of E^0 with $B \subseteq X$ there exists a $C \in \mathscr{F}$ such that $C \cap (A \cup B) = \emptyset$ and $p_B \precsim p_C$. Write $B = \{v_1, \ldots, v_n\}$. Since by hypothesis $M(v_1)$ is infinite there is a $w_1 \in M(v_1)$ such that $w_1 \notin A \cup B$. Then $v_1 \precsim w_1$. Assume that for $i \geq 1$ we have chosen distinct $w_1, \ldots, w_i$ in E^0 such that $\{w_1, \ldots, w_i\} \cap (A \cup B) = \emptyset$. Since $M(v_{i+1})$ is infinite, we can choose w_{i+1} in $M(v_{i+1})$ such that $w_{i+1} \notin \{w_1, \ldots, w_i\} \cup (A \cup B)$. Using this inductive procedure we get distinct $w_1, \ldots, w_n$ in E^0 so that, with $C = \{w_1, \ldots, w_n\}$, we have $C \cap (A \cup B) = \emptyset$. Note that

$$p_B = v_1 + \cdots + v_n \precsim w_1 + \cdots + w_n = p_C,$$

as desired.

Hence by Lemma 4.4.6, we get $\mathrm{sr}(L_K(E)) \leq 2$. Since all idempotents in a ring with stable rank one are finite (see [151, Theorems 2.6 and 3.9]), we conclude that $\mathrm{sr}(L_K(E)) = 2$. □

Lemma 4.4.9 *Let R be a ring, and let $I \lhd R$ be an ideal with local units. If there exists an ideal $J \lhd I$ such that I/J is a unital simple ring, then there exists an ideal $M \lhd R$ such that $R/M \cong I/J$.*

Proof Given $a \in J$, there exists an $x \in I$ such that $a = ax = xa$. Thus, $J \subseteq JI$, and $J \subseteq IJ$. Hence, $J \lhd R$.

By hypothesis, there exists an element $e \in I$ such that $\overline{e} \in I/J$ is the unit. Consider the set $\mathscr{C}$ of ideals L of R such that $J \subseteq L$ and $e \notin L$. If we order $\mathscr{C}$ by inclusion, it is easy to see that it is inductive. Thus, by Zorn's Lemma, there exists a maximal element of $\mathscr{C}$, say M. Then, $J \subseteq M \cap I \subsetneqq I$, whence $J = M \cap I$ by the maximality of J in I. Thus,

$$I/J = I/(M \cap I) \cong (I + M)/M \lhd R/M.$$

Suppose that $R \neq I + M$. Clearly, $\overline{e} \in (I + M)/M$ is a unit. Thus, $\overline{e}$ is a central idempotent of R/M generating $(I + M)/M$. So, $L = \{a - a\overline{e} \mid a \in R/M\}$ is an ideal of R/M, and

$$R/M = \overline{e}(R/M) + L,$$

the sum being an internal direct sum. If $\pi\colon R \twoheadrightarrow R/M$ is the natural projection map, then $\pi^{-1}(L) = M + \{a - ae \mid a \in R\}$ is an ideal of R containing M (and so J). If $e \in \pi^{-1}(L)$, then $L = R/M$, which is impossible. Hence, $\pi^{-1}(L) \in \mathscr{C}$, and strictly contains M, contradicting the maximality of M in $\mathscr{C}$. Thus, $I + M = R$, and so $R/M \cong I/J$, as desired. □

Corollary 4.4.10 *Let E be an arbitrary graph and K any field. Let $H \in \mathscr{H}_E$. If there exists a $J \lhd I(H)$ such that $I(H)/J$ is a unital simple ring, then there exists an ideal $M \lhd L_K(E)$ such that $L_K(E)/M \cong I(H)/J$.*

Proof By Theorem 2.5.19, $I(H) \cong L_K({}_HE)$, whence $I(H)$ has local units. Thus, the result holds by Lemma 4.4.9. □

Proposition 4.4.11 *Let E be a row-finite graph and K any field. Let J be a maximal ideal of $L_K(E)$. If $L_K(E)/J$ is a unital purely infinite simple ring, then J is a graded ideal of $L_K(E)$. Concretely, $J = I(J \cap E^0)$.*

Proof We show first that we may assume that E has a finite number of vertices. Let α be an element of $L_K(E)$ such that $\alpha + J$ is the unit element in $L_K(E)/J$. Let $v_1, \cdots, v_n \in E^0$ be such that $\alpha \in \left(\sum_{i=1}^n v_i\right) L_K(E) \left(\sum_{i=1}^n v_i\right)$. Since $\alpha v = v\alpha = 0$ for every $v \in E^0 \setminus \{v_1, \ldots, v_n\}$, it follows that the hereditary saturated set $H = J \cap E^0$ satisfies that E/H has a finite number of vertices, and so (by the row-finiteness of E) we get that E/H is finite. Since $L_K(E)/I(H) \cong L_K(E/H)$ by Corollary 2.4.13(i), and $(L_K(E)/I(H)) \,/\, (J/I(H)) \cong L_K(E)/J$, the Leavitt path algebra $L_K(E/H)$ has a unital purely infinite quotient. Passing to $L_K(E)/I(H) \cong L_K(E/H)$, we can assume that E is a finite graph and that $E^0 \cap J = \emptyset$.

Since E is finite, the lattice $\mathscr{L}_{\mathrm{gr}}(L_K(E))$ of graded ideals (equivalently, the lattice of idempotent-generated ideals, by Corollary 2.9.11) of $L_K(E)$ is finite by Theorem 2.5.9, so that there exists a nonempty $H \in \mathscr{H}_E$ such that $I = I(H)$ is minimal as a graded ideal. Since $I+J = L_K(E)$ by our assumption that $J \cap E^0 = \emptyset$, we have

$$I/(I \cap J) \cong L_K(E)/J,$$

so that I has a unital purely infinite simple quotient. Since $I \cong L_K({}_HE)$ (see Theorem 2.5.19) and $J \cap I$ does not contain nonzero idempotents, it follows from our previous argument that ${}_HE$ is finite and so I is unital. So $I = eL_K(E)$ for a central idempotent e in $L_K(E)$. Since I is (unital) graded-simple (by the minimality of I, together with Corollary 2.9.12), and E is finite, the Trichotomy Principle for graded simple Leavitt path algebras 3.1.14 implies that I is isomorphic to either $\mathrm{M}_n(K)$, or isomorphic to $\mathrm{M}_n(K[x, x^{-1}])$ for some $n \geq 1$, or is purely infinite simple. Since I has a quotient algebra which is purely infinite simple, it follows that $I \cap J = \{0\}$, and $J = (1-e)L_K(E)$ is a graded ideal. Indeed we get $e = 1$, because we are assuming that J does not contain nonzero idempotents. □

Next, we characterize in terms of graph conditions when a Leavitt path algebra has a unital purely infinite simple quotient (i.e., satisfies the hypotheses of Proposition 4.4.11).

Corollary 4.4.12 *Let E be an arbitrary graph and K any field. Then $L_K(E)$ has a unital purely infinite simple quotient if and only if there exists an $H \in \mathscr{H}_E$ such that the quotient graph E/H is nonempty, finite, cofinal, contains no sinks, and satisfies Condition (L).*

Proof Apply Proposition 4.4.11 and the Purely Infinite Simplicity Theorem 3.1.10. □

Lemma 4.4.13 *Let E be an arbitrary graph and K any field. If there exists a unital purely infinite simple quotient of $L_K(E)$, then the stable rank of $L_K(E)$ is ∞.*

Proof If there exists a maximal ideal $M \lhd L_K(E)$ such that $L_K(E)/M$ is a unital purely infinite simple ring, then $\mathrm{sr}(L_K(E)/M) = \infty$ (Examples 4.4.4(ii)). Since $\mathrm{sr}(L_K(E)/M) \leq \mathrm{sr}(L_K(E))$ (Theorem 4.4.3(iii)), we conclude that $\mathrm{sr}(L_K(E)) = \infty$. □

We adapt the following terminology from [72]: we say that a graph E has *isolated cycles* if whenever $e_1e_2\cdots e_n$ and $f_1f_2\cdots f_m$ are closed simple paths in E such that $s(e_i) = s(f_j)$ for some i, j, then $e_i = f_j$. Notice that, in particular, if E has isolated cycles, then the only closed simple paths E can contain are cycles.

Lemma 4.4.14 (cf. [72, Lemma 3.2]) *Let E be a row-finite graph and K any field. If $L_K(E)$ does not have any unital purely infinite simple quotients, then there exists a graded ideal $J \lhd L_K(E)$ with $\mathrm{sr}(J) \leq 2$ such that $L_K(E)/J$ is isomorphic to the Leavitt path algebra of a graph with isolated cycles. Moreover, $\mathrm{sr}(J) = 1$ if and only if $J = \{0\}$.*

Proof Set

$$X_0 = \{v \in E^0 \mid \exists\, e \neq f \in E^1 \text{ with } s(e) = s(f) = v,\ r(e) \geq v,\ r(f) \geq v\},$$

and let X be the hereditary saturated closure of X_0. Consider $J = I(X)$. Then J is a graded ideal of $L_K(E)$ and $L_K(E)/J \cong L_K(E/X)$ by Corollary 2.4.13(i). It is clear from the definition of X_0 that E/X is a graph with isolated cycles. Observe that $|CSP(v)| \geq 2$ for all $v \in X_0$. Assuming that $X_0 \neq \emptyset$, we will show that $\mathrm{sr}(J) = 2$.

By Proposition 2.5.19, $J \cong L_K({}_XE)$. We will show that every vertex lying on a closed simple path of ${}_XE$ is left infinite. Suppose that there exists a closed simple path α in ${}_XE$ such that the set Y of vertices of ${}_XE$ connecting to the vertices of α^0 is finite. It is not difficult to see that $\alpha^0 \cup Y$ is a maximal tail in ${}_XE$. Let M be a maximal tail of smallest cardinality contained in $\alpha^0 \cup Y$. Observe that $M \cap X_0 \neq \emptyset$; otherwise $X \setminus M$, which is a proper hereditary saturated subset of X, would contain X_0, which is impossible. Denote by $\widetilde{M}$ the quotient graph of ${}_XE$ by the hereditary saturated set $H = {}_XE^0 \setminus M$, i.e., $\widetilde{M} = {}_XE/H$. Then, since M is finite, $L_K(\widetilde{M})$ is a unital ring. Further, since M does not contain smaller maximal tails, $L_K(\widetilde{M})$ is graded-simple. If $v \in M \cap X_0$, then $|CSP_{\widetilde{M}}(v)| \geq 2$, and so, as $L_K(\widetilde{M})$ is graded-simple, it must be purely infinite simple. Thus, $L_K(\widetilde{M}) \cong L_K({}_XE)/I$ is a unital purely infinite simple ring, where I is the ideal of $L_K({}_XE)$ generated by H. By Corollary 4.4.10, $L_K(E)$ has a unital purely infinite simple quotient, contradicting the hypothesis. Hence, every vertex lying on a closed simple path in ${}_XE$ is left infinite. Thus, $\mathrm{sr}(J) = \mathrm{sr}(L_K({}_XE)) = 2$ by Proposition 4.4.8, as desired. □

Definition 4.4.15 Let A be a unital ring with stable rank n. We say that A has stable rank *closed by extensions* if for any unital ring extension

$$0 \longrightarrow I \longrightarrow B \longrightarrow A \longrightarrow 0$$

of A with $\mathrm{sr}(I) \leq n$, we have $\mathrm{sr}(B) = n$.

A unital ring R is said to have *elementary rank* n, written $\mathrm{er}(R) = n$, if, for every $t \geq n + 1$, the elementary group $E_t(R)$ acts transitively on the set $U_c(t, R)$ of t-unimodular columns with coefficients in R, see [116, 11.3.9]. In the next lemma we collect some properties of elementary rank that we will need in the sequel.

Lemma 4.4.16 *Let A be a unital ring. Assume that* $\mathrm{sr}(A) = n < \infty$.

(i) If $\mathrm{er}(A) < n$ *then* $\mathrm{M}_m(A)$ *has stable rank closed by extensions for every* $m \geq 1$.
(ii) Let D be any (commutative) Euclidean domain such that $\mathrm{sr}(D) > 1$ *and let m be a positive integer. Then* $\mathrm{sr}(\mathrm{M}_m(D)) = 2$ *and* $\mathrm{er}(\mathrm{M}_m(D)) = 1$. *In particular,* $\mathrm{M}_m(D)$ *has stable rank closed by extensions.*
(iii) Let

$$0 \longrightarrow I \longrightarrow B \longrightarrow A \longrightarrow 0$$

be a unital extension of A. If $\mathrm{er}(A) < n$ *and I has a set of local units* $\{g_i\}$ *such that* $\mathrm{sr}(g_iIg_i) \leq n$ *and* $\mathrm{er}(g_iIg_i) < n$ *for all i, then* $\mathrm{sr}(B) = n$ *and* $\mathrm{er}(B) < n$.

(iv) *For unital rings R and S, we have*

$$\mathrm{er}(R \times S) = \max\{\mathrm{er}(R), \mathrm{er}(S)\}.$$

Proof (i) This is essentially contained in [150]. We include a sketch of the proof for the convenience of the reader. Assume that we have a unital extension B of A with $\mathrm{sr}(I) \leq n$. Let $\mathbf{a} = (a_1, \dots, a_{n+1})^t \in U_c(n+1, B)$. Then $\overline{\mathbf{a}} = (\overline{a_1}, \dots, \overline{a_{n+1}})^t \in U_c(n+1, A)$. Since $\mathrm{sr}(A) = n$, there exists $b_1, \dots, b_n \in B$ such that $(\overline{a_1} + \overline{b_1}\overline{a_{n+1}}, \dots, \overline{a_n} + \overline{b_n}\overline{a_{n+1}})^t \in U_c(n, A)$. Replacing $\mathbf{a}$ with $(a_1 + b_1a_{n+1}, \dots, a_n + b_na_{n+1}, a_{n+1})$, we can assume that $(\overline{a_1}, \dots, \overline{a_n})^t \in U_c(n, A)$.

Since $\mathrm{er}(A) \leq n-1$, there exists an $E \in E(n, B)$ such that $\overline{E} \cdot (\overline{a_1}, \dots, \overline{a_n})^t = (1, 0, \dots, 0)^t$. Since $\mathbf{a}$ is reducible if and only if $\mathrm{diag}(E, 1) \cdot \mathbf{a}$ is reducible, we can assume that $(\overline{a_1}, \dots, \overline{a_n})^t = (1, 0, \dots, 0)^t$. Finally, replacing a_{n+1} with $a_{n+1} - a_1a_{n+1}$, we can assume that $\overline{\mathbf{a}} = (1, 0, \dots, 0)^t$, that is, $\mathbf{a} \in U_c(n+1, I)$ (using Lemma 4.4.2). Now, as $\mathrm{sr}(I) \leq n$, $\mathbf{a}$ is reducible in I, and so in B, as desired.

For any $m \in \mathbb{N}$, $\mathrm{sr}(\mathrm{M}_m(A)) = \lceil (\mathrm{sr}(A) - 1)/m \rceil + 1$ by Theorem 4.4.3(ii), and $\mathrm{er}(\mathrm{M}_m(A)) \leq \lceil \mathrm{er}(A)/m \rceil$ by McConnell and Robson [116, Theorem 11.5.15]. So, it is clear that $\mathrm{er}(A) < \mathrm{sr}(A)$ implies $\mathrm{er}(\mathrm{M}_m(A)) < \mathrm{sr}(\mathrm{M}_m(A))$. Hence, by the first part of the proof, $\mathrm{M}_m(A)$ has stable rank closed by extensions, as desired.

(ii) It is well known that a Euclidean domain has stable rank less than or equal to 2, and that it has elementary rank equal to 1, see e.g., [116, Proposition 11.5.3]. So, the result follows from part (i).

(iii) Since $\mathrm{sr}(I) \leq n$, the fact that $\mathrm{sr}(B) = n$ follows from part (i). Now, take $m \geq n$, and set $\mathbf{a} = (a_1, \dots, a_m)^t \in U_c(m, B)$. Since $\mathrm{er}(A) < n$, there exists an $E \in E(m, B)$ such that $\overline{E} \cdot \overline{\mathbf{a}} = (1, 0, \dots, 0)^t$. So, $\mathbf{b} := E \cdot \mathbf{a} \equiv (1, 0, \dots, 0)^t (\mathrm{mod}\, I)$. Let $g \in I$ be an idempotent in the local unit such that $b_1 - 1, b_2, \dots, b_m \in gIg$. Since $\mathrm{er}(gIg) < n$ by hypothesis, there exists a $G \in E(m, gIg)$ such that $(G + \mathrm{diag}(1-g, \dots, 1-g)) \cdot \mathbf{b} = (1, 0, \dots, 0)^t$.

(iv) This follows from the fact that $E_t(R \times S) = E_t(R) \times E_t(S)$, for $t \geq 2$. □

Given any K-algebra R, we define the *unitization* of R to be the ring $R^1 = R \times K$, with product given by

$$(r, a) \cdot (s, b) = (rs + as + rb, ab).$$

Corollary 4.4.17 *Let A be a unital K-algebra with* $\mathrm{sr}(A) = n \geq 2$ *and* $\mathrm{er}(A) < \mathrm{sr}(A)$. *Then, for any (not-necessarily-unital) K-algebra B and two-sided ideal I of B such that* $B/I \cong A$ *and* $\mathrm{sr}(I) \leq n$, *we have* $\mathrm{sr}(B) = n$.

Proof Consider the unital extension

$$0 \longrightarrow I \longrightarrow B^1 \longrightarrow A^1 \longrightarrow 0.$$

Notice that $A^1 \cong A \times K$, because A is unital. So, $\mathrm{sr}(A^1) = \mathrm{sr}(A)$ (Theorem 4.4.3(i)), and $\mathrm{er}(A^1) = \mathrm{er}(A)$ (Lemma 4.4.16(iv)). Now, by Lemma 4.4.16(i), $\mathrm{sr}(B^1) \le n$. Since $n \le \mathrm{sr}(B) \le \mathrm{sr}(B^1) \le n$, the conclusion follows. □

Proposition 4.4.18 *Let E be a finite graph with isolated cycles and K any field. Then $\mathrm{sr}(L_K(E)) \le 2$ and $\mathrm{er}(L_K(E)) = 1$. Moreover, $\mathrm{sr}(L_K(E)) = 1$ if and only if E is acyclic.*

Proof We proceed by induction on the number of cycles of E. If E has no cycles then $\mathrm{sr}(L_K(E)) = 1$ by Lemma 4.4.5, so that $\mathrm{er}(L_K(E)) = 1$ by McConnell and Robson [116, Proposition 11.3.11]. Assume that E has cycles $C_1, \dots, C_n$. Define a preorder on the set of cycles by setting $C_i \ge C_j$ iff there exists a finite path α such that $s(\alpha) \in C_i^0$ and $r(\alpha) \in C_j^0$. Since E is a graph with isolated cycles, $\ge$ is easily seen to be a partial order. Since the set of cycles is finite, there exists a maximal one, say C_1. Set $A = \{e \in E^1 \mid s(e) \in C_1 \text{ and } r(e) \notin C_1\}$, and define $B = \{r(e) \mid e \in A\} \cup \mathrm{Sink}(E) \cup \bigcup_{i=2}^n C_i^0$. Let H be the hereditary saturated closure of B. By construction of H, C_1 is the unique cycle in E/H, and it has no exits. Moreover, E/H coincides with the hereditary saturated closure of C_1. Note too that any vertex not connecting to C_1 must be in the hereditary saturated set generated by the sinks and C_i^0 for $i = 2, \dots, n$, which implies that there are no sinks in E/H. So we may apply Corollary 4.2.14 to get that $L_K(E/H) \cong \mathrm{M}_k(K[x, x^{-1}])$ for some $k \ge 1$.

Consider the extension

$$0 \longrightarrow I(H) \longrightarrow L_K(E) \longrightarrow L_K(E/H) \longrightarrow 0.$$

Now, by Lemma 4.4.16(ii), $\mathrm{sr}(L_K(E/H)) = 2$ and $\mathrm{er}(L_K(E/H)) = 1$. Consider the local unit (p_X) of $L_K({}_HE) \cong I(H)$ consisting of idempotents $p_X = \sum_{v \in X} v$ where X ranges over the set of vertices of ${}_HE$ containing H. Since these sets are hereditary in $({}_HE)^0$, we get that $p_X I(H) p_X = p_X L_K({}_HE) p_X = L_K(({}_HE)_X)$ is a Leavitt path algebra of a graph with isolated cycles, containing exactly $n - 1$ cycles. By the induction hypothesis, $\mathrm{sr}(p_X I(H) p_X) \le 2$ and $\mathrm{er}(p_X I(H) p_X) = 1$. So, by Lemma 4.4.16(iii), we conclude that $\mathrm{sr}(L_K(E)) = 2$ and $\mathrm{er}(L_K(E)) = 1$. Hence, the induction step works, so we are done. □

We are now ready to obtain our main result.

Theorem 4.4.19 *Let E be a row-finite graph and K any field. Then the values of the stable rank of $L_K(E)$ are determined as follows.*

(i) $\mathrm{sr}(L_K(E)) = 1$ *if E is acyclic.*
(ii) $\mathrm{sr}(L_K(E)) = \infty$ *if there exists an $H \in \mathcal{H}_E$ such that the quotient graph E/H is nonempty, finite, cofinal, contains no sinks, and satisfies Condition (L).*
(iii) $\mathrm{sr}(L_K(E)) = 2$ *otherwise.*

Proof (i) derives from Lemma 4.4.5, while (ii) derives from Corollary 4.4.12 and Lemma 4.4.13. We can thus assume that E contains cycles and, using Lemma 4.4.13, that $L_K(E)$ does not have any unital purely infinite simple quotients.

By Lemma 4.4.14, there exists a hereditary saturated subset X of E^0 such that $\mathrm{sr}(I(X)) \leq 2$, while E/X is a graph having isolated cycles. By Corollary 1.6.16 there is an upward directed set $\{E_i \mid i \in \Lambda\}$ of complete finite subgraphs of E/X such that $E/X = \bigcup_{i\in\Lambda} E_i$. So, by Theorem 1.6.10, $L_K(E/X) \cong \varinjlim_{i\in\Lambda} L_K(E_i)$. For each $i \in \Lambda$, there is a natural graded K-algebra homomorphism $\phi_i \colon L_K(E_i) \to L_K(E/X)$. The kernel of ϕ_i is a graded ideal of $L_K(E_i)$ whose intersection with E_i^0 is empty, so ϕ_i is injective by the Graded Uniqueness Theorem 2.2.15, and thus the image L_i of $L_K(E_i)$ through ϕ_i is isomorphic to $L_K(E_i)$. It follows from Proposition 4.4.18 that, for every $i \in \Lambda$, $\mathrm{sr}(L_i) \leq 2$ and $\mathrm{er}(L_i) = 1$. If $\pi : L_K(E) \to L_K(E/X)$ denotes the natural epimorphism (see Corollary 2.4.13(i)), then given any $i \in \Lambda$, we have

$$0 \longrightarrow I(X) \longrightarrow \pi^{-1}(L_i) \longrightarrow L_i \longrightarrow 0.$$

If $\mathrm{sr}(L_i) = 1$, then $\mathrm{sr}(\pi^{-1}(L_i)) \leq 2$ by Vaseršteĭn [150, Theorem 4].

If $\mathrm{sr}(L_i) = 2$, then it follows from Corollary 4.4.17 that $\mathrm{sr}(\pi^{-1}(L_i)) = 2$. Since $L_K(E) = \bigcup_{i\in\Lambda} \pi^{-1}(L_i)$ we get that $\mathrm{sr}(L_K(E)) \leq 2$. Since E contains cycles we have that either $I(X) \neq 0$ or E/X contains cycles. If $I(X) \neq 0$ then $\mathrm{sr}(I(X)) = 2$ by Lemma 4.4.14 and so $\mathrm{sr}(L_K(E)) = 2$ by Vaseršteĭn [150, Theorem 4]. If $I(X) = 0$, then E has isolated cycles. Take a vertex v in a cycle C of E and let H be the hereditary subset of E generated by v. Then $L_K(E_H) = pL_K(E)p$ for the idempotent $p = \sum_{w\in H^0} w \in \mathscr{M}(L_K(E))$, where $\mathscr{M}(L_K(E))$ denotes the multiplier algebra of $L_K(E)$; see [36]. Let I be the ideal of $pL_K(E)p$ generated by all the idempotents of the form $r(e)$, where $e \in E^1$ is such that $s(e) \in C$ and $r(e) \notin C$. Since E has isolated cycles it follows that I is a proper ideal of $pL_K(E)p$ and moreover $pL_K(E)p/I \cong \mathrm{M}_k(K[x, x^{-1}])$, where k is the number of vertices in C. We get

$$\mathrm{sr}(pL_K(E)p) \geq \mathrm{sr}(pL_K(E)p/I) = 2.$$

It follows that $1 < \mathrm{sr}(L_K(E)) \leq 2$ and thus $\mathrm{sr}(L_K(E)) = 2$, as desired. □

Remark 4.4.20 The result in Theorem 4.4.19 remains valid for arbitrary graphs, as was shown by Larki and Riazi in [111].

Some remarks on the relationship between the stable rank of Leavitt path algebras and the stable rank of graph C^*-algebras will be given at the end of Sect. 5.6.

We present below several examples of Leavitt path algebras, and we compute their stable rank by using Theorem 4.4.19.

Examples 4.4.21 The basic examples to illustrate Theorem 4.4.19 coincide with those given in Chap. 1.

(i) The Leavitt path algebra associated with the acyclic graph A_n

$$\bullet^{v_1} \longrightarrow \bullet^{v_2} \longrightarrow \bullet^{v_3} \cdots\cdots \bullet^{v_{n-1}} \longrightarrow \bullet^{v_n}$$

satisfies $L_K(A_n) \cong \mathrm{M}_n(K)$ (see Proposition 1.3.5). Thus, $\mathrm{sr}(L_K(A_n)) = 1$ by Theorem 4.4.19(i). (This is of course well known, see Examples 4.4.4(i).)

(ii) If E is a finite graph for which $L_K(E)$ is purely infinite simple, then by the Purely Infinite Simplicity Theorem 3.1.10 we have that $L_K(E)$ satisfies the conditions of Theorem 4.4.19(ii) (with $H = \emptyset$), so that $\mathrm{sr}(L_K(E)) = \infty$. In particular, for $n \geq 2$, the Leavitt path algebra of the graph R_n

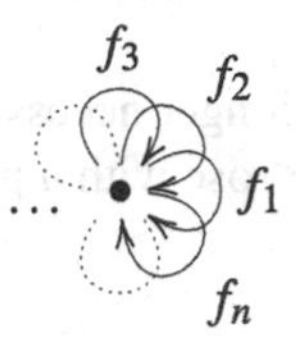

has $\mathrm{sr}(L_K(R_n)) = \infty$, i.e., $\mathrm{sr}(L_K(1,n)) = \infty$. (In this particular case, one can also recover this conclusion using [133, Proposition 6.5].) For additional examples of purely infinite simple unital Leavitt path algebras (and so, additional examples of Leavitt path algebras having infinite stable rank), see Examples 3.2.7.

(iii) Finally, the Leavitt path algebra of the graph G

has $\mathrm{sr}(L_K(G)) = 2$ by Theorem 4.4.19(iii). In other words, by Proposition 1.3.4, we recover the fact that $\mathrm{sr}(K[x,x^{-1}]) = 2$. In a similar manner, we get that $\mathrm{sr}(\mathscr{T}_K) = 2$ as well (where $\mathscr{T}_K$ is the Toeplitz algebra of Example 1.3.6), which in turn yields by Proposition 1.3.7 that $\mathrm{sr}(K\langle X, Y | XY = 1\rangle) = 2$.

Example 4.4.22 We present an additional example that illustrates an interesting phenomenon arising in the discussion of stable rank in the context of Leavitt path algebras. On one hand, stable rank 2 examples can be obtained (more or less) as extensions of the ring of Laurent polynomials, as we can see with the Leavitt path algebra of the graph E

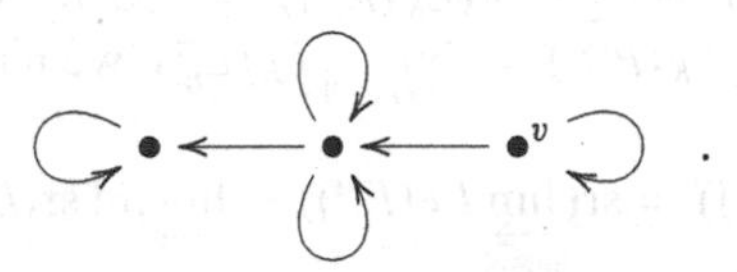

Here the ideal I in Lemma 4.4.14 is $I = I(E^0 \setminus \{v\})$; we see $L_K(E)/I \cong K[x,x^{-1}]$. Notice that, because of Lemma 4.4.14, $\mathrm{sr}(I) = 2$, while $\mathrm{sr}(L_K(E)) = \mathrm{sr}(L_K(E)/I) = 2$ as well by Theorem 4.4.19(iii). The remarkable fact behind Theorem 4.4.19 is that in the context of Leavitt path algebras, extensions of stable rank 2 rings by stable rank 2 ideals cannot attain stable rank 3. (This statement is not true for more general algebras.)

Remark 4.4.23 Stable rank is not a Morita invariant in general, but in the case of Leavitt path algebras some interesting phenomena arise. Suppose that E, F are finite graphs such that $L_K(E)$ and $L_K(F)$ are Morita equivalent. Thus, $L_K(E) \cong P \cdot \mathrm{M}_n(L_K(F)) \cdot P$ for some $n \in \mathbb{N}$ and some full idempotent $P \in \mathrm{M}_n(L_K(F))$. Since the values 1 and ∞ of stable rank are preserved by passing to matrices [150, Theorem 4] and full corners [26, Theorem 7 and Theorem 8], Theorem 4.4.19 implies that $\mathrm{sr}(L_K(E)) = \mathrm{sr}(L_K(F))$. So, stable rank is a Morita invariant for unital Leavitt path algebras.

However, this conclusion no longer necessarily follows if E and/or F is infinite. For instance, let R_n be the usual "rose with n petals" graph, and let F^∞ be the graph

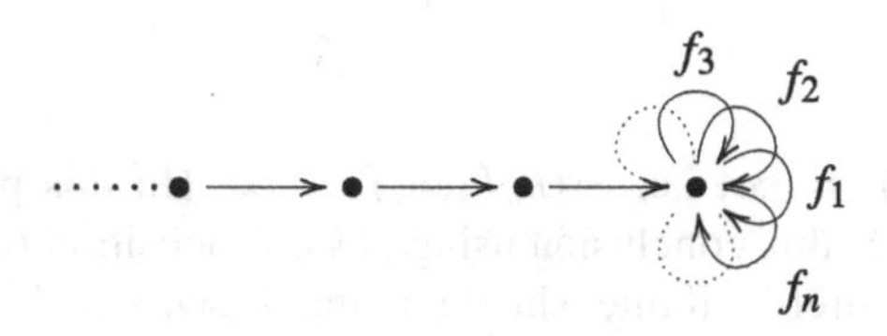

i.e., R_n with an infinite tail added. As we noted in Examples 4.4.21, $\mathrm{sr}(L_K(R_n)) = \infty$. On the other hand, an easy induction argument using [6, Proposition 13] shows that $L_K(F^\infty) \cong \mathrm{M}_{\mathbb{N}}(L_K(R_n))$, which gives in particular that $L_K(F^\infty)$ and $L_K(R_n)$ are Morita equivalent. Now observe that $L_K(F^\infty)$ is simple by the Simplicity Theorem 2.9.1, and non-unital (because the graph F^∞ has infinitely many vertices), so that $L_K(F^\infty)$ has no unital purely infinite simple quotients, and thus $\mathrm{sr}(L_K(F^\infty)) = 2$ by Theorem 4.4.19(iii).

Moreover, the graph F^∞ is a direct limit of the graphs E_n^m

$$\bullet^{v_m} \longrightarrow \bullet^{v_{m-1}} \cdots\cdots \bullet^{v_3} \longrightarrow \bullet^{v_2} \longrightarrow \bullet^{v_1} \quad f_3 \; f_2 \; f_1 \; f_n$$

i.e., R_n with a tail of length $m-1$ added. Since $L_K(E_n^m) \cong \mathrm{M}_m(L_K(1,n))$ (see Proposition 2.2.19), we get $\mathrm{sr}(L_K(E_n^m)) = \infty$ by Examples 4.4.21(i) and Theorem 4.4.3(ii). Since $L_K(F^\infty) \cong \varinjlim_{m\in\mathbb{N}} L_K(E_n^m)$, we have

$$2 = \mathrm{sr}(L_K(F^\infty)) = \mathrm{sr}(\varinjlim_{m\in\mathbb{N}} L_K(E_n^m)) < \liminf_{m\in\mathbb{N}} \mathrm{sr}(L_K(E_n^m)) = \infty.$$

In particular, the inequality invoked in Theorem 4.4.3(iv) may indeed be strict.

Chapter 5
Graph C^*-Algebras, and Their Relationship to Leavitt Path Algebras

There is a close, fundamental relationship between the Leavitt path algebra $L_{\mathbb{C}}(E)$ and an analytic structure known as a *graph* C^**-algebra*. In the first section of this chapter we give a very general overview of some of the basic properties of C^*-algebras. In the following section we then present the definition of the graph C^*-algebra $C^*(E)$, and establish that $L_{\mathbb{C}}(E)$ embeds (as a dense $*$-subalgebra) inside $C^*(E)$ (Theorem 5.2.9). Further into that section, we present the appropriate analogs of the two "Uniqueness Theorems", namely, the Gauge-Invariant Uniqueness Theorem and Cuntz–Krieger Uniqueness Theorem for graph C^*-algebras. In Sect. 5.3 we investigate what was, historically, the first structural connection established between $L_{\mathbb{C}}(E)$ and $C^*(E)$, to wit, that the $\mathscr{V}$-monoids of these two $\mathbb{C}$-algebras are in fact isomorphic (Theorem 5.3.5). The remainder of Sect. 5.3, as well as all of Sects. 5.4 and 5.5, are taken up with a description of the closed ideals of a graph C^*-algebra. As we will see, their structure mimics to a great extent (but not completely) the structure of the ideals of $L_{\mathbb{C}}(E)$ established in Sect. 2.8. Finally, in Sect. 5.6, we present a number of results which bring to the foreground the extremely tight (but not yet completely understood) relationships between the complex algebra $L_{\mathbb{C}}(E)$, the graph C^*-algebra $C^*(E)$, and the directed graph E.

5.1 A Brief Overview of C^*-Algebras

In this section, we present some basic material on C^*-algebras and pre-C^*-algebras. The presentation is biased by our interest in paying close attention to the relationship between Leavitt path algebras and graph C^*-algebras. The reader is referred to the book [129] by Raeburn for the theory of graph C^*-algebras, and to Murphy's book [118] and Goodearl's book [85] for the general basic theory of C^*-algebras. Section 1.2 in Ara–Mathieu's book [30] can serve as a guide to the most important facts on operator algebras.

G. Abrams et al., *Leavitt Path Algebras*, Lecture Notes in Mathematics 2191,
DOI 10.1007/978-1-4471-7344-1_5

Most of the results in this section can be generalized to semi-pre-C^*-algebras, see [123, 139, 140].

A *complex $*$-algebra* is an algebra A over $\mathbb{C}$ endowed with an involution $*$ such that $(\lambda x)^* = \overline{\lambda} x^*$, for all $\lambda \in \mathbb{C}$ and $x \in A$, where $\overline{\lambda}$ denotes the complex conjugate of λ.

A *C^*-seminorm* on a complex $*$-algebra A is a function $\|\cdot\|: A \to \mathbb{R}_+$ satisfying the following properties for all $a, b \in A$ and $\lambda \in \mathbb{C}$:

(i) $\|ab\| \leq \|a\| \cdot \|b\|$,
(ii) $\|a + b\| \leq \|a\| + \|b\|$,
(iii) $\|aa^*\| = \|a\|^2 = \|a^*\|^2$, and
(iv) $\|\lambda a\| = |\lambda| \|a\|$ for $\lambda \in \mathbb{C}$.

If, in addition, $\|a\| = 0$ implies $a = 0$, then we say that $\|\cdot\|$ is a *C^*-norm*. It follows easily that if $\|\cdot\|$ is a nonzero C^*-seminorm then $\|0\| = 0$, and $\|1_A\| = 1$ if A is unital. A *pre-C^*-algebra* is a complex $*$-algebra A endowed with a C^*-norm $\|\cdot\|$. A *C^*-algebra* is a pre-C^*-algebra A such that A is complete with respect to the norm topology. If A is a pre-C^*-algebra, then the completion $\overline{A}$ of A with respect to the norm topology is a C^*-algebra and the natural map $A \longrightarrow \overline{A}$ is an isometry. In this chapter we are interested in studying the Leavitt path algebra $L_{\mathbb{C}}(E)$ from an analytic point of view. Note that $L_{\mathbb{C}}(E)$ is a complex $*$-algebra.

For a locally compact subset X of $\mathbb{C}$, $C(X)$ denotes the C^*-algebra of bounded continuous functions from X to $\mathbb{C}$, while $C_0(X)$ denotes the C^*-algebra of continuous functions from X to $\mathbb{C}$ that *disappear at* ∞, i.e., those functions f for which for all $\epsilon > 0$ there is a compact subset S of X such that $|f| < \epsilon$ outside S. If X is compact then $C(X) = C_0(X)$.

A complex $*$-algebra may admit more than one C^*-norm, contrasting with the fact that a C^*-algebra admits only one C^*-norm, complete or not. As an illustrative example, consider the $*$-algebra $\mathbb{C}[z, z^{-1}] \cong \mathbb{C}[\mathbb{Z}]$ of Laurent polynomials. If K is a compact subset of

$$\mathbb{T} := \{z \in \mathbb{C} : |z| = 1\}$$

having nonempty interior, then the composition $\mathbb{C}[z, z^{-1}] \to C(\mathbb{T}) \to C(K)$, where $\mathbb{C}[z, z^{-1}] \to C(\mathbb{T})$ is the natural embedding, and $C(\mathbb{T}) \to C(K)$ is the natural projection, is an injective $*$-homomorphism, and so induces a C^*-norm $\|\cdot\|_K$ on $\mathbb{C}[z, z^{-1}]$. Observe that $\|\cdot\|_K \neq \|\cdot\|_{K'}$ if $K \neq K'$.

It may also happen that a complex $*$-algebra admits no nonzero C^*-seminorms. This is the case for the Weyl algebra, which is the $\mathbb{C}$-algebra $\mathscr{W}$ generated by Q, P subject to the relation $QP - PQ = 1$. It follows easily by induction that $Q^nP - PQ^n = nQ^{n-1}$ in $\mathscr{W}$. Suppose that $\|\cdot\|$ is a nonzero submultiplicative seminorm on $\mathscr{W}$. Then, using the above equations one gets $\|Q^n\| \neq 0$ for all positive integers n. Moreover,

$$n\|Q^{n-1}\| \leq 2\|Q\|\|Q^{n-1}\|\|P\|.$$

Since $\|Q^{n-1}\| \neq 0$ we get $\|P\|\|Q\| \geq \frac{n}{2}$ for all n, which is impossible.

Definitions 5.1.1 Let A be a complex $*$-algebra. An element $b \in A$ is called a *partial isometry* if $b = bb^*b$, and a *projection* if $b = b^2 = b^*$. (Clearly then any projection is a partial isometry, and b is a partial isometry precisely when bb^* is a projection.) If A is unital, then $b \in A$ is called an *isometry* if $b^*b = 1_A$, and a *unitary* if $bb^* = b^*b = 1_A$. (Clearly then any unitary is an isometry.)

An obvious necessary algebraic condition for a complex $*$-algebra to admit a C^*-norm is that A is positive definite, as defined here.

Definitions 5.1.2 ([90, 152]) We say that a complex $*$-algebra A is *positive definite* if, for $x_1, \ldots, x_n \in A$, if $\sum_{i=1}^n x_i^* x_i = 0$ then $x_i = 0$ for all $i = 1, \ldots, n$. If A is positive definite, we define the *positive cone* A_{++} of A as the set of elements of A of the form $\sum_{i=1}^n x_i^* x_i$. In this situation, there is a partial order defined on A, where for $a, b \in A$,

$$a \leq b \text{ if } b - a \in A_{++}.$$

Assume now that A is unital. An element a in A is said to be *bounded* if $a^*a \leq \lambda 1_A$ for some positive real number λ. The set A_b of bounded elements of A is a $*$-subalgebra of A, cf. [152], [90, p. 339], or [49, Proposition 54.1]. A C^*-seminorm $\|\cdot\|$ is defined on A_b by using the partially ordered structure on A, namely

$$\|x\| = \inf\{\lambda \in \mathbb{R}_+ : x^*x \leq \lambda^2 \cdot 1\};$$

cf. [152], or [90, p. 342]. Observe that A_b contains all the partial isometries of A, so that $A = A_b$ if A is generated as a $*$-algebra by its partial isometries.

Let A be a positive definite unital complex $*$-algebra. A *state* of A is a linear functional $\phi: A \to \mathbb{C}$ such that $\phi(x^*x) \geq 0$ for any $x \in A$, and $\phi(1) = 1$. Given a state ϕ of A, one has automatically that $\phi(x^*y) = \overline{\phi(y^*x)}$ for all $x, y \in A$, and that the Cauchy–Schwarz inequality holds: for all $x, y \in A$,

$$|\phi(x^*y)|^2 \leq \phi(x^*x)\phi(y^*y).$$

It follows that $L_\phi := \{x \in A \mid \phi(x^*x) = 0\}$ is a left ideal of A and that the quotient A/L_ϕ has the structure of a pre-Hilbert space given by $\langle x + L_\phi, y + L_\phi \rangle = \phi(x^*y)$. There is a $*$-representation $\pi_\phi: A \to \mathscr{L}(A/L_\phi)$, the GNS-representation associated to the state ϕ. Here $\mathscr{L}(A/L_\phi)$ denotes the $*$-algebra of adjointable operators on the pre-Hilbert space A/L_ϕ. The above representation π_ϕ extends to a $*$-representation $\pi_\phi: A \to B(\mathscr{H}_\phi)$, where $\mathscr{H}_\phi := \overline{A/L_\phi}$ is the Hilbert space completion of A/L_ϕ, if and only if for each $a \in A$ there is a positive constant $K(a)$ such that

$$\phi(b^*a^*ab) \leq K(a)\phi(b^*b)$$

for all $b \in A$. We call a state as above a *bounded state*. Let $S(A)$ be the set of bounded states of A. Define

$$\|a\|_{\max} = \sup\{\phi(a^*a)^{1/2} \mid \phi \in S(A)\} \in [0, +\infty].$$

Lemma 5.1.3 *There is a maximal C^*-seminorm on A if and only if $\|a\|_{max} < \infty$ for all $a \in A$, and in this case the maximal C^*-seminorm on A is precisely $\| \cdot \|_{max}$.*

Proof Assume first that $\|a\|_{\max} < \infty$ for all $a \in A$. Since each $\phi \in S(A)$ defines a $*$-representation

$$\pi_\phi : A \longrightarrow B(\mathcal{H}_\phi)$$

such that $\phi(a^*a) = \|\pi_\phi(a)\|^2$, it follows that $\| \cdot \|_{\max}$ is a C^*-seminorm on A. Now let ρ be an arbitrary C^*-seminorm on A. Let I be the set of elements a in A such that $\rho(a) = 0$. Then I is a $*$-ideal of A, and A/I is a pre-C^*-algebra under $\overline{\rho}(a + I) = \rho(a)$. Let B be the completion of A/I with respect to the norm $\overline{\rho}$. Then B is a C^*-algebra, and there is a canonical $*$-homomorphism $\tau : A \to B$. Let a be an element of A. By Kadison and Ringrose [101, Theorem 4.3.4(iv)], there is a state ϕ' of B such that $\phi'(\tau(a^*a)) = \rho(a)^2$. Then $\phi := \phi' \circ \tau$ is a bounded state of A, and so

$$\rho(a)^2 = \phi(a^*a) \le \|a\|^2_{\max}.$$

This shows that $\rho \le \| \cdot \|_{\max}$ and so $\| \cdot \|_{\max}$ is the maximal C^*-seminorm on A.

Conversely, assume there is a maximal C^*-seminorm ρ on A. For $a \in A$ we have

$$\begin{aligned}
\|a\|^2_{\max} &= \sup\{\phi(a^*a) \mid \phi \in S(A)\} \\
&= \sup\{\|\pi_\phi(a^*a)\| \mid \phi \in S(A)\} \\
&\le \sup\{\|\pi(a^*a)\| \mid \pi \text{ is a } * \text{-representation on a Hilbert space}\} \\
&= \rho(a)^2 .
\end{aligned}$$

This shows that $\|a\|_{\max} < \infty$ for all $a \in A$. □

Definition 5.1.4 We say that a complex $*$-algebra A is a *universal pre-C^*-algebra* if there is a C^*-norm ρ on A such that each $*$-homomorphism $\psi : A \to B$ from A to a C^*-algebra B extends to a $*$-homomorphism $\overline{\psi} : \overline{A} \to B$, where $\overline{A}$ is the completion of A with respect to ρ.

We establish the following useful result.

Proposition 5.1.5 *Let A be a unital complex $*$-algebra.*

(i) A is a universal pre-C^*-algebra if and only if A is positive definite and

$$0 < \|a\|_{max} < \infty$$

for all nonzero a in A. In this case, A is a universal pre-C^*-algebra with respect to the norm $\| \cdot \|_{max}$.

(ii) If A is positive definite and $A = A_b$ then all the states of A are automatically bounded and the C^*-seminorm $\| \cdot \|$ on A defined by using the partially ordered

structure as in Definition 5.1.2 coincides with the maximal C^*-seminorm, so that

$$\|x\| = \|x\|_{max} = \sup\{\|\pi(x)\| \mid \pi: A \to B(\mathscr{H}) \text{ is a } *\text{-representation}\}.$$

In particular, there is a maximal C^*-seminorm on A.

Proof (i) follows from Lemma 5.1.3.

(ii) Assume that A is positive definite and $A = A_b$. If ϕ is a state of A and $a \in A$, then since $a \in A_b$ there is a positive constant $K(a)$ and elements $x_1, \ldots, x_n$ in A such that

$$a^*a + \sum_{i=1}^{n} x_i^* x_i = K(a) \cdot 1_A.$$

For $b \in A$ we have $b^*a^*ab + \sum_{i=1}^{n}(x_i b)^*(x_i b) = K(a)b^*b$, and since ϕ is positive we get

$$\phi(b^*a^*ab) \leq K(a)\phi(b^*b).$$

This shows that all states of A are bounded.

If a is an element of A and $a^*a + \sum_{i=1}^{n} x_i^* x_i = \lambda \cdot 1_A$ for some $x_i \in A$ and $\lambda \in \mathbb{R}_+$, then we have $\phi(a^*a) \leq \lambda$ for all $\phi \in S(A)$. It follows that

$$\|a\|_{\max} \leq \|a\|.$$

In particular, we see from Lemma 5.1.3 that $\|\cdot\|_{\max}$ is the maximal C^*-seminorm on A. Since $\|\cdot\|$ is a C^*-seminorm on A we get that $\|\cdot\| \leq \|\cdot\|_{\max}$ as well, and so $\|\cdot\| = \|\cdot\|_{\max}$, as desired. □

Remark 5.1.6 Assume that $(A, \|\cdot\|_{\max})$ is a (unital) universal pre-C^*-algebra, so that $\|\cdot\|_{\max}$ is a C^*-norm on A. In this situation it is not necessarily true that the positive cone A_{++} described in Definitions 5.1.2 coincides with the positive cone $A_+ = \overline{A}_+ \cap A$ obtained by considering A as an operator system. The positive cone A_+ can be intrinsically described in terms of A in each one of the following equivalent alternative ways:

(1) $A_+ = \{x \in A : \phi(x) \geq 0 \ \forall \ \phi \in S(A)\}$;

(2) For $x \in A$, we have

$$x \in A_+ \iff x = \lim_{n\to\infty} x_n \text{ for a sequence } (x_n) \text{ in } A_{++}; \text{ or}$$

(3) $x = x^*$ and $x + \epsilon 1 \in A_{++}$ for all $\epsilon > 0$ (see e.g, [123, Theorem 1]).

Remark 5.1.7 Now assume that A is a non-unital complex $*$-algebra. We denote by $\widetilde{A}$ the $\mathbb{C}$-algebra unitization of A (see the proof of Corollary 4.4.17), and observe that it is a unital $*$-algebra, with $(a, \lambda)^* = (a^*, \overline{\lambda})$, for $a \in A$ and $\lambda \in \mathbb{C}$. Now every $*$-representation $\pi: A \to B(\mathscr{H})$ on a Hilbert space $\mathscr{H}$ can be uniquely extended to a unital $*$-representation $\widetilde{\pi}: \widetilde{A} \to B(\mathscr{H})$. It is easy to check that A is positive definite if and only if so is $\widetilde{A}$. Moreover, if A is generated by partial isometries, then clearly so is $\widetilde{A}$. Therefore if A is positive definite and generated by partial isometries and if $\|a\|_{\max} > 0$ for all nonzero $a \in A$ then it follows from Proposition 5.1.5 that A is a universal pre-C^*-algebra with respect to the norm

$$\|a\|_{\max} = \|a\| = \inf\{\lambda \in \mathbb{R}_+ \mid a^* a \leq \lambda^2 \cdot 1_{\widetilde{A}}\},$$

where $\leq$ is the order induced by the positive cone $\widetilde{A}_{++}$ on $\widetilde{A}$.

The following corollary uses a result of Tomforde [147], which we will prove in Sect. 5.2.

Corollary 5.1.8 *Let E be an arbitrary graph. Then the Leavitt path algebra $L_{\mathbb{C}}(E)$ is a universal pre-C^*-algebra with respect to the C^*-norm*

$$\|a\|_{\max} = \|a\| = \inf\{\lambda \in \mathbb{R}_+ \mid a^* a \leq \lambda^2 \cdot 1_{\widetilde{L_{\mathbb{C}}(E)}}\},$$

where $\leq$ is the order induced by the positive cone $\widetilde{L_{\mathbb{C}}(E)}_{++}$ on $\widetilde{L_{\mathbb{C}}(E)}$.

Proof By Tomforde [147, Theorem 7.3], $L_{\mathbb{C}}(E)$ is a pre-C^*-algebra. Therefore $L_{\mathbb{C}}(E)$ is positive definite and $\|a\|_{\max} > 0$ for every nonzero $a \in L_{\mathbb{C}}(E)$. Moreover, it is clear by definition that $L_{\mathbb{C}}(E)$ is generated by partial isometries. The result follows from Remark 5.1.7. □

We note that, more generally, it has been proved in [42, Proposition 3.4] that if K is a field with positive definite involution, then the induced involution on $L_K(E)$ is positive definite.

5.2 Graph C^*-Algebras, and Connections to Leavitt Path Algebras

Let E be an arbitrary graph. We are going to consider an enveloping C^*-algebra of the complex $*$-algebra $L_{\mathbb{C}}(E)$. Following the standard procedure (cf. [129, p. 13] and Sect. 5.1), we consider $*$-representations π of $L_{\mathbb{C}}(E)$ into Hilbert spaces $\mathscr{H}$. Observe that for $a \in L_{\mathbb{C}}(E)$ there exists a $K = K(a)$ such that for any $*$-representation π we have

$$\|\pi(a)\| \leq K.$$

Indeed, if $a = \sum a_{\lambda,\mu} \lambda \mu^* \in L_{\mathbb{C}}(E)$ with $a_{\lambda,\mu} \in \mathbb{C}$ and $\lambda, \mu \in \text{Path}(E)$, then

$$\|\pi(a)\| \leq \sum |a_{\lambda,\mu}| \|\pi(\lambda\mu^*)\| \leq \sum |a_{\lambda,\mu}|,$$

because $\pi(\lambda\mu^*)$ is a partial isometry, and $\|w\| \leq 1$ for any partial isometry w. Therefore $\|\cdot\|_1$ is an algebra seminorm satisfying $\|a^*a\|_1 = \|a\|_1^2$. This puts us in position to define the central focus of this chapter.

Definition 5.2.1 Let J be the $*$-ideal of $L_{\mathbb{C}}(E)$ consisting of those elements a such that $\|a\|_1 = 0$. Then $L_{\mathbb{C}}(E)/J$ is a $*$-algebra, and the quotient norm $\|\cdot\|_0$ defined by $\|a+J\|_0 = \inf\{\|a+j\|_1 \mid j \in J\}$ is a C^*-norm. So the completion of $L_{\mathbb{C}}(E)/J$ is a C^*-algebra, which we denote by

$$C^*(E),$$

and which we call *the graph C^*-algebra* of E.

Observe that any $*$-representation of $L_{\mathbb{C}}(E)$ extends uniquely to a representation of $C^*(E)$. We will show later that the $*$-ideal J is trivial (Theorem 5.2.9). As shown in Corollary 5.1.8, this implies that $L_{\mathbb{C}}(E)$ is a universal pre-C^*-algebra (in the sense of Definition 5.1.4).

Remark 5.2.2 There are two distinct notations which have arisen when describing the multiplication in the graph C^*-algebra $C^*(E)$. The notation we have chosen to use here is consistent with the notation used in describing multiplication within Leavitt path algebras; it is the "left to right" notation. On the other hand, viewing the elements of $C^*(E)$ as operators on a Hilbert space, it also makes sense to view multiplication in $C^*(E)$ as function composition, in which case the "right to left" notation is quite natural. The "left to right" notation is almost universally utilized in the context of Leavitt path algebras, while the notation used in the graph C^*-algebra literature is not universally agreed-upon; for example, Tomforde [147] uses left-to-right, while Raeburn [129] uses right-to-left.

Example 5.2.3 Let $E = R_1$ (i.e., E has exactly one vertex and one edge). Then $C^*(E) = C(\mathbb{T})$ (recall $\mathbb{T} = \{z \in \mathbb{C} \mid |z| = 1\}$). In this case the canonical map $L_{\mathbb{C}}(R_1) \to C^*(R_1)$ is precisely the canonical inclusion $\mathbb{C}[x, x^{-1}] \to C(\mathbb{T})$ which sends x to z, where z denotes the inclusion mapping $\mathbb{T} \to \mathbb{C}$.

Example 5.2.4 Let $E = R_n$ ($n \geq 2$) be the rose with n-petals. A central role is played by the *Cuntz algebra*

$$\mathscr{O}_n = C^*(R_n).$$

The Cuntz algebras $\{\mathscr{O}_n \mid n \geq 2\}$ were first introduced in Cuntz' seminal paper [68].

Definitions 5.2.5 Let E be an arbitrary graph. A *Cuntz–Krieger E-family* on a Hilbert space $\mathscr{H}$ consists of a set $\{P_v \mid v \in E^0\}$ of mutually orthogonal projections of $\mathscr{H}$ and a set $\{S_e \mid e \in E^1\}$ of partial isometries on $\mathscr{H}$ satisfying relations:

(CK1) $S_e^* S_{e'} = \delta_{e,e'} P_{r(e)}$ for all $e, e' \in E^1$,
(CK2) $P_v = \sum_{\{e \in E^1 | s(e) = v\}} S_e S_e^*$ for every regular vertex $v \in E^0$, and
(CK3) $S_e S_e^* \leq P_{s(e)}$ for all $e \in E^1$.

We note that the (CK3) condition can be shown to follow from (CK2) if $s(e)$ is a regular vertex. We will denote by p_v and s_e the images of $v \in E^0$ and $e \in E^1$ through $L_{\mathbb{C}}(E) \to C^*(E)$, respectively.

Clearly $C^*(E)$ is the C^*-algebra generated by a *universal* Cuntz–Krieger E-family. In addition, we note that the (CK1) and (CK2) conditions given here mimic exactly the identically-named conditions in a Leavitt path algebra, see Definition 1.2.3.

There is a natural $*$-representation of $L_{\mathbb{C}}(E)$, as follows.

Example 5.2.6 Let E be an arbitrary graph. Select (nonzero) Hilbert spaces $\mathscr{H}_v$, all of the same Hilbertian dimension. For each $v \in E^0 \setminus \mathrm{Sink}(E)$, define $\mathscr{H}_v = \bigoplus_{e \in s^{-1}(v)} \mathscr{H}_e$, where $\mathscr{H}_e$ is a Hilbert space of the same dimension as $\mathscr{H}_v$. Of course, if E is countable we can select all the Hilbert spaces to be separable infinite-dimensional. If E is uncountable we can always select a big enough cardinality for the dimension of the Hilbert spaces, so that the above decomposition exists. Set

$$\mathscr{H} = \bigoplus_{v \in E^0} \mathscr{H}_v.$$

For $v \in E^0$, let P_v be the orthogonal projection onto $\mathscr{H}_v$, and for each $e \in E^1$ select a partial isometry S_e with initial space $\mathscr{H}_{r(e)}$ and final space $\mathscr{H}_e$. This collection of projections and partial isometries defines a representation $\pi \colon C^*(E) \to B(\mathscr{H})$. In particular, this shows that $p_v \neq 0$ in $C^*(E)$ for all $v \in E^0$.

We are now going to describe the *gauge action*, which is a fundamental tool in the theory of graph C^*-algebras. See [129, Proposition 2.1] for a proof of the following result.

Proposition 5.2.7 *Let E be an arbitrary graph. Then there is an action γ of $\mathbb{T}$ on $C^*(E)$ such that $\gamma_z(s_e) = zs_e$ and $\gamma_z(p_v) = p_v$ for all $z \in \mathbb{C}$, $v \in E^0$, and $e \in E^1$.*

Lemma 5.2.8 *Let c be a cycle without exits and let $a = a_1 \cdots a_r$ be a representative of c, with $s(a) = r(a) = v$. Then $p_v C^*(E) p_v = C^*(a) \cong C(\mathbb{T})$. In particular, the map $vL_{\mathbb{C}}(E)v \to p_v C^*(E) p_v$ is injective.*

Proof Observe that a is a unitary element in $p_v C^*(E) p_v$. By Murphy [118, Theorem 2.1.13], to show that $C^*(a) \cong C(\mathbb{T})$, it is enough to show that the spectrum of a is $\mathbb{T}$. Choose λ in the spectrum of a. Then for each $z \in \mathbb{T}$, $\gamma_z(a - \lambda p_v)$ is not invertible in $p_v C^*(E) p_v$. Since

$$\gamma_z(a - \lambda p_v) = z^r a - \lambda p_v = z^r(a - z^{-r}\lambda p_v),$$

we see that any element in $\mathbb{T}$ belongs to the spectrum of a.

Recall that $\mathbb{C}[x, x^{-1}] \cong vL_{\mathbb{C}}(E)v$ (by Lemma 2.2.7); then the map $\mathbb{C}[x, x^{-1}] \cong vL_{\mathbb{C}}(E)v \to p_v C^*(E) p_v$ has dense image, and its image is contained in $C^*(a) \cong C(\mathbb{T})$. Since $C^*(a)$ is complete it follows that $p_v C^*(E) p_v = C^*(a)$. This proves the result. □

We are now ready to obtain the following result, whose original proof is due to Tomforde [147]. The proof presented here is different.

Theorem 5.2.9 *For any graph E the natural map $L_{\mathbb{C}}(E) \to C^*(E)$ is injective.*

Proof The result follows from Example 5.2.6, Lemma 5.2.8, and the Reduction Theorem 2.2.11. □

From now on we will identify $L_{\mathbb{C}}(E)$ with its image in $C^*(E)$. In particular, we will write $\lambda\mu^*$ instead of $S_\lambda S_\mu^*$, when λ and μ are paths of E with $r(\lambda) = r(\mu)$, and we will write v instead of p_v for $v \in E^0$.

Following, e.g., [77, Definition 16.2], given a discrete group Γ, we define a *Γ-graded C^*-algebra* as a C^*-algebra B with a family $\{B_t \mid t \in \Gamma\}$ of closed subspaces B_t of B such that: $B_tB_s \subseteq B_{ts}$ and $B_t^* = B_{t^{-1}}$ for all $t, s \in \Gamma$; the sum $\bigoplus_{t\in\Gamma} B_t$ is a direct sum; and

$$B = \overline{\bigoplus_{t\in\Gamma} B_t}\,.$$

Let $B \subseteq A$ be an inclusion of C^*-algebras. A *conditional expectation* from A onto B is a map $\phi\colon A \to B$ such that

(i) ϕ is a positive map, that is, $\phi(a) \geq 0$ for $a \geq 0$,
(ii) $\|\phi(a)\| \leq \|a\|$ for $a \in A$,
(iii) $\phi(b) = b$ for $b \in B$, and
(iv) $\phi(ba) = b\phi(a)$ and $\phi(ab) = \phi(a)b$ for all $a \in A$ and $b \in B$.

A conditional expectation $\phi\colon A \to B$ is *faithful* if, for all $a \geq 0$ in A, $\phi(a) = 0 \implies a = 0$.

The following result gives the relationship between actions of $\mathbb{T}$ on a C^*-algebra and $\mathbb{Z}$-gradings.

Lemma 5.2.10 *Let $\alpha\colon \mathbb{T} \to A$ be an action of $\mathbb{T}$ on a C^*-algebra A. Then there is a faithful conditional expectation $\Phi\colon A \to A^\alpha$, given by*

$$\Phi(a) = \int_{\mathbb{T}} \alpha_z(a)dz.$$

For $n \in \mathbb{Z}$, define

$$A_n = \{a \in A \mid \int_{\mathbb{T}} z^{-n}\alpha_z(a)dz = a\}.$$

Then $\{A_n \mid n \in \mathbb{Z}\}$ are closed subspaces of A such that $A_nA_m \subseteq A_{n+m}$ and $A_n^* = A_{-n}$ for all $m, n \in \mathbb{Z}$. Moreover, $\sum_{n\in\mathbb{Z}} A_n$ is a direct sum, so that the C^*-subalgebra $\overline{\bigoplus_{n\in\mathbb{Z}} A_n}$ of A is a $\mathbb{Z}$-graded C^*-algebra.

Proof The proof that Φ is a faithful conditional expectation is given in [129, Proposition 3.2]. Let $a_n \in A_n$ and $a_m \in A_m$. We have

$$\begin{aligned} a_n a_m &= \int_{\mathbb{T}} z^{-n}\alpha_z(a_n)dz \cdot \int_{\mathbb{T}} w^{-m}\alpha_w(a_m)dw \\ &= \int_{\mathbb{T}} z^{-n-m}\alpha_z(a_n)\alpha_z\Big(\int_{\mathbb{T}} (z^{-1}w)^{-m}\alpha_{z^{-1}w}(a_m)dw\Big)dz \\ &= \int_{\mathbb{T}} z^{-n-m}\alpha_z(a_n)\alpha_z(a_m)dz \\ &= \int_{\mathbb{T}} z^{-(n+m)}\alpha_z(a_n a_m)dz, \end{aligned}$$

showing that $a_n a_m \in A_{n+m}$. Similarly one checks that $A_n^* = A_{-n}$.

Assume now that $a \in A_n$ with $n \neq 0$. Then we have

$$\begin{aligned} \Phi(a) = \int_{\mathbb{T}} \alpha_z(a)dz &= \int_{\mathbb{T}} \alpha_z\Big(\int_{\mathbb{T}} w^{-n}\alpha_w(a)dw\Big)dz \\ &= \int_{\mathbb{T}} z^n\Big(\int_{\mathbb{T}} (zw)^{-n}\alpha_{zw}(a)dw\Big)dz \\ &= \int_{\mathbb{T}} z^n a dz = 0. \end{aligned}$$

Thus $\Phi(a) = 0$ for $a \in A_n$ with $n \neq 0$. Now if $\sum_{i=-N}^{N} a_i = 0$, with $a_i \in A_i$, we have

$$0 = \Phi(\sum_{i=-N}^{N} a_j^* a_i) = \sum_{i=-N}^{N} \Phi(a_j^* a_i) = \Phi(a_j^* a_j)$$

using the previous observation, so that $a_j = 0$ because Φ is faithful. It follows that $\sum_{n\in\mathbb{Z}} A_n$ is a direct sum, as claimed. □

Proposition 5.2.11 *Let E be an arbitrary graph. Then $C^*(E)$ is a $\mathbb{Z}$-graded C^*-algebra:*

$$C^*(E) = \overline{\bigoplus_{n\in\mathbb{Z}} C^*(E)_n}\,,$$

with $C^*(E)_0 = C^*(E)^\gamma$, the fixed-point algebra with respect to the gauge action. Moreover, there is a faithful conditional expectation $\Phi\colon C^*(E) \to C^*(E)_0$, and $C^*(E)_n = \overline{L_{\mathbb{C}}(E)_n}$ for all $n \in \mathbb{Z}$.

Proof For $n \in \mathbb{Z}$ define

$$C^*(E)_n = \{a \in C^*(E) \mid \int_{\mathbb{T}} z^{-n}\gamma_z(a)dz = a\}.$$

Everything follows from Lemma 5.2.10 once we show that $C^*(E)_n = \overline{L_{\mathbb{C}}(E)_n}$ for all $n \in \mathbb{Z}$. Observe that if $\lambda\mu^* \in L_{\mathbb{C}}(E)$ with $|\lambda| - |\mu| = n$ then $\gamma_z(\lambda\mu^*) = z^n\lambda\mu^*$ so that $\int_{\mathbb{T}} z^{-n}\gamma_z(\lambda\mu^*)dz = \int_{\mathbb{T}} \lambda\mu^* dz = \lambda\mu^*$. So we get that $L_{\mathbb{C}}(E)_n \subseteq C^*(E)_n$. Observe also that $C^*(E)_n$ is a closed subspace of $C^*(E)$ and that the map $\Phi_n: C^*(E) \to C^*(E)_n$ given by $\Phi_n(a) = \int_{\mathbb{T}} z^{-n}\gamma_z(a)dz$ is norm-decreasing and projects onto $C^*(E)_n$. Moreover, $\Phi_n(\overline{L_{\mathbb{C}}(E)_m}) = 0$ for $m \neq n$. Since $L_{\mathbb{C}}(E)$ is dense in $C^*(E)$, it follows that $C^*(E)_n = \overline{L_{\mathbb{C}}(E)_n}$. □

We are now in position to obtain the two so-called *Uniqueness Theorems*; these are the graph C^*-algebra analogs of Theorems 2.2.15 and 2.2.16.

Theorem 5.2.12 (The Gauge-Invariant Uniqueness Theorem) *Let E be an arbitrary graph. Let $\pi: C^*(E) \to B$ be a $*$-homomorphism from $C^*(E)$ to a C^*-algebra B such that $\pi(v) \neq 0$ for all $v \in E^0$. Assume that there is an action $\beta: \mathbb{T} \to \mathrm{Aut}(B)$ such that $\beta_z(\pi(a)) = \pi(\gamma_z(a))$ for all a in $C^*(E)$ and all $z \in \mathbb{T}$. Then π is injective.*

Proof By Lemma 5.2.10, $B' := \overline{\bigoplus_{n\in\mathbb{Z}} B_n}$ is a $\mathbb{Z}$-graded C^*-algebra. Replacing B with B', we may assume that B is $\mathbb{Z}$-graded. Moreover, π is a graded $*$-homomorphism and $\pi(\Phi(a)) = \Phi'(\pi(a))$ for all $a \in C^*(E)$, where $\Phi': B \to B_0$ is the conditional expectation corresponding to $\beta: \mathbb{T} \to \mathrm{Aut}(B)$. It follows from Theorem 2.2.15 that $\pi_{|L_{\mathbb{C}}(E)}$ is injective. In particular, $\pi_{|L_{\mathbb{C}}(E)_0}$ is injective. By Proposition 2.1.14 we have that $L_{\mathbb{C}}(E)_0$ is an algebraic direct limit of finite-dimensional C^*-algebras. Hence, since any injective $*$-homomorphism between C^*-algebras is isometric, we get that π is isometric on $L_{\mathbb{C}}(E)_0$ and so on $C^*(E)_0 = C^*(E)^\gamma = \overline{L_{\mathbb{C}}(E)_0}$.

Let a be a positive element in the kernel of π. Then

$$\pi(\Phi(a)) = \Phi'(\pi(a)) = 0,$$

and so $\Phi(a) = 0$. Since Φ is faithful we get that $a = 0$. This shows that π is injective, as desired. □

Lemma 5.2.13 *Let E be an arbitrary graph. Assume that λ, ν, τ are paths of E with $s(\lambda) = r(\tau) = r(\nu)$, with $|\tau| < |\nu|$ and $|\lambda| > |\nu| - |\tau|$. Then $\lambda^*\nu^*\tau\lambda \neq 0$ in $L_{\mathbb{C}}(E)$ if and only if there is a decomposition $\lambda = (\nu')^r \cdot \lambda'$, $r \geq 1$, such that $\nu = \tau\nu'$ and $\nu' = \lambda'\lambda''$.*

Proof Assume first that $\lambda = (\nu')^r\lambda'$, with $\nu = \tau\nu'$ and $\nu' = \lambda'\lambda''$. Then

$$\lambda^*\nu^*\tau\lambda = \lambda^*(\nu')^*\lambda = (\lambda')^*(\nu'^*)^{r+1}(\nu')^r\lambda' = (\lambda')^*(\nu')^*\lambda' = (\lambda')^*(\lambda'')^*(\lambda')^*\lambda' = (\lambda''\lambda')^* \neq 0.$$

Conversely, assume that $\lambda^*\nu^*\tau\lambda \neq 0$. Write $\nu = \tau\nu'$. Then $\lambda^*\nu^*\tau\lambda = \lambda^*(\nu')^*\lambda$. Since $|\lambda| > |\nu'|$, we can write $\lambda = (\nu')^r\lambda'$ with $r \geq 1$ and such that λ' does not start with ν'. Now $0 \neq (\lambda')^*(\nu')^*\lambda'$ and since ν' is not an initial segment of λ', we get $\nu' = \lambda'\lambda''$, as desired. □

Lemma 5.2.14 *Let E be an arbitrary graph satisfying Condition (L), and let $v \in E^0$. Then for any $n \in \mathbb{N}$ there is a path λ of E such that $s(\lambda) = v$, and $\lambda^*\nu^*\tau\lambda = 0$ for all ν, τ such that $0 < \big|\,|\nu| - |\tau|\,\big| < n$.*

Proof Assume first that all paths starting at v have length $< n$. Then $v\nu^*\tau v = 0$ for all paths ν, τ such that $|\nu| \neq |\tau|$ so we can take $\lambda = v$. Now assume that there is a path λ of length n such that $s(\lambda) = v$. If $\lambda^*\nu^*\tau\lambda \neq 0$ for some ν, τ such that, say, $0 < |\nu| - |\tau| < n$, then by Lemma 5.2.13 we have

$$\lambda = (\lambda'\lambda'')^r\lambda',$$

with $r \geq 1$, where $\nu = \tau\nu'$ and $\nu' = \lambda'\lambda''$. Note that either $|\lambda'| \geq 1$ or $r \geq 2$. It follows from this that $\nu' = \lambda'\lambda''$ is a closed path in E based at v, of positive length. Thus there is a cycle β in E based at v. By the Condition (L) hypothesis there is an exit for β, call it e, and denote by β' the path from v to $s(e)$. Then it follows from Lemma 5.2.13 that $\rho = \beta^n\beta' e$ satisfies $\rho^*(\nu')^*\tau'\rho = 0$ for all ν', τ' such that $0 < \big|\,|\nu'| - |\tau'|\,\big| < n$. □

It is worth remarking that only relations (V), (E1), (E2) and (CK1) have been used in the proof of the above two lemmas, which are purely algebraic.

We are now ready to establish the second of the two Uniqueness Theorems.

Theorem 5.2.15 (The Cuntz–Krieger Uniqueness Theorem) *Let E be a graph satisfying Condition (L). Let $\pi\colon C^*(E) \to B$ be a $*$-homomorphism from $C^*(E)$ to a C^*-algebra B such that $\pi(v) \neq 0$ for every $v \in E^0$. Then π is injective.*

Proof By Theorem 2.2.16, we know that $\pi_{|L_{\mathbb{C}}(E)}$ is injective, and in particular that $\pi_{|L_{\mathbb{C}}(E)_0}$ is injective. As in the proof of the Gauge-Invariant Uniqueness Theorem 5.2.12, we get that $\pi_{|C^*(E)^\gamma}$ is an isometry.

We will show that

$$\|\pi(\Phi(a))\| \leq \|\pi(a)\| \qquad \forall a \in C^*(E). \tag{5.1}$$

From this inequality and the faithfulness of Φ we will conclude that π is injective, just as in the proof of Theorem 5.2.12. By continuity it suffices to show (5.1) for elements of the form $a = \sum_{(\mu,\nu)\in F} c_{\mu,\nu}\mu\nu^*$ in $L_{\mathbb{C}}(E)$, where F is a finite set of pairs (μ, ν) in Path(E) with $r(\mu) = r(\nu)$. We will find a projection Q in $L_{\mathbb{C}}(E)_0$ which satisfies

$$\|Q\Phi(a)Q\| = \|\Phi(a)\|, \qquad \text{and} \tag{5.2}$$

$$Q\mu\nu^*Q = 0 \quad \text{when } (\mu, \nu) \in F \text{ and } |\mu| \neq |\nu|. \tag{5.3}$$

Let F' be the set of paths that appear as a first or second component of a pair in F. Let $k = \max\{|\gamma| : \gamma \in F'\}$, and let X be a finite complete subset of Path(E) consisting of paths of length $\leq k$ such that $\Phi(a) \in \mathscr{F}(X)$ (see Definition 2.1.11). By Proposition 2.1.12, we can find a finite complete subgraph E' of E and a subset of vertices V of E' such that X is precisely the set of all paths of length k from E' which start at a vertex in V, together with the set of all paths in E' of length $< k$ starting at a vertex in V and ending in a sink of E. Observe that by the construction of X, the elements $\tau_i(v)$ considered in the proof of Proposition 2.1.14 do not depend on i, only on v, and we will denote this element by $\tau(v)$. Indeed, we have

$$\tau(v) = v - \sum_{e \in (E')^1, s(e)=v} ee^*,$$

and so $\tau(v)\gamma = 0 = \gamma^*\tau(v)$ for every path γ in E' of positive length.

Since $\mathscr{F}(X)$ is a matricial $\mathbb{C}$-algebra there is a simple component $\mathscr{S}$ of it such that $\|\Phi(a)\| = \|a'\|$, where a' is the projection of $\Phi(a)$ onto the simple component $\mathscr{S}$. There are various possibilities for this component. If $\mathscr{S} = \mathscr{G}_{i,v}(X)$ where v is a sink, then let $Q = \sum_{\tau \in G} \tau\tau^*$ be the unit of $\mathscr{G}_{i,v}(X)$, where G is the set of paths in E' of length i starting at V and ending at v. Then we have that (5.2) is obviously satisfied. To show (5.3) we consider any element of the form $\mu\nu^*$, where $|\mu| > |\nu|$ and μ, ν are paths in E' starting at V, of length $\leq k$. Assume that $\tau\tau^*\mu\nu^*\tau_1\tau_1^* \neq 0$ for some $\tau, \tau_1 \in G$. Then $|\tau| \geq |\mu|$, because τ ends in a sink, and similarly $|\tau_1| \geq |\nu|$. Writing $\tau = \mu\tau'$ and $\tau_1 = \nu\tau_1'$, we get $0 \neq \tau(\tau')^*\tau_1'(\tau_1)^*$, and since

$$|\tau'| = i - |\mu| < i - |\nu| = |\tau_1'|,$$

we get that $0 \neq \tau\rho(\tau_1)^*$ for a path ρ of positive length starting at the sink v, which is a contradiction. This shows (5.3) in this case.

Assume now that $\mathscr{S} = \mathscr{F}_{i,v}(X)$, where $i \leq k$, and let $Q = \sum_{\lambda \in G} \lambda\tau(v)\lambda^*$ be the unit of $\mathscr{F}_{i,v}(X)$, where G is the set of paths in E' of length $i-1$ starting at V and ending at v, and v is an infinite emitter in E such that $s_{E'}^{-1}(v) \neq \emptyset$.

In this case we obviously have (5.2). By using that $\gamma\tau(v) = 0 = \tau(v)\gamma^*$ for every path γ in E' of positive length, we easily obtain (5.3) as well.

Finally, assume that $\mathscr{S} = \mathscr{G}_{k,v}(X)$, and that v is not a sink in E. By Lemma 5.2.14 there is a path λ in E such that $s(\lambda) = v$ and $\lambda^*\mu^*\nu\lambda = 0$ for all paths μ, ν in E' of length $\leq k$ such that $|\mu| \neq |\nu|$. Let G be the set of paths in E' of length k starting at V and ending at v. Consider

$$Q = \sum_{\tau \in G} \tau\lambda\lambda^*\tau^*.$$

Observe that the set $\{\tau\lambda\lambda^*\tau_1^* \mid \tau, \tau_1 \in G\}$ is a set of matrix units, generating a matrix algebra $\mathscr{G}_{i,v}^{\lambda}(X)$, and that $x \mapsto QxQ$ defines a $*$-isomorphism (and thus an isometry) from $\mathscr{G}_{i,v}(X)$ onto $\mathscr{G}_{i,v}^{\lambda}(X)$. Hence (5.2) is also established in this case.

We now note that if $\tau, \tau_1 \in G$ and $\tau\lambda\lambda^*\tau^*\mu\nu^*\tau_1\lambda\lambda^*(\tau_1)^* \neq 0$ for $(\mu, \nu) \in F$ with $|\mu| \neq |\nu|$, then $\tau = \mu\tau'$ and $\tau_1 = \nu\tau_1'$, with $|\tau'| \neq |\tau_1'|$, so that

$$\lambda^*\tau^*\mu\nu^*\tau_1\lambda = \lambda^*(\tau')^*\tau_1'\lambda = 0,$$

because τ' and τ_1' are paths in E' of length $\leq k$, with different lengths. This gives a contradiction, and so we must have that

$$Q\mu\nu^*Q = \sum_{\tau,\tau_1 \in G} \tau\lambda\lambda^*\tau^*\mu\nu^*\tau_1\lambda\lambda^*(\tau_1)^* = 0,$$

showing (5.3) in this case.

Observe that (5.2) and (5.3) together give

$$\|\Phi(a)\| = \|Q\Phi(a)Q\| = \|QaQ\|,$$

so that, recalling that $\pi_{|L_{\mathbb{C}}(E)_0}$ is an isometry, we get

$$\|\pi(\Phi(a))\| = \|\Phi(a)\| = \|QaQ\| = \|\pi(QaQ)\| \leq \|\pi(a)\|,$$

as desired. □

Later we will need the following immediate consequence of Theorem 5.2.15.

Corollary 5.2.16 *Let E be a graph satisfying Condition (L) and let J be a nonzero closed two-sided ideal of $C^*(E)$. Then there exists some $v \in E^0$ such that the projection v belongs to J.*

Proof Apply Theorem 5.2.15 to the $*$-homomorphism $\pi: C^*(E) \to C^*(E)/J$ given by the canonical projection map. □

5.3 Projections in Graph C^*-Algebras

The goal of this section is to establish an isomorphism between the monoids $\mathcal{V}(L_{\mathbb{C}}(E))$ and $\mathcal{V}(C^*(E))$ for a row-finite graph E. This gives in particular that every projection P in $\mathrm{M}_n(C^*(E))$ is equivalent to a diagonal projection $\mathrm{diag}(v_1, \dots, v_n)$ for some vertices $v_1, \dots, v_n$ in E. We draw a nice consequence of this relationship in Corollary 5.3.7. (The proof of Lemma 5.4.1 will also make use of this fact.)

In order to downsize the level of technicalities, we will work in the rest of this chapter only with row-finite graphs; however, many of the results presented here have an adaptation to arbitrary graphs.

We start by establishing the relationship between hereditary saturated subsets of E^0 and gauge-invariant ideals of the graph C^*-algebra $C^*(E)$. This is completely analogous to what we know for Leavitt path algebras about the relationship between hereditary saturated subsets of E^0 and graded ideals of $L_{\mathbb{C}}(E)$ (see Chap. 2).

For a subset X of E^0, denote by $\mathfrak{I}(X)$ the closed ideal of $C^*(E)$ generated by X. Consistent with the notation employed elsewhere in this book, we will denote by $I(X)$ the ideal of $L_{\mathbb{C}}(E)$ generated by X.

Lemma 5.3.1 *Let E be a row-finite graph, and let H be a hereditary saturated subset of E. Then $C^*(E)/\mathfrak{I}(H) \cong C^*(E/H)$.*

Proof This only uses the universal property of the graph C^*-algebra. The proof is left to the reader. □

A closed ideal I of $C^*(E)$ is said to be *gauge-invariant* if $\gamma_z(I) = I$ for all $z \in \mathbb{T}$. In that case the quotient C^*-algebra $C^*(E)/I$ admits a gauge action: $\gamma_z(a + I) = \gamma_z(a) + I$.

Theorem 5.3.2 *Let E be a row-finite graph. Then there is a bijective correspondence between the set of hereditary saturated subsets of E and the set of gauge-invariant ideals of $C^*(E)$. This correspondence sends $H \in \mathscr{H}_E$ to $\mathfrak{I}(H)$, and sends I to $E^0 \cap I$ (for a gauge-invariant ideal I).*

Proof We have that $I(H)$ is a graded ideal of $L_{\mathbb{C}}(E)$, with $I(H)_n = I(H)_0 L_{\mathbb{C}}(E)_n = L_{\mathbb{C}}(E)_n I(H)_0$. Since $\gamma_z(a_n) = z^n a_n$ for $a \in C^*(E)_n$, we get that $I(H)$ is γ_z-invariant for all $z \in \mathbb{T}$, and so $\mathfrak{I}(H) = \overline{I(H)}$ is gauge-invariant. Note also that Lemma 5.3.1 gives that the set of vertices v such that $v \in \mathfrak{I}(H)$ is exactly H.

Let I be a gauge-invariant ideal of $C^*(E)$, and let H be the set of vertices v in E^0 such that $v \in I$. Then H is hereditary and saturated, as in Lemma 2.4.3. We obviously have $\mathfrak{I}(H) \subseteq I$. Consider the quotient map $\pi: C^*(E)/\mathfrak{I}(H) \to C^*(E)/I$. By Lemma 5.3.1 we have $C^*(E)/\mathfrak{I}(H) \cong C^*(E/H)$ in the natural way, so we get a quotient map $\pi': C^*(E/H) \to C^*(E)/I$. Observe that $\pi'(v) \neq 0$ for all $v \in E^0 \setminus H$ and that $\gamma_z(\pi'(a)) = \pi'(\gamma_z(a))$ for all $a \in C^*(E/H)$. It follows from the Gauge-Invariant Uniqueness Theorem 5.2.12 that π' is an isomorphism, and so we get $I = \mathfrak{I}(H)$. □

We note the following, which is an immediate consequence of Theorems 2.5.9 and 5.3.2.

Corollary 5.3.3 *Let E be a row-finite graph and K any field. Then there is a lattice isomorphism between the lattice of graded ideals of $L_K(E)$ and the lattice of gauge-invariant ideals of $C^*(E)$.*

We need a lemma whose proof is similar to that of Corollary 1.6.16, and so is omitted.

Lemma 5.3.4 *The assignment $E \mapsto C^*(E)$ can be extended to a continuous functor from the category $\mathscr{RG}$ of row-finite graphs and complete graph homomorphisms to the category of C^*-algebras and $*$-homomorphisms. Every graph C^*-algebra $C^*(E)$ is the direct limit of graph C^*-algebras associated with finite graphs.*

In order to obtain the main result of this section (Theorem 5.3.5), we will need to utilize two ideas which will not be formally introduced until Chap. 6. The first is the general notion of the Grothendieck group $K_0(A)$ of an algebra A; this can be defined as the universal group of the monoid $\mathcal{V}(A)$, see Definition 6.1.4. The second is a specific result about $K_0(A)$ when A is purely infinite simple. In this situation $K_0(A) = \mathcal{V}(A) \setminus \{[0]\}$, see Proposition 6.1.3.

Theorem 5.3.5 ([31, Theorem 7.1]) *Let E be a row-finite graph. Then the natural inclusion $\psi: L_{\mathbb{C}}(E) \to C^*(E)$ induces a monoid isomorphism $\mathcal{V}(\psi): \mathcal{V}(L_{\mathbb{C}}(E)) \to \mathcal{V}(C^*(E))$.*

In particular, the monoid $\mathcal{V}(C^*(E))$ is naturally isomorphic to the monoid M_E.

Proof The algebra homomorphism $\psi: L_{\mathbb{C}}(E) \to C^*(E)$ induces the following commutative square.

$$\begin{array}{ccc} \mathcal{V}(L_{\mathbb{C}}(E)) & \xrightarrow{\mathcal{V}(\psi)} & \mathcal{V}(C^*(E)) \\ \downarrow{\scriptstyle \varphi_1} & & \downarrow{\scriptstyle \varphi_2} \\ K_0(L_{\mathbb{C}}(E)) & \xrightarrow{K_0(\psi)} & K_0(C^*(E)) \end{array}$$

The map $K_0(\psi)$ is an isomorphism, by Theorem 6.1.9 together with [130, Theorem 3.2]. Using Lemma 1.6.16 and Lemma 5.3.4, we see that it is enough to show that $\mathcal{V}(\psi)$ is an isomorphism for a finite graph E.

Assume that E is a finite graph. We first show that the map $\mathcal{V}(\psi): \mathcal{V}(L_{\mathbb{C}}(E)) \to \mathcal{V}(C^*(E))$ is injective. Suppose that P and Q are idempotents in $M_{\mathbb{N}}(L_{\mathbb{C}}(E))$ such that $P \sim Q$ in $C^*(E)$. By Theorem 3.2.5, we can assume that each of P and Q are equivalent in $M_{\mathbb{N}}(L_{\mathbb{C}}(E))$ to direct sums of "basic" projections, that is, projections of the form v, with $v \in E^0$. Let J be the closed ideal of $C^*(E)$ generated by the entries of P. Since $P \sim Q$, the closed ideal generated by the entries of P agrees with the closed ideal generated by the entries of Q, and indeed it agrees with the closed ideal generated by the projections of the form w, where w ranges over the hereditary saturated subset H of E^0 generated by $\{v \in E^0 \mid P = \oplus v\}$ (see Theorem 5.3.2). It follows from Theorem 2.5.9 that P and Q generate the same ideal I_0 in $L_{\mathbb{C}}(E)$. There is a projection $e \in L_{\mathbb{C}}(E)$, which is the sum of the basic projections w, where w ranges over H, such that $I_0 = L_{\mathbb{C}}(E)eL_{\mathbb{C}}(E)$ and $eL_{\mathbb{C}}(E)e = L_{\mathbb{C}}(H)$ is also a Leavitt path algebra. Note that P and Q are full projections in $L_{\mathbb{C}}(H)$, and so $[1_H] \leq m[P]$ and $[1_H] \leq m[Q]$ in $\mathcal{V}(L_{\mathbb{C}}(H))$ for some $m \geq 1$. Now consider the map $\psi_H: L_{\mathbb{C}}(H) \to C^*(H)$. Since $\mathcal{V}(\psi_H)([P]) = \mathcal{V}(\psi_H)([Q])$ in $\mathcal{V}(C^*(H))$ we get $K_0(\psi_H)(\varphi_1([P])) = K_0(\psi_H)(\varphi_1([Q]))$, and since $K_0(\psi_H)$ is an isomorphism we get $\varphi_1([P]) = \varphi_1([Q])$. This means that there exists a $k \geq 0$ such that $[P] + k[1_H] = [Q] + k[1_H]$. But since $\mathcal{V}(L_{\mathbb{C}}(E))$ is separative (Theorem 3.6.12) and $[1_H] \leq m[P]$ and $[1_H] \leq m[Q]$, we get $[P] = [Q]$ in $\mathcal{V}(L_{\mathbb{C}}(E))$.

Now we will establish that the map $\mathcal{V}(\psi): \mathcal{V}(L_{\mathbb{C}}(E)) \to \mathcal{V}(C^*(E))$ is surjective. It suffices to show that any projection P in $M_{\mathbb{N}}(C^*(E))$ is equivalent to a finite sum of basic projections (corresponding to vertices of E).

By Theorem 5.3.2, there is a natural isomorphism between the lattice of hereditary saturated subsets of E^0 and the lattice of closed gauge-invariant ideals of $C^*(E)$. Thus, since E is finite, the number of closed gauge-invariant ideals of

$C^*(E)$ is finite, and there is a finite chain $I_0 = \{0\} \leq I_1 \leq \cdots \leq I_n = C^*(E)$ of closed gauge-invariant ideals such that each quotient I_{i+1}/I_i is gauge-simple. We proceed by induction on n.

If $n = 1$ we have the case in which $C^*(E)$ is gauge-simple. So by Bates et al. [47], we conclude that $C^*(E)$ is either purely infinite simple, or AF, or Morita-equivalent to $C(\mathbb{T})$. In any of the three cases the result follows. Note that in the purely infinite case, we use that $\mathcal{V}(C^*(E)) = K_0(C^*(E)) \sqcup \{[0]\} = K_0(L_{\mathbb{C}}(E)) \sqcup \{[0]\} = \mathcal{V}(L_{\mathbb{C}}(E))$; see also Proposition 6.1.3.

Now assume that the result is true for graph C^*-algebras of (gauge) length $n-1$ and let $A = C^*(E)$ be a graph C^*-algebra of length n. Let H be the hereditary saturated subset of E^0 corresponding to the gauge-simple ideal I_1. Note that H is a minimal hereditary saturated subset of E^0, and thus H is cofinal. Set $B = A/I_1$. By Lemma 5.3.1, we have $B \cong C^*(E/H)$. Observe that by the induction hypothesis we know that every projection in B is equivalent to a finite orthogonal sum of basic projections of the form v, where v ranges over $(E/H)^0 = E^0 \setminus H$. Let $\pi: A \to B$ denote the canonical projection. Since I_1 is the closed ideal generated by its projections, there is an embedding $\mathcal{V}(A)/\mathcal{V}(I_1) \to \mathcal{V}(B)$. (This follows from [24, Proposition 5.3(c)], taking into account that every closed ideal generated by projections is an almost trace ideal.) By the induction hypothesis, $\mathcal{V}(B) = \mathcal{V}(C^*(E/H))$ is generated as a monoid by $\{[v] \mid v \in E^0 \setminus H\}$, and so the map $\mathcal{V}(A)/\mathcal{V}(I_1) \to \mathcal{V}(B)$ is also surjective, so that $\mathcal{V}(B) \cong \mathcal{V}(A)/\mathcal{V}(I_1)$. In particular, $\pi(P) \sim \pi(Q)$ for two projections $P, Q \in M_{\mathbb{N}}(A)$ if and only if there are projections $P', Q' \in M_{\mathbb{N}}(I_1)$ such that $P \oplus P' \sim Q \oplus Q'$ in $M_{\mathbb{N}}(A)$.

We now deal simultaneously with the two cases where I_1 is either AF or Morita equivalent to $C(\mathbb{T})$; these correspond to the case where I_1 has stable rank 1. Note that in this case either H contains a sink v, or we have a unique cycle without exits, in which case we select v as a vertex in this cycle. Note that, by the cofinality of H, any projection in I_1 is equivalent to a projection of the form $k \cdot v$ for some $k \geq 0$. Now take any projection P in $M_{\mathbb{N}}(A)$. Since $\pi(P) \sim \pi(v_1 \oplus \cdots \oplus v_r)$ for some vertices $v_1, \ldots, v_r$ in $E^0 \setminus H$, there exist $a, b \in \mathbb{Z}^+$ such that

$$P \oplus a \cdot v \sim v_1 \oplus \cdots \oplus v_r \oplus b \cdot v.$$

Since the stable rank of vAv is 1, the projection v cancels in direct sums [133], and so, if $b \geq a$, we get

$$P \sim v_1 \oplus \cdots \oplus v_r \oplus (b-a)v,$$

so that P is equivalent to a finite orthogonal sum of basic projections. If $b < a$, then we have $P \oplus (a-b)v \sim v_1 \oplus \cdots \oplus v_r$. We claim that there is some $1 \leq i \leq r$ such that v is in $T(v_i)$, the tree of v_i. For, assume to the contrary that $v \notin \bigcup_{i=1}^r T(v_i)$. We will see that v is not in the hereditary saturated subset of E generated by $v_1, \ldots, v_r$. Note that the set $D = \bigcup_{i=1}^r T(v_i)$ is hereditary, and that the hereditary saturated subset of E generated by $v_1, \ldots, v_r$ is $\overline{D} = \bigcup_{j\in\mathbb{N}} S^j(D)$ (see Lemma 2.0.7). Observe also that, since v is either a sink or belongs to a cycle without exits, and H is cofinal,

v belongs to the tree of any vertex in H, whence $D \cap H = \emptyset$. Let v' be a vertex in H. If $v' \in S(D)$ then $s^{-1}(v') \neq \emptyset$ and $r(s^{-1}(v')) \subseteq D \cap H$. Since $D \cap H = \emptyset$, this is impossible. So $S(D) \cap H = \emptyset$. Indeed, an easy induction shows that $S^i(D) \cap H = \emptyset$ for all i, and so $\overline{D} \cap H = \emptyset$. But as v is equivalent to a subprojection of $v_1 \oplus \cdots \oplus v_r$, the projection v belongs to the closed ideal of A generated by $v_1, \ldots, v_r$, and so v belongs to $\overline{D}$. This contradiction shows that v belongs to the tree of some v_i, as claimed.

Now, the fact that v belongs to the tree of v_i implies that there is a projection Q which is a finite orthogonal sum of basic projections such that $v_i \sim v \oplus Q$. Therefore we get

$$P \oplus (a-b)v \sim (v_1 \oplus \cdots \oplus v_{i-1} \oplus v_{i+1} \oplus \cdots \oplus v_r \oplus Q) \oplus v.$$

Since v can be cancelled in direct sums, we get

$$P \oplus (a-b-1)v \sim (v_1 \oplus \cdots \oplus v_{i-1} \oplus v_{i+1} \oplus \cdots \oplus v_r \oplus Q),$$

and so, using induction, we obtain that P is equivalent to a finite orthogonal sum of basic projections.

We consider now the third possibility for I_1, namely, the case where I_1 is a purely infinite simple C^*-algebra. Recall that in this case I_1 has real rank zero [56], and that $\mathscr{V}(I_1)\setminus\{0\}$ is a group. So there is a nonzero projection e in I_1 such that $e \sim e \oplus e$, and such that for every nonzero projection p in $M_{\mathbb{N}}(I_1)$ there exists a nonzero projection $q \in I_1$ such that $p \oplus q \sim e$. Let P be a nonzero projection in $M_k(A)$, for some $k \geq 1$, and denote by I the closed ideal of A generated by the entries of P. If $I \cdot I_1 = 0$, then $I \cong (I + I_1)/I_1$, so that I is a closed ideal in the quotient C^*-algebra $B = A/I_1$. It then follows by our assumption on B that P is equivalent to a finite orthogonal sum of basic projections. Assume now that $I \cdot I_1 \neq 0$. Then there is a nonzero column $C = (a_1, a_2, \ldots, a_k)^t \in A^k$ such that $C = PCe$. Consider the positive element $c = C^*C$, which belongs to eAe. Since $e \in I_1$ and I_1 has real rank zero, the C^*-algebra eAe also has real rank zero, so that we can find a nonzero projection $p \in cAc$. Take $x \in A$ such that $p = cxc$. By using standard techniques (see e.g., [136]), we can now produce a projection $P' \leq P$ such that $p \sim P'$. Namely, consider the idempotent $F = CpC^*CxC^*$ in $PM_k(A)P$. Then p and F are equivalent as idempotents, and F is equivalent to some projection P' in $PM_k(A)P$; see [136, Exercise 3.11(i)]. Since p and P' are equivalent as idempotents, they are also Murray-von Neumann equivalent (see e.g., [136, Exercise 3.11(ii)]), as desired. We have thus shown that there is a nonzero projection p in I_1 such that p is equivalent to a subprojection of P. Since I_1 is purely infinite simple, every projection in I_1 is equivalent to a subprojection of p, and so every projection in I_1 is equivalent to a subprojection of P.

Now we are ready to conclude the proof. There is a projection q in I_1 such that $P \oplus q$ is equivalent to a finite orthogonal sum of basic projections. Let q' be a nonzero projection in I_1 such that $q \oplus q' \sim e$, and observe that

$$P \oplus e \sim (P \oplus q) \oplus q',$$

so that $P \oplus e$ is also a finite orthogonal sum of basic projections. By the above argument, there is a projection e' such that $e' \leq P$ and $e \sim e'$. Write $P = e' + P'$. Then we have

$$P \oplus e \sim P' \oplus e' \oplus e \sim P' \oplus e \oplus e \sim P' \oplus e \sim P.$$

It follows that $P \sim P \oplus e$ and so P is equivalent to a finite orthogonal sum of basic projections. This completes the proof. □

We note that an extension of Theorem 5.3.5 to countable (but not necessarily row-finite) graphs has been achieved in [91].

Corollary 5.3.6 *Let E be a row-finite graph. Then the monoid $\mathcal{V}(C^*(E))$ is a refinement monoid, and $C^*(E)$ is separative. Moreover, $\mathcal{V}(C^*(E))$ is an unperforated monoid and $K_0(C^*(E))$ is an unperforated group.*

Proof By Theorem 5.3.5, $\mathcal{V}(C^*(E)) \cong M_E$, and so $\mathcal{V}(C^*(E))$ is a refinement monoid by Proposition 3.6.8. It follows from Theorem 3.6.12 that $\mathcal{V}(C^*(E))$ is a separative monoid. The statements about unperforation follow from Proposition 3.6.14. □

By Ara et al.[31, Proposition 2.1], a C^*-algebra is separative if and only if it has stable weak cancellation, a property studied by Brown [55] and Brown and Pedersen [57]. The interested reader can consult these articles for more information about this class of C^*-algebras.

Corollary 5.3.7 *Let E be a row-finite graph and let I be a closed ideal of $C^*(E)$ generated by projections. Then I is a gauge-invariant ideal and $I = \overline{I \cap L_{\mathbb{C}}(E)}$. Moreover, if I and J are closed ideals of $C^*(E)$ generated by projections then*

$$I \cap L_{\mathbb{C}}(E) = J \cap L_{\mathbb{C}}(E) \implies I = J.$$

Proof If p is a projection in $C^*(E)$ then it follows from Theorem 5.3.5 that $p \sim v_1 \oplus \cdots \oplus v_n$ for some $v_1, \ldots, v_n \in E^0$. This yields that the ideal generated by p coincides with the ideal generated by $v_1, \ldots, v_n$. So I is a closed ideal generated by a set of vertices of E, and thus it is a gauge-invariant ideal. Since I is generated by a family of elements of $L_{\mathbb{C}}(E)$, we see that $I = \overline{I \cap L_{\mathbb{C}}(E)}$.

If I and J are closed ideals of $C^*(E)$ generated by projections and $I \cap L_{\mathbb{C}}(E) = J \cap L_{\mathbb{C}}(E)$ then $I = \overline{I \cap L_{\mathbb{C}}(E)} = \overline{J \cap L_{\mathbb{C}}(E)} = J$, as desired. □

The above property does not necessarily hold for arbitrary closed ideals of $C^*(E)$. An example can be found by considering $C^*(R_1) = C(\mathbb{T})$, with $L_{\mathbb{C}}(R_1)$ being the algebra $\mathbb{C}[z, z^{-1}]$ of Laurent polynomials. Indeed, for any proper compact subset K of $\mathbb{T}$ with nonempty interior, we have that the ideal

$$I = \{f \in C(\mathbb{T}) : f(K) = 0\}$$

satisfies $I \cap L_{\mathbb{C}}(R_1) = \{0\}$, but of course $I \neq \{0\}$. There are also different $*$-ideals I and J of $L_{\mathbb{C}}(E)$ such that $\overline{I} = \overline{J}$, for instance take $I = L_{\mathbb{C}}(R_1)$ and J the ideal generated by any polynomial $p(z)$ for which the ideals generated by $p(z)$ and by $p(z^{-1})$ are equal, e.g., $p(z) = 1 + 4z + z^2$.

5.4 Closed Ideals of Graph C^*-Algebras Containing No Vertices

In this section we consider ideals of $C^*(E)$ which stand on the opposite end of the spectrum from the ideals considered in the previous section. To wit, we investigate the structure of the closed ideals of $C^*(E)$ which contain no nonzero projections. We again remind the reader of our standing hypothesis throughout this chapter that E is always assumed to be a row-finite graph.

We recall that for a subset X of E^0, $\mathfrak{I}(X)$ denotes the closed ideal of $C^*(E)$ generated by X.

Lemma 5.4.1 *Let H be a hereditary saturated subset of E^0. Then projections from $C^*(E)/\mathfrak{I}(H)$ lift to $C^*(E)$, that is, if $x \in C^*(E)$ is such that $x + \mathfrak{I}(H)$ is a projection then there is a projection p in $C^*(E)$ such that $x - p \in \mathfrak{I}(H)$.*

Proof Assume that $\overline{x} := x + \mathfrak{I}(H)$ is a projection in $C^*(E)/\mathfrak{I}(H) \cong C^*(E/H)$. Then there are (not necessarily distinct) vertices $v_1, \ldots, v_m$ such that

$$\overline{x} \sim v_1 \oplus \cdots \oplus v_m.$$

Assume first that $m = 1$, so that $\overline{x} \sim v$ for $v \in E^0$. Then we can write

$$v = z^* z + \alpha, \qquad x = zz^* + \beta,$$

where $z, \alpha, \beta \in C^*(E)$, $zv = z$, $\alpha = v\alpha v$, and $\alpha, \beta \in \mathfrak{I}(H)$. Since $\mathfrak{I}(H)$ is the norm closure of the ideal $I(H)$ of $L_{\mathbb{C}}(E)$, it follows that there are distinct paths $\tau_1, \ldots, \tau_n \in F_E(H)$ (see Definition 2.5.16 and Theorem 2.5.19) such that $s(\tau_i) = v$ for all i and

$$\|\alpha - \sum_{i,j=1}^{n} \tau_i a_{ij} \tau_j^* \| < 1,$$

where $a_{ij} \in L_{\mathbb{C}}(E_H)$, that is, a_{ij} belong to the subalgebra of $L_{\mathbb{C}}(E)$ generated by all the terms $\gamma\nu^*$, where γ, ν are paths of E starting at vertices of H. Obviously we can assume that $v \notin H$.

Define $g := \sum_{i=1}^{n} \tau_i\tau_i^*$. By the structure of $F_E(H)$ we have $\tau_i^*\tau_j = \delta_{ij}r(\tau_i)$, so that g is a projection. Observe that $g \leq v$ and that $g \in \mathfrak{I}(H)$. Moreover,

$$g \cdot (\sum_{i,j=1}^{n} \tau_i a_{ij}\tau_j^*) = \sum_{i,j=1}^{n} \tau_i a_{ij}\tau_j^* = (\sum_{i,j=1}^{n} \tau_i a_{ij}\tau_j^*) \cdot g,$$

and thus

$$\|(v-g)\alpha(v-g)\| = \|(v-g)(\alpha - \sum_{i,j=1}^{n} \tau_i a_{ij}\tau_j^*)(v-g)\| < 1. \tag{5.4}$$

Now, multiplying the equation $v = z^*z + \alpha$ by $v - g$ on both sides and using (5.4) we get

$$\|(v-g) - (v-g)z^*z(v-g)\| < 1,$$

from which we conclude that $(v-g)z^*z(v-g)$ is invertible in $(v-g)C^*(E)(v-g)$ (see e.g., [118, Theorem 1.2.2]); denote its inverse by w. Now $p := z(v-g)w(v-g)z^*$ is a projection, and since $w + \mathfrak{I}(H) = v + \mathfrak{I}(H)$, it follows that $x - p \in \mathfrak{I}(H)$ as desired.

Now assume that $m > 1$. Write $\overline{x} = \overline{x}_1 + \cdots + \overline{x}_m$, where $\overline{x}_i$ are orthogonal projections with $\overline{x}_i \sim v_i$. By the case $m = 1$ there exists a projection p_1 in $C^*(E)$ such that $p_1 + \mathfrak{I}(H) = x_1 + \mathfrak{I}(H)$. Now assume that $1 \leq i < m$ and that we have constructed orthogonal projections $p_1, \ldots, p_i$ in $C^*(E)$ such that $p_j + \mathfrak{I}(H) = x_j + \mathfrak{I}(H)$ for $j = 1, \ldots, i$. Write $P_i := p_1 + \cdots + p_i$. Then there are elements $z_{i+1}, \alpha_{i+1}, \beta_{i+1}$ such that

$$v_{i+1} = z_{i+1}^* z_{i+1} + \alpha_{i+1}, \qquad x_{i+1} = z_{i+1}z_{i+1}^* + \beta_{i+1},$$

with $(1 - P_i)z_{i+1}v_{i+1} = z_{i+1}$, $\alpha_{i+1} = v_{i+1}\alpha_{i+1}v_{i+1}$, and $\alpha_{i+1}, \beta_{i+1} \in \mathfrak{I}(H)$. The same proof as in the $m = 1$ case allows us to build a projection $p_{i+1} = z_{i+1}(v_{i+1} - g_{i+1})w_{i+1}(v_{i+1} - g_{i+1})z_{i+1}^*$ such that $p_{i+1} + \mathfrak{I}(H) = x_{i+1} + \mathfrak{I}(H)$. Observe that $p_1, \ldots, p_{i+1}$ are orthogonal projections.

Following this procedure, we eventually arrive at a sequence $p_1, \ldots, p_m$ of orthogonal projections in $C^*(E)$ such that $p_i + \mathfrak{I}(H) = x_i + \mathfrak{I}(H)$ for all i. So $p := p_1 + \cdots + p_m$ is a projection in $C^*(E)$ such that $p + \mathfrak{I}(H) = x + \mathfrak{I}(H)$, and the result is proved. □

The following result follows from Theorems 5.3.2 and 2.7.3, together with Lemma 5.2.8.

Proposition 5.4.2 *Let E be a row-finite graph. Let K be the closed ideal of $C^*(E)$ generated by the vertices in cycles without exits. Then K is isomorphic to the C^*-algebraic direct sum*

$$\bigoplus_{C} \mathscr{K}_C \otimes C(\mathbb{T}),$$

where $\mathscr{K}_C$ is the algebra of compact operators on some (finite-dimensional or separable infinite-dimensional) Hilbert space $\mathscr{H}_C$, and C ranges over all the cycles without exits in E.

Proposition 5.4.3 *Let E be a finite graph and let J be a closed ideal of $C^*(E)$ containing no nonzero projections. Then J is contained in the closed ideal of $C^*(E)$ generated by the vertices in cycles without exits.*

Proof We proceed by induction on the number of cycles of E. If E is acyclic then $C^*(E)$ is a finite-dimensional (and hence matricial) C^*-algebra, and it is well known that every closed ideal in such a C^*-algebra is generated by its projections.

Now let $n \geq 1$ and assume that the result is true for finite graphs with fewer than n cycles. Let E be a finite graph containing exactly n cycles. If every cycle of E has an exit (i.e., if E satisfies Condition (L)), then every nonzero closed ideal contains a nonzero projection by Corollary 5.2.16. Thus we may assume that there is at least one cycle without exits in E. Let H be the hereditary saturated closure of the set of vertices of E belonging to cycles without exits. It is clear that the only cycles contained in H are the cycles without exits. Consider the quotient C^*-algebra $C^*(E)/\mathfrak{I}(H) \cong C^*(E/H)$ (Lemma 5.3.1).

Claim The ideal $J + \mathfrak{I}(H)/\mathfrak{I}(H)$ does not contain nonzero projections.

Proof of Claim We proceed by way of contradiction. Assume that $x + \mathfrak{I}(H)$ is a nonzero projection in $C^*(E)/\mathfrak{I}(H) \cong C^*(E/H)$, where $x \in J$. By Lemma 5.4.1 there is a projection p in $C^*(E)$ such that $p + \alpha = x$ for some $\alpha \in \mathfrak{I}(H)$. Now there are vertices $v_1, \dots, v_r \in E^0$ with $v_1 \notin H$ such that

$$p \sim v_1 \oplus \cdots \oplus v_r.$$

Write $p = p_1 + \cdots + p_r$, where p_i are orthogonal projections with $p_i \sim v_i$ for $i = 1, \dots, r$. Then

$$p_1 + p_1 \alpha p_1 = p_1 x p_1 \in J. \tag{5.5}$$

Set $p_1 = ww^*$, $v_1 = w^*w$ for some partial isometry w in $C^*(E)$. Multiplying (5.5) on the left by w^* and on the right by w we get

$$v_1 + w^*\alpha w = w^*xw \in v_1 J v_1.$$

As in the proof of Lemma 5.4.1, there is a projection g in $\mathfrak{I}(H)$ such that $g \leq v_1$ and

$$\|(v_1 - g)w^*\alpha w(v_1 - g)\| < 1.$$

Now since

$$\|(v_1 - g) - (v_1 - g)w^*xw(v_1 - g)\| = \|(v_1 - g)w^*\alpha w(v_1 - g)\| < 1,$$

by again invoking [118, Theorem 1.2.2] we get that $(v_1 - g)w^*xw(v_1 - g)$ is invertible in $(v_1 - g)C^*(E)(v_1 - g)$, and hence $v_1 - g \in J$, contradicting the fact that J does not contain nonzero projections. This proves the Claim.

Since E/H has fewer than n cycles, it follows from the induction hypothesis that $J + \mathfrak{I}(H)/\mathfrak{I}(H)$ is contained in the ideal K of $C^*(E/H)$ generated by the vertices in E/H belonging to cycles without exits in E/H. By Proposition 5.4.2, the ideal K is isomorphic (as a C^*-algebra) to the C^*-algebraic direct sum

$$\bigoplus_C \mathcal{K}_C \otimes C(\mathbb{T}),$$

where $\mathcal{K}_C$ is the algebra of compact operators on a finite-dimensional or a separable infinite-dimensional Hilbert space, and C ranges over all the cycles without exits in E/H. It follows that $J + \mathfrak{I}(H)/\mathfrak{I}(H)$ contains a nonzero positive element of the form $f(c)$, where $f \in C(\mathbb{T})_+$ and $c = e_1e_2\cdots e_r$ is a cycle without exits in E/H.

Let us now consider the cycle c as being in E, and the element $f(c)$ as an element in $C^*(E)$. Note that c is a cycle with exits in E, and that if e is an exit of c then $r(e) \in H$. Since

$$e_1^*f(c)e_1 = f(e_2\cdots e_re_1)$$

we may assume without loss of generality that there is an exit $\tilde{e}$ of c such that $s(\tilde{e}) = v_0 := s(e_1)$. Observe that $\Phi(f(c)) = \lambda \cdot v_0$, where $\lambda > 0$ (and where Φ is as defined in Lemma 5.2.10).

Since $f(c) \in J + \mathfrak{I}(H)$, there exists an $\alpha \in \mathfrak{I}(H)$ such that

$$f(c) - \alpha \in J,$$

and we may assume that $v_0\alpha v_0 = \alpha$. As before, since $\mathfrak{I}(H)$ is the norm closure of the ideal $I(H)$ of $L_{\mathbb{C}}(E)$, it follows that there are distinct paths $\tau_1, \ldots, \tau_n \in F_E(H)$ such that $s(\tau_i) = v_0$ for all i and

$$\|\alpha - \sum_{i,j=1}^{n} \tau_i a_{ij} \tau_j^*\| < \lambda,$$

where $a_{ij} \in L_{\mathbb{C}}(E_H)$. Observe that $\{\tau_i\}_{i=1}^n$ must be a subset of the set of edges $e \in s_E^{-1}(v_0)$ such that $e \neq e_1$, because $r(e) \in H$ for all such edges e.

Set $\beta := \sum_{i,j=1}^n \tau_i a_{ij} \tau_j^*$. Consider

$$g := \sum_{e \in s^{-1}(v_0), e \neq e_1} ee^*.$$

Observe that $v_0 - g = e_1 e_1^*$ by relation (CK2) at v_0. Since $(v_0 - g)\beta(v_0 - g) = 0$ we get $e_1^* \beta e_1 = 0$ and so

$$\|e_1^* \alpha e_1\| < \lambda.$$

We have $c^* f(c) c = f(c)$ and $\|c^* \alpha c\| < \lambda$. Furthermore,

$$\tilde{e}^* f(c) \tilde{e} = \lambda \cdot r(\tilde{e}).$$

We get

$$\lambda \cdot r(\tilde{e}) - \tilde{e}^* c^* \alpha c \tilde{e} = \tilde{e}^* c^* (f(c) - \alpha) c \tilde{e} = y,$$

where $y \in r(\tilde{e}) J r(\tilde{e})$, so that $r(\tilde{e}) - \lambda^{-1} y = \lambda^{-1} \tilde{e}^* c^* \alpha c \tilde{e}$. Since

$$\|r(\tilde{e}) - \lambda^{-1} y\| = \|\lambda^{-1} \tilde{e}^* c^* \alpha c \tilde{e}\| \leq \lambda^{-1} \|c^* \alpha c\| < \lambda^{-1} \cdot \lambda = 1,$$

we conclude (yet again by Murphy [118, Theorem 1.2.2]) that $\lambda^{-1} y$ is invertible in $r(\tilde{e}) C^*(E) r(\tilde{e})$ and so $r(\tilde{e}) \in J$. But this contradicts the fact that J does not contain nonzero projections, and thus concludes the proof of the proposition. □

Proposition 5.4.3 now puts us in position to prove the following result, which is the main goal of this section. The result is the analog for graph C^*-algebras of Proposition 2.7.9.

Theorem 5.4.4 *Let E be a row-finite graph and let J be a closed ideal of $C^*(E)$ containing no nonzero projections. Then J is contained in the closed ideal of $C^*(E)$ generated by the vertices in cycles without exits.*

Proof We have $C^*(E) = \varinjlim C^*(F)$, where F runs over the family of finite complete subgraphs of E (see Lemma 5.3.4). Consequently $J = \varinjlim (J \cap C^*(F))$, and by Proposition 5.4.3, $J \cap C^*(F)$ is contained in the closed ideal of $C^*(F)$ generated by the vertices in cycles without exits in F. Since F is a complete subgraph of E, the cycles without exits of F cannot have exits in E, and so $J \cap C^*(F)$ is contained in the closed ideal of $C^*(E)$ generated by the vertices in cycles without exits in E. This shows that J is contained in the closed ideal generated by the cycles without exits in E. □

5.5 The Structure of Closed Ideals of Graph C^*-Algebras of Row-Finite Graphs

In this section we will describe all the closed ideals of $C^*(E)$ for a row-finite graph E, Theorem 5.5.3. The description is similar to that given in the Structure Theorem for Ideals 2.8.10, and its consequence for row-finite graphs, Proposition 2.8.11. We use here terminology analogous to that given in Sect. 2.8.

Definition 5.5.1 For any graph E, define $\mathscr{L}_{HSK}(E)$ as the set of all triples of the form $(H, \mathscr{S}, \{K_c\}_{c\in\mathscr{S}})$, where H is a hereditary saturated subset of E^0, $\mathscr{S}$ is a subset of $C_u(E)$ such that $\mathscr{S}^0 \cap H = \emptyset$ and $\mathscr{S}^{<<} \subseteq H$, and for each $c \in \mathscr{S}$, K_c is a compact nonempty proper subset of $\mathbb{T}$.

Lemma 5.5.2 *Let E be a row-finite graph. Then $\mathscr{L}_{HSK}(E) := \{(H, \mathscr{S}, \{K_c\}_{c\in\mathscr{S}})\}$ is a lattice, with the order $\leq$ given by:*

$$(H_1, \mathscr{S}_1, \{K_c^{(1)}\}_{c\in\mathscr{S}^1}) \leq (H_2, \mathscr{S}_2, \{K_c^{(2)}\}_{c\in\mathscr{S}^2}) \text{ if:}$$

(i) $H_1 \subseteq H_2$,
(ii) $\mathscr{S}_1^0 \subseteq H_2 \cup \mathscr{S}_2^0$, and
(iii) $K_c^{(2)} \subseteq K_c^{(1)}$ for every $c \in \mathscr{S}_1 \cap \mathscr{S}_2$.

Proof It is very easy to see that $\leq$ is reflexive and antisymmetric. To prove the transitivity, take three triples in $\mathscr{L}_{HSK}(E)$ such that $(H_1, \mathscr{S}_1, \{K_c^{(1)}\}_{c\in\mathscr{S}^1}) \leq (H_2, \mathscr{S}_2, \{K_c^{(2)}\}_{c\in\mathscr{S}^2})$ and $(H_2, \mathscr{S}_2, \{K_c^{(2)}\}_{c\in\mathscr{S}^2}) \leq (H_3, \mathscr{S}_3, \{K_c^{(3)}\}_{c\in\mathscr{S}^3})$.

Since $H_1 \subseteq H_2$ and $H_2 \subseteq H_3$, it follows $H_1 \subseteq H_3$. On the other hand, $\mathscr{S}_1^0 \subseteq H_2 \cup \mathscr{S}_2^0$ and $\mathscr{S}_2^0 \subseteq H_3 \cup \mathscr{S}_3^0$ implies $\mathscr{S}_1^0 \subseteq H_2 \cup \mathscr{S}_2^0 \subseteq H_3 \cup \mathscr{S}_3^0$.

Finally, let $c \in \mathscr{S}_1 \cap \mathscr{S}_3$. Note that $c \in \mathscr{S}_3$ implies $c^0 \cap H_3 = \emptyset$, hence $c \in \mathscr{S}_2$ because otherwise $c^0 \subseteq H_2 \cup \mathscr{S}_2^0$ would imply $c^0 \subseteq H_2 \subseteq H_3$, a contradiction. Therefore $c \in \mathscr{S}_1 \cap \mathscr{S}_2 \cap \mathscr{S}_3$ and by the relations $K_c^{(2)} \subseteq K_c^{(1)}$ and $K_c^{(3)} \subseteq K_c^{(2)}$ we get $K_c^{(3)} \subseteq K_c^{(1)}$. Hence $(H_1, \mathscr{S}_1, \{K_c^{(1)}\}_{c\in\mathscr{S}^1}) \leq (H_3, \mathscr{S}_3, \{K_c^{(3)}\}_{c\in\mathscr{S}^3})$. □

Now we describe how to attach to every closed ideal of a graph C^*-algebra $C^*(E)$ of a row-finite graph E an element of $\mathscr{L}_{HSK}(E)$. Let I be a closed ideal of $C^*(E)$. Define $H := I \cap E^0$. Since $H \subseteq I$ then $\mathfrak{I}(H) \subseteq I$. Let us consider the ideal $I/\mathfrak{I}(H)$ of $C^*(E)/\mathfrak{I}(H)$. By Lemma 5.3.1, there is a natural isomorphism $C^*(E)/\mathfrak{I}(H) \cong C^*(E/H)$. Let J denote the image of $I/\mathfrak{I}(H)$ through this isomorphism. We claim that J is a closed ideal of $C^*(E/H)$ which does not contain vertices. Indeed, if $v \in J \cap (E/H)^0$, then $\overline{v} \in I/\mathfrak{I}(H)$, where $\overline{v}$ denotes the image of v by the canonical map $\pi : C^*(E) \to C^*(E)/\mathfrak{I}(H)$. But this gives $v \in I$, a contradiction. Hence, by Theorem 5.4.4, J is contained in the closed ideal of $C^*(E/H)$ generated by the vertices in cycles without exits.

Let $\overline{\mathscr{S}_I}$ be the set of cycles in E/H such that for every cycle $d \in E/H$ we have $J \cap \mathfrak{I}(d^0) \neq \{0\}$, and define $\mathscr{S}_I$ as the set of cycles of $\overline{\mathscr{S}_I}$ viewed as cycles of E. Then $\mathscr{S}_I$ is a subset of $C_u(E)$ such that $\mathscr{S}_I{}^0 \cap H = \emptyset$ and $\mathscr{S}_I{}^{<<} \subseteq H$.

For every $c \in \overline{\mathscr{S}_I}$ we have $J \cap \mathfrak{I}(c^0)$ is a nonzero proper ideal of $\mathfrak{I}(c^0)$, which is isomorphic to $\mathscr{K}_c \otimes C(\mathbb{T})$. It follows that there is a unique nonempty proper compact subset K_c of $\mathbb{T}$ such that

$$J \cap \mathfrak{I}(c^0) \cong \mathscr{K}_c \otimes C_0(\mathbb{T} \setminus K_c).$$

We associate to the closed ideal I the triple $(H, \mathscr{S}_I, \{K_c\}_{c \in \mathscr{S}_I})$, and will prove in Theorem 5.5.3 that we can recover the ideal I from this data.

Before stating the theorem that gives the correspondence between closed ideals of a graph C^*-algebra of a row-finite graph and elements of $\mathscr{L}_{HSK}(E, K)$, we introduce the following notation: $\mathscr{L}(C^*(E))$ will stand for the lattice of closed ideals of $C^*(E)$, with the order given by inclusion.

Theorem 5.5.3 *Let E be a row-finite graph. Then the following maps are mutually inverse lattice isomorphisms.*

$$\begin{array}{llll} \varphi: & \mathscr{L}_{HSK}(E) & \longrightarrow & \mathscr{L}(C^*(E)) \\ & (H, \mathscr{S}, \{K_c\}_{c \in \mathscr{S}}) & \mapsto & < \{H, \{f(c) \mid f \in C_0(\mathbb{T} \setminus K_c)\}_{c \in \mathscr{S}}\} >, \end{array}$$

$$\begin{array}{lll} \varphi' : \mathscr{L}(C^*(E)) & \longrightarrow & \mathscr{L}_{HSK}(E) \\ I & \mapsto & (I \cap E^0, \mathscr{S}_I, \{K_c\}_{c \in \mathscr{S}_I}). \end{array}$$

Proof We start by showing that $\varphi' \circ \varphi = id_{\mathscr{L}_{HSK}(E)}$. Take $(H, \mathscr{S}, \{K_c\}_{c \in \mathscr{S}}) \in \mathscr{L}_{HSK}(E)$ and denote by I its image under φ. We show that $I \cap E^0 = H$. Clearly $H \subseteq I \cap E^0$. To show the reverse containment, consider $I/\mathfrak{I}(H) =< \{f(\overline{c}) \mid f \in C_0(\mathbb{T} \setminus K_c)\}_{\overline{c} \in \overline{\mathscr{S}}} >$. We thus obtain

$$I/\mathfrak{I}(H) \cong \bigoplus_{c \in \mathscr{S}} \mathscr{K}_c \otimes C_0(\mathbb{T} \setminus K_c), \tag{5.6}$$

and since $K_c \neq \mathbb{T}$ for all $c \in \mathscr{S}$, we get that $I/\mathfrak{I}(H)$ does not contain nonzero projections. Hence $I \cap E^0 = H$ and we have shown our claim.

Now, for each $c \in C_{ne}(E/H)$ we have $\mathfrak{I}(c^0) \cong \mathscr{K}_c \otimes C(\mathbb{T})$, and it follows from (5.6) that $\mathfrak{I}(c^0) \cap I/\mathfrak{I}(H) \neq \{0\}$ only if $c \in \mathscr{S}$. On the other hand, since $K_c \neq \mathbb{T}$ for $c \in \mathscr{S}$ we see that $I/\mathfrak{I}(H) \cap \mathfrak{I}(c^0) \neq \{0\}$ for all $c \in \mathscr{S}$. This implies that $\mathscr{S}_I = \mathscr{S}$. Finally (5.6) shows that we also recover the compact sets K_c for $c \in \mathscr{S}$.

We now establish that $\varphi \circ \varphi' = id_{\mathscr{L}(C^*(E))}$. To this end, let $I \in \mathscr{L}(C^*(E))$. Then we have $\varphi \circ \varphi'(I) = \varphi(\ (I \cap E^0, \mathscr{S}_I, \{K_c\}_{c \in \mathscr{S}_I})\ )$, where $\mathscr{S}_I$ and K_c are defined as above. Define $H := I \cap E^0$, which is a hereditary saturated subset of E^0.

Set $J = \varphi(\varphi'(I))$. Then $\mathfrak{I}(H) \subseteq I$ and $\mathfrak{I}(H) \subseteq J$. Moreover, by construction we have

$$J/\mathfrak{I}(H) = I/\mathfrak{I}(H) \cong \bigoplus \mathscr{K}_c \otimes C_0(\mathbb{T} \setminus K_c).$$

Using these facts we get that $I = J$, and the proof is complete. □

We conclude this section by discussing aspects of the relationship between ideals of $C^*(E)$ and those of $L_{\mathbb{C}}(E)$. For emphasis, for a graph E we define

$$\mathscr{L}_{HSP}(E, \mathbb{C})$$

to be the lattice $\mathscr{Q}_E$ given in Definition 2.8.6, where the field K is the complex numbers $\mathbb{C}$.

Corollary 5.5.4 *Let E be a row-finite graph, and let $I = \mathfrak{I}(H)$ be a gauge-invariant closed ideal of $C^*(E)$, where H is a hereditary saturated subset of E^0. Then $I \cap L_{\mathbb{C}}(E) = I(H)$, the ideal of $L_{\mathbb{C}}(E)$ generated by H.*

Proof Write $J = \mathfrak{I}(H) \cap L_{\mathbb{C}}(E)$, and observe that $\overline{J} = \mathfrak{I}(H)$. Clearly $I(H) \subseteq J$ and $J \cap E^0 = H = I(H) \cap E^0$. If $J \neq I(H)$ then by Proposition 2.8.11, there is a nonempty set $\mathscr{S}$ and Laurent polynomials $\{p_c\}_{c\in\mathscr{S}}$ such that $(H, \mathscr{S}, \{p_c\}_{c\in\mathscr{S}})$ belongs to $\mathscr{L}_{HSP}(E, \mathbb{C})$ and $J = I(H \cup \{p_c\}_{c\in\mathscr{S}})$. Now $\overline{J}$ is a closed ideal of $C^*(E)$ which strictly contains $\mathfrak{I}(H)$, because it will contain more vertices than $\mathfrak{I}(H)$ if some $p_c(z)$ does not have zeros in $\mathbb{T}$, or otherwise it will correspond to a triple in $\mathscr{L}_{HSK}(E)$ of the form $(H, \mathscr{S}, \{K_c\}_{c\in\mathscr{S}})$ where K_c is the (finite) set of zeros of $p_c(z)$ in $\mathbb{T}$. This would contradict the fact that $\overline{J} = \mathfrak{I}(H)$. □

We can indeed generalize the above corollary in the following way. We consider the extension map

$$e\colon \mathscr{L}_{\mathrm{id}}(L_{\mathbb{C}}(E)) \to \mathscr{L}(C^*(E)), \qquad e(I) = \overline{I}$$

and the restriction map

$$r\colon \mathscr{L}(C^*(E)) \to \mathscr{L}_{\mathrm{id}}(L_{\mathbb{C}}(E)), \qquad r(I) = I \cap L_{\mathbb{C}}(E).$$

These maps define a (monotone) Galois connection, and one can determine the effect of them on the corresponding isomorphic lattices $\mathscr{L}_{HSP}(E, \mathbb{C})$ and $\mathscr{L}_{HSK}(E)$ respectively, as follows.

Corollary 5.5.5 *Let E be a row-finite graph, and denote by*

$$e\colon \mathscr{L}_{HSP}(E, \mathbb{C}) \to \mathscr{L}_{HSK}(E) \quad \text{and} \quad r\colon \mathscr{L}_{HSK}(E) \to \mathscr{L}_{HSP}(E, \mathbb{C})$$

the maps induced by extension and restriction of ideals (closed ideals).

(i) *For* $(H, \mathcal{S}, \{p_c\}_{c\in\mathcal{S}}) \in \mathcal{L}_{HSP}(E,\mathbb{C})$, *we have* $e(H, \mathcal{S}, \{p_c\}_{c\in\mathcal{S}}) = (H', \mathcal{S}', \{K_c\}_{c\in\mathcal{S}'})$, *where* $\mathcal{S}'$ *is the set of those elements* c *in* $\mathcal{S}$ *such that* p_c *has at least one root in* $\mathbb{T}$, H' *is the hereditary saturated closure of* $H \cup \{c^0 \mid c \in \mathcal{S} \setminus \mathcal{S}'\}$, *and* K_c *is the finite nonempty subset of* $\mathbb{T}$ *consisting of the roots of* p_c *lying in* $\mathbb{T}$, *for* $c \in \mathcal{S}'$.
(ii) *For* $(H, \mathcal{S}, \{K_c\}_{c\in\mathcal{S}}) \in \mathcal{L}_{HSK}(E)$, *we have* $r(H, \mathcal{S}, \{K_c\}_{c\in\mathcal{S}}) = (H, \mathcal{S}'', \{p_c\}_{c\in\mathcal{S}''})$, *where* $\mathcal{S}''$ *is the set of those elements* c *in* $\mathcal{S}$ *such that* K_c *is a finite subset of* $\mathbb{T}$, *and for each* $c \in \mathcal{S}''$, p_c *is the unique admissible polynomial having as roots all the elements of* K_c *(with multiplicity one).*

In particular, e *and* r *restrict to an isomorphism between these two partially ordered sets: the poset of triples* $(H, \mathcal{S}, \{p_c\}_{c\in\mathcal{S}}) \in \mathcal{L}_{HSP}(E,\mathbb{C})$ *such that each* p_c, *for* $c \in \mathcal{S}$, *is a polynomial with simple roots, all of them in* $\mathbb{T}$; *and the poset of triples* $(H, \mathcal{S}, \{K_c\}_{c\in\mathcal{S}}) \in \mathcal{L}_{HSK}(E)$ *such that each* K_c, *for* $c \in \mathcal{S}$, *is a finite nonempty subset of* $\mathbb{T}$.

Proof This follows easily from the correspondences we have established previously, together with Corollary 5.5.4. □

5.6 Comparing Properties of Leavitt Path Algebras and Graph C^*-Algebras

Now that a description and discussion of graph C^*-algebras is in hand, we conclude this chapter by describing a clearly very strong, but still very mysterious, connection between various structural properties of the algebras $L_{\mathbb{C}}(E)$ and $C^*(E)$. Our goal here is not to present all currently-known connections, but rather only enough such connections to convince the reader that additional investigation would be both interesting and merited. Specifically, we will focus on these connections in the context of finite graphs. In particular, both $L_{\mathbb{C}}(E)$ and $C^*(E)$ are thereby unital algebras, and, in addition, $C^*(E)$ is separable.

We have already seen an example of one such connection in Theorem 5.3.5, namely, that the monoids $\mathcal{V}(L_K(E))$ and $\mathcal{V}(C^*(E))$ are isomorphic, and that each is isomorphic to the graph monoid M_E. Moreover, by Corollary 5.3.3, the lattice of graded ideals of a Leavitt path algebra is identical to the lattice of closed gauge-invariant ideals of the corresponding graph C^*-algebra. On the other hand, the relationship between the corresponding lattices of all ideals is not as tight, and this is mainly due to the differences arising in the case of the graph $E = R_1$; see Corollary 5.5.5.

Some historical perspective is in order here. While there were a number of articles predating it which discussed structures of a similar nature, the article [47] is generally recognized as the starting point for the study of graph C^*-algebras. By the time Leavitt path algebras made their appearance in the literature in 2005, many of the structural properties of graph C^*-algebras had already been established, including the properties discussed below. One of the two foundational articles on Leavitt path algebras [31] included the description of the $\mathscr{V}$-monoid of both $L_{\mathbb{C}}(E)$ and $C^*(E)$; the information about $C^*(E)$ had theretofore been unknown. On the other hand, the second of the two foundational articles on Leavitt path algebras [5] included the Simplicity Theorem for Leavitt path algebras; both algebraists and analysts took note that the conditions for the simplicity of $L_{\mathbb{C}}(E)$ were identical to the conditions for the simplicity of $C^*(E)$ given in [47]. Indeed, some (but definitely not all) of the subsequent results established for Leavitt path algebras had as their bases the analogous previously-proved results for graph C^*-algebras. It is fair to say that there has been a very satisfactory and fruitful exchange of ideas between the algebraists and analysts, owing to the similarities of some of these results. On the other hand, although many of the results are similar in appearance, there is currently no known vehicle by which the results about one of the structures directly implies the results about the other.

Any C^*-algebra A can be considered from two different points of view: not only is A a ring, but A comes equipped with a topology as well, so that one may also view the ring-theoretic structure of A from a topological/analytic viewpoint. For instance, one may define the (algebraic) simplicity of a C^*-algebra either as a ring (no nontrivial two-sided ideals), or the (topological) simplicity as a topological ring (no nontrivial closed two-sided ideals). In general, the algebraic and topological properties of a given C^*-algebra A need not coincide.

Parts of the following discussion appear in [1].

Isomorphism and Morita Equivalence Perhaps the most basic possible connection between Leavitt path algebras and graph C^*-algebras is this: if E and F are graphs for which the two Leavitt path algebras $L_{\mathbb{C}}(E)$ and $L_{\mathbb{C}}(F)$ are "the same", must the corresponding graph C^*-algebras $C^*(E)$ and $C^*(F)$ also be "the same"? Here "the same" could potentially take on many meanings, for example: isomorphic as rings, isomorphic as $\mathbb{C}$-algebras, isomorphic as $*$-algebras, isomorphic as $\mathbb{Z}$-graded algebras, Morita equivalent, etc. As a first attempt to answer a question of this type, it was established [13, Theorem 8.6] that if E and F are row-finite graphs such that $L_{\mathbb{C}}(E)$ and $L_{\mathbb{C}}(F)$ are simple rings, and if $L_{\mathbb{C}}(E) \cong L_{\mathbb{C}}(F)$ as rings, then $C^*(E) \cong C^*(F)$ as $*$-algebras. It was conjectured in [13] both that a similar conclusion should hold for all Leavitt path algebras over countable graphs, and that a similar conclusion should hold with *isomorphic* replaced by *Morita equivalent*. Five years after the appearance of [13], using deep, powerful tools from both symbolic dynamics and ordered, filtered K-theory, the following significant advance was achieved in this regard.

Theorem ([76, Theorem 14.7]) *Let E and F be graphs with finitely many vertices and at most countably many edges.*

(i) *If $L_{\mathbb{C}}(E) \cong L_{\mathbb{C}}(F)$ as rings, then $C^*(E) \cong C^*(F)$ as $*$-algebras.*
(ii) *If $L_{\mathbb{C}}(E)$ and $L_{\mathbb{C}}(F)$ are Morita equivalent, then $C^*(E)$ and $C^*(F)$ are (strongly) Morita equivalent.*

Simplicity A ring A is called *simple* if A has no nontrivial two-sided ideals; a topological ring is called *simple* if A has no nontrivial closed two-sided ideals.

By the Simplicity Theorem 2.9.7, $L_{\mathbb{C}}(E)$ is simple if and only if E is cofinal and has Condition (L). On the other hand, by Bates et al. [47, Proposition 5.1] (for the case without sources), and [129] (for the general case), $C^*(E)$ is (topologically) simple if and only if E is cofinal and has Condition (L). (It is worth noting here that, by Cuntz [69, p. 215], for any unital C^*-algebra A, A is topologically simple if and only if A is algebraically simple.) Consequently, we get that these are equivalent for any finite graph E:

(1) $L_{\mathbb{C}}(E)$ is simple.
(2) $C^*(E)$ is (topologically) simple.
(3) $C^*(E)$ is (algebraically) simple.
(4) E is cofinal and satisfies Condition (L).

Purely Infinite Simplicity A simple ring R is called *purely infinite simple* if every nonzero left ideal of R contains an infinite idempotent; a simple C^*-algebra A is called purely infinite (simple) if for every positive $x \in A$, the subalgebra $\overline{xAx}$ contains an infinite projection.

By Ara et al. [29, Theorem 1.6] (see also Remark 3.8.4), (algebraic) purely infinite simplicity for unital rings is equivalent to: R is not a division ring, and for all nonzero $x \in R$ there exist $\alpha, \beta \in R$ for which $\alpha x \beta = 1$. On the other hand, by Blackadar [52, Proposition 6.11.5], (topological) purely infinite simplicity for unital C^*-algebras is equivalent to: $A \neq \mathbb{C}$ and for every $x \neq 0$ in A there exist $\alpha, \beta \in A$ for which $\alpha x \beta = 1$. (Remark: Blackadar *defines* purely infinite simplicity this way, and then shows this definition is equivalent to Cuntz' definition given in [70].) Easily, for any graph E, $C^*(E)$ is a division ring if and only if E is a single vertex, in which case $C^*(E) = \mathbb{C}$. Thus we have, for graph C^*-algebras, $C^*(E)$ is (algebraically) purely infinite simple if and only if $C^*(E)$ is (topologically) purely infinite simple.

By the Purely Infinite Simplicity Theorem 3.1.10, $L_{\mathbb{C}}(E)$ is purely infinite simple if and only if $L_{\mathbb{C}}(E)$ is simple, and E has the property that every vertex connects to a cycle. On the other hand, by Bates et al. [47, Proposition 5.3], $C^*(E)$ is (topologically) purely infinite simple if and only if $C^*(E)$ is simple, and E has the property that every vertex connects to a cycle. Consequently, we get that these are equivalent for any finite graph E:

(1) $L_{\mathbb{C}}(E)$ is purely infinite simple.
(2) $C^*(E)$ is (topologically) purely infinite simple.
(3) $C^*(E)$ is (algebraically) purely infinite simple.
(4) E is cofinal, every cycle in E has an exit, and every vertex in E connects to a cycle.

The Exchange Property A ring R is an *exchange ring* if for any $a \in R$ there exists an idempotent $e \in R$ for which $e \in Ra$ and $1 - e \in R(1 - a)$. (Note: The original definition of *exchange ring* was given by Warfield, in terms of a property on direct sum decomposition of modules; this property clarifies the genesis of the name *exchange*. The definition given here is equivalent to Warfield's; this equivalence was shown independently by Goodearl and Warfield in [88, discussion on p. 167], and by Nicholson in [121, Theorem 2.1].) On the other hand, a $*$-ring A with positive definite involution is said to have the *exchange property* in case for every $x > 0$ there exists a projection p such that $p \in Ax$ and $1 - p \in A(1 - x)$. Here $x > 0$ means that x is a finite sum of elements of the form a^*a for $a \in A$, and a projection is a self-adjoint idempotent. (We call this condition "$*$-exchange"; there does not seem to be an explicit definition of "$*$-exchange ring" in the literature.)

By Theorem 3.3.11, $L_{\mathbb{C}}(E)$ is an exchange ring if and only if E satisfies Condition (K). On the other hand, by Jeong and Park [100, Theorem 4.1] $C^*(E)$ has real rank zero if and only if E satisfies Condition (K). Furthermore, by Ara et al. [28, Theorem 7.2], for a unital C^*-algebra A, A has real rank zero if and only if A is a $*$-exchange ring if and only if A is an exchange ring. Consequently, we get that these are equivalent for any finite graph E:

(1) $L_{\mathbb{C}}(E)$ is an exchange ring.
(2) $C^*(E)$ is a $*$-exchange ring.
(3) $C^*(E)$ is an (algebraic) exchange ring.
(4) E satisfies Condition (K).

Primitivity A ring R is (left) primitive if there exists a simple faithful left R-module; a topological ring A is (topologically) primitive if there exists an irreducible faithful $*$-representation of A. (That is, there is a faithful irreducible representation $\pi : A \to B(\mathscr{H})$ for a Hilbert space $\mathscr{H}$.) It is shown in Theorem 4.1.10 that (for row-finite E) $L_{\mathbb{C}}(E)$ is left (and/or right) primitive if and only if E is downward directed and satisfies Condition (L). On the other hand, it is shown in [45, Proposition 4.2] that $C^*(E)$ is (topologically) primitive if and only if E is downward directed and satisfies Condition (L). It is shown in [74, Corollary to Theorem 2.9.5] that a C^*-algebra is algebraically primitive if and only if it is topologically primitive. Consequently, we get that these are equivalent for any finite graph E:

(1) $L_{\mathbb{C}}(E)$ is primitive.
(2) $C^*(E)$ is (topologically) primitive.
(3) $C^*(E)$ is (algebraically) primitive.
(4) E is downward directed and satisfies Condition (L).

(We note that the first three properties have been shown to be equivalent for arbitrary graphs as well, with the fourth condition being replaced by: E is downward directed, satisfies Condition (L), and has the Countable Separation Property. See Theorems 7.2.5 and 7.2.7 below.)

It is interesting to observe that for the properties discussed above (simplicity, purely infinite simplicity, exchange, and primitivity), the algebraic and topological conditions on $C^*(E)$ are identical. Perhaps there is something in this observation which will lead to a deeper understanding of why there seems to be such a strong relationship between these properties of $L_{\mathbb{C}}(E)$ and $C^*(E)$.

There are situations where the analogies between the Leavitt path algebras and graph C^* algebras are not as tight as those presented above. We have already mentioned one: the comparison of the ideal lattice of $L_{\mathbb{C}}(E)$ with the (closed) ideal lattice of $C^*(E)$. We discuss two more of those now: primeness and stable rank.

We will discuss others in Chap. 6, including questions about tensor products (see Sect. 6.4).

Even in these situations, much similarity between the algebras $L_{\mathbb{C}}(E)$ and $C^*(E)$ remains. Indeed, oftentimes the only differences in the relationships occur with respect to graphs containing cycles without exits, e.g., the graph R_1.

Primeness A ring R is called *prime* if $\{0\}$ is a prime ideal of R; that is, if for any two-sided ideals I, J of R, $I \cdot J = \{0\}$ if and only if $I = \{0\}$ or $J = \{0\}$. A C^*-algebra A is *prime* if $\{0\}$ is a prime ideal of A; that is, if for any *closed* two-sided ideals I, J of R, $I \cdot J = \{0\}$ if and only if $I = \{0\}$ or $J = \{0\}$.

By Proposition 4.1.5, $L_{\mathbb{C}}(E)$ is prime if and only if E is downward directed. But by Dixmier [73, Corollaire 1], any separable C^*-algebra is (topologically) prime if and only if it is (topologically) primitive. So (for finite E) $C^*(E)$ is prime if and only if $C^*(E)$ is primitive, which as mentioned directly above happens if and only if E is downward directed *and* satisfies Condition (L). (We note that since $I \cdot J = \{0\}$ implies $\overline{I} \cdot \overline{J} = \{0\}$, it is straightforward to show that A is algebraically prime if and only if A is analytically prime.)

So, for example, if $E = R_1$ is the graph with one vertex and one loop, then $L_{\mathbb{C}}(E) \cong \mathbb{C}[x, x^{-1}]$ is prime (clearly, as it is an integral domain), but $C^*(E) \cong C(\mathbb{T})$ is not prime (it is not difficult to write down nonzero orthogonal continuous functions on the unit circle $\mathbb{T}$.)

Specifically, we see that in situations where E satisfies Condition (L), then primeness of $L_{\mathbb{C}}(E)$ is equivalent to primeness of $C^*(E)$ (because in each case these are equivalent to primitivity).

Stable Rank The definition of the *stable rank* $\mathrm{sr}(R)$ of a ring R is given in Definitions 4.4.1. The *topological stable rank* of Banach algebras was introduced by Rieffel in his seminal paper [133]. It was shown by Herman and Vaserstein [95] that the topological stable rank $\mathrm{tsr}(A)$ coincides with the ring-theoretic (a.k.a. 'Bass') stable rank $\mathrm{sr}(A)$.

The value of the stable rank of $L_{\mathbb{C}}(E)$ for all possible configurations of the graph E is given in Theorem 4.4.19. On the other hand, the value of the stable rank of $C^*(E)$ for all possible configurations of the graph E is given in [72, Proposition 3.1 and Theorem 3.4], to wit:

- $\mathrm{sr}(C^*(E)) = 1$ if no cycle in E has an exit (i.e., has Property (NE));
- $\mathrm{sr}(C^*(E)) = \infty$ if there exists an $H \in \mathscr{H}_E$ such that the quotient graph E/H is nonempty, finite, cofinal, contains no sinks and each cycle has an exit; and
- $\mathrm{sr}(C^*(E)) = 2$ otherwise.

Consequently, if E is not acyclic and has property (NE), then $\mathrm{sr}(L_{\mathbb{C}}(E)) = 2$ by Theorem 4.4.19, but $\mathrm{sr}(C^*(E)) = 1$ by the above-quoted result from [72]. As in the primeness discussion above, the simplest example of this situation is the graph $E = R_1$. As noted in Example 4.4.21(iii), $L_{\mathbb{C}}(R_1) \cong \mathbb{C}[z, z^{-1}]$ has $\mathrm{sr}(L_{\mathbb{C}}(R_1)) = 2$. To explicitly show why $\mathrm{sr}(L_{\mathbb{C}}(R_1)) > 1$, observe that $(1+z)\mathbb{C}[z, z^{-1}] + (1+z^2)\mathbb{C}[z, z^{-1}] = \mathbb{C}[z, z^{-1}]$. It is straightforward to see that there is no element $v \in \mathbb{C}[z, z^{-1}]$ such that $(1+z) + v(1+z^2)$ is invertible in $\mathbb{C}[z, z^{-1}]$, i.e., that there is a 2-unimodular row which is not reducible. On the other hand, the completion $C^*(R_1) \cong C(\mathbb{T})$ of $L_{\mathbb{C}}(R_1)$ has stable rank 1. So necessarily there exists a $v \in C^*(E)$ such that $(1+z) + v(1+z^2)$ is invertible in $C(\mathbb{T})$. Since a (continuous) function in $C(\mathbb{T})$ is invertible if and only if it has no zeros in $\mathbb{T}$, we see that we can take $v = 1$.

Chapter 6
K-Theory

In this chapter we focus on a number of K-theoretic properties of $L_K(E)$. In Sect. 6.1 we focus on the relationship between the monoid $\mathscr{V}(L_K(E))$ and the Grothendieck group $K_0(L_K(E))$; in particular, we realize $K_0(L_K(E))$ as the cokernel of an appropriate linear transformation between free abelian groups. In the subsequent Sect. 6.2 we describe the Whitehead group $K_1(L_K(E))$, and show that its description is quite closely related to the description of $K_0(L_K(E))$. In Sect. 6.3 we present in great detail the Restricted Algebraic Kirchberg–Phillips Theorem, and the still open (as of 2017) Algebraic Kirchberg–Phillips Question. It is not hyperbolic to say that this question has been, and continues to be, at the heart of a substantial portion of the research effort in the subject. We finish the chapter with Sect. 6.4, in which we describe various properties of tensor products of Leavitt path algebras in the wider context of Hochschild homology.

For additional background on K-theoretic concepts, see e.g., [159].

6.1 The Grothendieck Group $K_0(L_K(E))$

In this first section of Chap. 6 we completely describe the group $K_0(L_K(E))$, where E is a row-finite graph and K is any field. We start the section by giving an overview of the groups $K_0(R)$ for a general ring R, and then subsequently focus on the case $R = L_K(E)$.

Let M be an abelian monoid, written additively; that is, $(M, +)$ is a set with an associative, commutative binary operation $+$, for which there is an element 0 having $0+m = m+0 = m$ for all $m \in M$. The goal is to associate M with an abelian group $G = G(M)$ in a natural, universal way. Intuitively, this should be done by "adding inverses when necessary"; for instance, if $M = \mathbb{Z}^+$, then the appropriate group G is simply $\mathbb{Z}$. Moreover, if M is already a group, then $G(M)$ should just be M itself. The main issue that arises in this process is in the situation where M is not cancellative

G. Abrams et al., *Leavitt Path Algebras*, Lecture Notes in Mathematics 2191,
DOI 10.1007/978-1-4471-7344-1_6

(i.e., there exist $a, b, c \in M$ for which $a \neq b$ but $a + c = b + c$). In this situation M clearly cannot be embedded in a group.

Formally, for any abelian monoid M there exists a universal (abelian) group $G(M)$, having the following property: there exists a monoid homomorphism $\varphi : M \to G(M)$ such that, for every abelian group G', and every monoid homomorphism $\psi : M \to G'$, there exists a unique group homomorphism $\delta : G(M) \to G'$ for which $\psi = \delta \circ \varphi$. In general φ need not be an injection. There is an explicit construction of $G(M)$, as follows. Define an equivalence relation $\sim$ on $M \times M$ by setting $(m_1, m_2) \sim (n_1, n_2)$ if there exists a $k \in M$ for which $m_1 + n_2 + k = m_2 + n_1 + k$. Let $G(M)$ denote the equivalence classes in $M \times M$ under $\sim$ (we denote an individual class by $[\ ]_0$), and define $+$ on $G(M)$ as expected: $[(m_1, m_2)]_0 + [(n_1, n_2)]_0 = [(m_1 + n_1, m_2 + n_2)]_0$. It is straightforward to check that this operation is well defined, and that $(G(M), +)$ is indeed an abelian group. Specifically, the identity of $G(M)$ is $[(m, m)]_0$ (for any $m \in M$); the inverse of $[(m, n)]_0 \in G(M)$ is $[(n, m)]_0$; and the monoid homomorphism $\varphi : M \to G(M)$ is given by $\varphi(m) = [(m, 0)]_0$. The image of φ is called the *positive cone* of $G(M)$. Effectively, the construction of the group $G(M)$ takes care of any lack of cancellation in M by ensuring that if $a + c = b + c$ in M for $a \neq b$, then $\varphi(a) = \varphi(b)$ in $G(M)$.

Example 6.1.1 If $M = (\mathbb{Z}^+)^n$ (the direct sum of n copies of $\mathbb{Z}^+$), then it is easy to show directly that $G(M) \cong \mathbb{Z}^n$.

Remark 6.1.2 Of special importance in the context of Leavitt path algebras is the following example (cf. Examples 3.2.2(i)). Consider the monoid $M = \{0, x, 2x, \ldots, (n-1)x\}$ with obvious operation $+$ and relation $nx = x$. Then the subset $S = M \setminus \{0\}$ is closed under $+$. But, since $x + (n-1)x = nx = x$ in S, we see that S is indeed a *group* under $+$ (with identity element $(n-1)x$), specifically, $S \cong \mathbb{Z}/(n-1)\mathbb{Z}$. In this situation, it is not hard to show that $S \cong G(M)$.

This phenomenon happens more generally.

Proposition 6.1.3 *Suppose $(M, +)$ is an abelian monoid with the property that $S = M \setminus \{0\}$ is a group (under the same operation $+$). Then $S \cong G(M)$.*

Proof Let $e \in S$ denote the presumed identity element in S.

We claim that for each element $(m_1, m_2) \in M \times M$ there exists a unique element $x \in S$ for which $[(m_1, m_2)]_0 = [(x, 0)]_0$. There are three cases to establish the existence. First, if $m_2 \in S$ then, because S is a group, there exists an $s \in S$ with $s + m_2 = e$. Defining $x = s + m_1 \in S$, we have $x + m_2 = s + m_1 + m_2 = e + m_1 = m_1$ in S, so that $(m_1, m_2) \sim (x, 0)$ (using any element k of M in the definition of $\sim$). Second, if $m_2 = 0$ and $m_1 \neq 0$ then the result is clear. Finally, $(0, 0) \sim (e, 0)$, as $0 + 0 + e = 0 + e + e$ in S (since e is the identity element of S).

For uniqueness, if $x, y \in S$ with $(x, 0) \sim (y, 0)$ then $x + 0 + k = y + 0 + k$ for some $k \in M$. If $k = 0$ then $x = y$; if on the other hand $k \in S$ then by hypothesis there exists an $\ell \in S$ with $k + \ell = e$; and by adding ℓ to both sides we get $x + 0 + e = y + 0 + e$, so that again $x = y$. □

We note that in the above situation the monoid homomorphism $\varphi : M \to G(M) = M \setminus \{0\}$ is given by $\varphi(m) = [(m, 0)]_0$ for $m \in S$, and $\varphi(0) = [(e, 0)]_0$.

Definition 6.1.4 Recall that, as noted in Sect. 3.2, for any unital ring R we denote by $\mathscr{V}(R)$ the monoid of isomorphism classes of finitely generated projective left R-modules, with operation $\oplus$. The *Grothendieck group* $K_0(R)$ of a unital ring R is the universal group $G(\mathscr{V}(R))$.

As is standard, if R is unital we denote the equivalence class of the left regular module R in $K_0(R)$ by $[1_R]_0$.

Notation 6.1.5 In the construction of $K_0(R)$ as the universal group of the monoid $\mathscr{V}(R)$ there are two equivalence relations in play: the isomorphism relation in $\mathscr{V}(R)$, and the relation described above which yields the equivalence classes in $K_0(R)$. We distinguish these two types of equivalence classes notationally, by writing [ ] to denote elements of $\mathscr{V}(R)$, and writing $[\ ]_0$ to denote elements of $K_0(R)$.

Combining the previous observations with Example 3.2.6, we get

Corollary 6.1.6 *Let K be any field, and $2 \leq n \in \mathbb{N}$. Then $K_0(L_K(R_n)) \cong \mathbb{Z}/(n-1)\mathbb{Z}$. Moreover, under this isomorphism, $[1_{L_K(R_n)}]_0 \mapsto \overline{1}$.*

In general, from a slightly different point of view, when R is unital then $K_0(R)$ is the group F/S, where F is the free abelian group (written additively) generated by the isomorphism classes of finitely generated projective left R-modules, and S is the subgroup of F generated by symbols of the form $[P \oplus Q]_0 - [P]_0 - [Q]_0$. From this perspective, one can show that $[A]_0 = [B]_0$ as elements of $K_0(R)$ (where A and B are finitely generated projective left R-modules) precisely when A and B are *stably isomorphic*, i.e, when there exists a positive integer n for which $A \oplus R^n \cong B \oplus R^n$.

We briefly remind the reader of some basic properties of K_0 for general (unital) rings. (See [86, Chap. 15] for a discussion of these, and additional properties.)

(i) $K_0(K) \cong \mathbb{Z}$ for any division ring K.
(ii) If R is a unital ring with Jacobson radical J, then the maps $\mathscr{V}(R) \to \mathscr{V}(R/J)$ and $K_0(R) \to K_0(R/J)$ are both injective. This follows from Bass' Theorem [107, Lemma 19.27]. In particular, it follows that $\mathscr{V}(R) \cong \mathbb{Z}^+$ and $K_0(R) \cong \mathbb{Z}$ for any local ring R.
(iii) K_0 is preserved under Morita equivalence; that is, if R and S are Morita equivalent rings, then $K_0(R) \cong K_0(S)$.
(iv) K_0 preserves direct sums: for rings $\{R_i | i \in I\}$, $K_0(\oplus_{i\in I} R_i) \cong \oplus_{i\in I} K_0(R_i)$.
(v) Generalizing the previous item, if $(\{R_i | i \in I\}, \{\varphi_{i,j}\})$ is a direct system of rings and ring homomorphisms, then $K_0(\varinjlim(R_i, \varphi_{i,j})) = \varinjlim(K_0(R_i), K_0(\varphi_{i,j}))$.

If I is a non-unital ring, then $K_0(I)$ is defined as the kernel of the canonical map $K_0(\pi): K_0(I^1) \to K_0(\mathbb{Z})$, where $\pi: I^1 \to \mathbb{Z}$ is the projection from the unitization $I^1 = I \oplus \mathbb{Z}$ of I onto $\mathbb{Z}$. In this case, the monoid $\mathscr{V}(I)$ has already been defined in Sect. 3.2 as the monoid of isomorphism classes of finitely generated projective modules in $FP(I, R)$, where R is any unital ring containing I as an ideal. (For instance, we can take $R = I^1$.) There is a natural map $G(\mathscr{V}(I)) \to K_0(I)$, which is neither injective nor surjective in general. However, as already remarked in Sect. 3.2, this map is an isomorphism if I is a ring with local units. In particular,

$K_0(L_K(E)) \cong G(\mathcal{V}(L_K(E)))$ for any graph E. Property (v) above also holds in the context of non-unital rings.

Remark 6.1.7 We now remind the reader of some additional properties of K_0, especially those that are most relevant in the current context.

(i) For unital K-algebras T and T', if T and T' are isomorphic then there exists an induced isomorphism $\varphi : K_0(T) \to K_0(T')$ for which $\varphi([1_T]_0) = [1_{T'}]_0$.
(ii) Let S denote the K-algebra $L_K(R_n)$. We note that in the isomorphism $K_0(S) \cong \mathbb{Z}/(n-1)\mathbb{Z}$ established in Corollary 6.1.6, the identity element of the group $K_0(S)$ is the element $[S^{(n-1)}]_0$. On the other hand, the element $[S]_0$ of $K_0(S)$ corresponds to the generator $\overline{1}$ of $\mathbb{Z}/(n-1)\mathbb{Z}$.
(iii) More generally, let d be any positive integer, and T any unital ring. Let T' denote the matrix ring $\mathrm{M}_d(T)$. Since T is Morita equivalent to T' we necessarily have $K_0(T) \cong K_0(T')$. However, in general this isomorphism need not take $[T]_0$ to $[T']_0$; indeed, the element $[T]_0$ of $K_0(T)$ is taken to the element $[T'e_{1,1}]_0$ of $K_0(T')$, while $[T']_0$ corresponds to $d[T'e_{1,1}]_0$ in $K_0(T')$.
(iv) In a situation which will be of interest in the sequel, we consider $T = L_K(1,n)$ and $T' = \mathrm{M}_d(L_K(1,n))$. Then $K_0(L_K(1,n)) \cong K_0(\mathrm{M}_d(L_K(1,n)))$, via an isomorphism which takes $\overline{1}$ of $\mathbb{Z}/(n-1)\mathbb{Z}$ to $\overline{d}$ of $\mathbb{Z}/(n-1)\mathbb{Z}$. From this, we see that there exists *some* group isomorphism from $K_0(L_K(1,n))$ to $K_0(\mathrm{M}_d(L_K(1,n)))$ which takes $[L_K(1,n)]_0$ to $[\mathrm{M}_d(L_K(1,n))]_0$ if and only if $\overline{d}$ is a generator of $\mathbb{Z}/(n-1)\mathbb{Z}$, i.e., if and only if g.c.d.$(d, n-1) = 1$.

In Theorem 3.2.5 we established the following. Let E be a row-finite graph. Let M_E be the abelian monoid with generators $\{a_v \mid v \in E^0\}$, and with relations given by setting, for each non-sink v of E, $a_v = \sum_{e \in s^{-1}(v)} a_{r(e)}$. Then $\mathcal{V}(L_K(E)) \cong M_E$. Specifically, we have an explicit description of the monoid $\mathcal{V}(L_K(E))$ as the monoid $\oplus_{v \in E^0} \mathbb{Z}^+$, modulo the indicated relations, where $a_v \in \mathcal{V}(L_K(E))$ corresponds to the element z_v of $\oplus_{v \in E^0} \mathbb{Z}^+$ consisting of 1 in the v-coordinate, 0 elsewhere. We consider now the factor group $(\oplus_{v \in E^0} \mathbb{Z})/T$, where T is the subgroup of $\oplus_{v \in E^0} \mathbb{Z}$ generated by the elements $\{z_v - \sum_{e \in s^{-1}(v)} z_{r(e)} \mid v \in \mathrm{Reg}(E)\}$. It is clear that whenever we have an abelian monoid defined by a presentation with generators and relations, then its universal group is the group defined by the same presentation. Using this observation, we immediately obtain the following result.

Theorem 6.1.8 *Let E be a row-finite graph and K any field. Let T denote the subgroup of $\oplus_{v \in E^0} \mathbb{Z}$ generated by the set $\{z_v - \sum_{e \in s^{-1}(v)} z_{r(e)} \mid v \in \mathrm{Reg}(E)\}$. Consider the monoid homomorphism*

$$\varphi : \mathcal{V}(L_K(E)) \to (\oplus_{v \in E^0} \mathbb{Z})/T$$

given by sending a_v to $z_v + T$ for each $v \in E^0$. Then

$$(\oplus_{v \in E^0} \mathbb{Z})/T \cong G(\mathcal{V}(L_K(E))) = K_0(L_K(E)),$$

and φ can be identified with the canonical map $\mathcal{V}(L_K(E)) \to K_0(L_K(E))$.

We now present the standard matrix interpretation of Theorem 6.1.8. As usual, let $A_E = (a_{i,j})$ denote the adjacency matrix of E. We let $A_{ns}(E)$ (or more compactly A_{ns} when E is clear) denote the matrix A_E with the zero-rows removed; that is, A_{ns} is the (non-square) matrix obtained from A_E by removing the rows corresponding to the sinks of E. Similarly, we denote by I_{ns} the matrix obtained by taking the $E^0 \times E^0$ identity matrix I and deleting the rows corresponding to the sinks of E. A moment's reflection yields that, for each element of $\oplus_{v \in E^0}\mathbb{Z}$ (viewed as a column vector) of the form $z_v - \sum_{e \in s^{-1}(v)} z_{r(e)}$ where $v \in \mathrm{Reg}(E)$, we have

$$z_v - \sum_{e \in s^{-1}(v)} z_{r(e)} = (I_{ns} - A_{ns})^t z_v ,$$

where as usual $(\ )^t$ denotes the transpose of a matrix. The upshot is that the subgroup T may be realized as the image of the linear transformation $(I_{ns} - A_{ns})^t : \oplus_{v \in \mathrm{Reg}(E)}\mathbb{Z} \to \oplus_{v \in E^0}\mathbb{Z}$, viewed as left multiplication on columns. In other words, we may restate Theorem 6.1.8 as follows.

Theorem 6.1.9 *Let E be a row-finite graph. Then*

$$K_0(L_K(E)) \cong \mathrm{Coker}((I_{ns} - A_{ns})^t : \oplus_{v \in \mathrm{Reg}(E)}\mathbb{Z} \to \oplus_{v \in E^0}\mathbb{Z}).$$

Corollary 6.1.10 *Let E be a finite graph containing no sinks. (As a specific case, by Theorem 3.1.10, we may suppose $L_K(E)$ is purely infinite simple unital.) Let $|E^0| = n$. Then*

$$K_0(L_K(E)) \cong \mathrm{Coker}(I_n - A_E^t : \oplus_{v \in E^0}\mathbb{Z} \to \oplus_{v \in E^0}\mathbb{Z}).$$

Examples 6.1.11

(i) Let E be a finite acyclic graph having s sinks. In Theorem 2.6.17 it is shown that $L_K(E)$ is isomorphic to a direct sum of s rings, each of which is a full matrix ring over K. Thus, using the aforementioned basic properties of K_0, we get that $K_0(L_K(E)) \cong \mathbb{Z}^s$.

(ii) It is well known that $K_0(K[x, x^{-1}]) \cong \mathbb{Z}$. (Indeed, this result also follows from an application of Theorem 3.2.5, since in this case we get $\mathcal{V}(K[x, x^{-1}]) = \mathbb{Z}^+$.) Similarly, let E be a finite graph having Condition (NE). Let m denote the number of (necessarily disjoint) cycles of E, and let s denote the number of sinks of E. Then $K_0(L_K(E)) \cong \mathbb{Z}^{m+s}$.

(iii) Let E_3 denote the (sink-free) graph 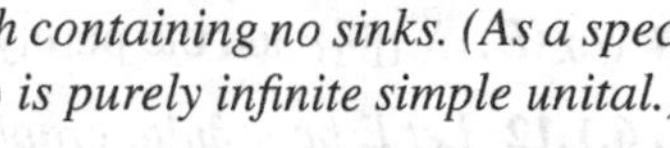 . Then $A_E = \begin{pmatrix} 1 & 1 & 0 \\ 1 & 0 & 1 \\ 0 & 1 & 1 \end{pmatrix}$, so that $I_3 - A_{E_3}^t = \begin{pmatrix} 0 & -1 & 0 \\ -1 & 1 & -1 \\ 0 & -1 & 1 \end{pmatrix}$. Using this description, it is easy to show that the image of the linear transformation from $\mathbb{Z}^3$ to $\mathbb{Z}^3$ given by left multiplication by $I_3 - A^t_{E_3}$ is generated by the column vectors $\begin{pmatrix}0\\1\\0\end{pmatrix}$ and $\begin{pmatrix}1\\0\\1\end{pmatrix}$, which in turn easily yields that the cokernel of this transformation is isomorphic to $\mathbb{Z}$. By Corollary 6.1.10 this gives that $K_0(L_K(E_3)) \cong \mathbb{Z}$. Moreover, under this isomorphism, $[1_{L_K(E_3)}]_0 \in K_0(L_K(E_3)) \mapsto 0 \in \mathbb{Z}$.

(iv) Let E_4 denote the (sink-free) graph $\bullet \rightleftarrows \bullet \rightleftarrows \bullet \rightleftarrows \bullet$ (each vertex with a loop). Then $A_{E_4} = \begin{pmatrix}1\,1\,0\,0\\1\,1\,1\,0\\0\,1\,1\,1\\0\,0\,1\,1\end{pmatrix}$, so that $I_4 - A^t_{E_4} = \begin{pmatrix}0&-1&0&0\\-1&0&-1&0\\0&-1&0&-1\\0&0&-1&0\end{pmatrix}$. Using this description, it is easy to show that the image of the linear transformation from $\mathbb{Z}^4$ to $\mathbb{Z}^4$ given by left multiplication by $I_4 - A^t_{E_4}$ is all of $\mathbb{Z}^4$, so that the cokernel of the transformation is $\{0\}$, which gives by Corollary 6.1.10 that $K_0(L_K(E_4)) \cong \{0\}$.

We conclude this section by demonstrating a close connection between the semigroup $\mathcal{V}(L_K(E)) \setminus \{[0]\}$ and the purely infinite simplicity of $L_K(E)$.

Proposition 6.1.12 *Let E be a finite graph and K any field. Then $L_K(E)$ is purely infinite simple if and only if $\mathcal{V}(L_K(E)) \setminus \{[0]\}$ is a group. Moreover, in this situation we have $\mathcal{V}(L_K(E)) \setminus \{[0]\} = K_0(L_K(E))$.*

Proof ($\Rightarrow$) This follows for any unital ring R by Ara et al. [29, Proposition 2.1]. Indeed, the result is proved by observing that for any purely infinite simple ring R, given any two elements x, y in $\mathcal{V}(R) \setminus \{[0]\}$, there exist a, b in $\mathcal{V}(R) \setminus \{[0]\}$ such that $x = y + a$ and $y = x + b$. It is easy to show that this implies that $\mathcal{V}(R) \setminus \{[0]\}$ is a group.

($\Leftarrow$) Let g be a nonzero idempotent in $L_K(E)$. Then g is infinite, as follows. We have $[L_K(E)g] \in \mathcal{V}(L_K(E)) \setminus \{[0]\}$, and by hypothesis there exists a nonzero finitely generated projective left $L_K(E)$-module P (specifically, the identity element of the presumed group $\mathcal{V}(L_K(E)) \setminus \{[0]\}$), for which $[L_K(E)g] = [L_K(E)g] \oplus [P]$, so that $L_K(E)g \cong L_K(E)g \oplus P$ as left R-modules.

In particular, this shows that every vertex v of E^0 is infinite, so by Corollary 3.5.5 we conclude that E satisfies Condition (L). So Proposition 2.9.13 gives that every nonzero left ideal of $L_K(E)$ contains a nonzero idempotent. But every nonzero idempotent of $L_K(E)$ is infinite by the previous observation. So every nonzero left ideal of $L_K(E)$ contains an infinite idempotent.

To conclude the proof we need only show that $L_K(E)$ is a simple ring. Pick any nonzero two-sided ideal I of $L_K(E)$; we show that $I = L_K(E)$. Arguing as above, we get that I contains a nonzero idempotent (call it g), and that there exists a nonzero finitely generated projective left $L_K(E)$-module Q for which $L_K(E)g \cong L_K(E) \oplus Q$.

In particular, there is an element $r \in L_K(E)$ and a left $L_K(E)$-module homomorphism $\varphi : L_K(E)g \to L_K(E)$ for which $(rg)\varphi = 1_{L_K(E)}$. But then by standard arguments this yields that there exists an $x \in gL_K(E)$ for which $rgx = 1_{L_K(E)}$. So we have $RgR = R$, so that R is simple, as desired.

The final statement follows from Proposition 6.1.3. □

Remark 6.1.13 Suppose R is a purely infinite simple ring. Then Proposition 6.1.12 together with Proposition 6.1.3 imply that we may view $K_0(R)$ as a submonoid of $\mathcal{V}(R)$. In particular, in this situation, if $[A]_0 = [B]_0$ as elements of $K_0(R)$, and neither A nor B is the zero module, then $[A] = [B]$ as elements of $\mathcal{V}(R) \setminus \{[0]\}$, i.e., $A \cong B$ as (nonzero) left R-modules. (In other words, in this situation, "stable isomorphism implies isomorphism".)

Thus when R is a purely infinite simple ring and A is a nonzero finitely generated projective left R-module, we have the choice to denote the element $[A]_0$ of $K_0(R)$ either using the $[A]_0$ notation, or the $[A]$ notation. For convenience we will typically use the latter.

We note that the 'only if' part of Proposition 6.1.12 does not hold for general rings. For instance, consider the ring $B(\mathcal{H})$ of bounded operators on a separable Hilbert space $\mathcal{H}$, and let R be the ring $B(\mathcal{H})/F(\mathcal{H})$, where $F(\mathcal{H})$ denotes the ideal of finite rank operators. Then R is not simple, because the Jacobson radical of R is the nonzero ideal $K(\mathcal{H})/F(\mathcal{H})$, where $K(\mathcal{H})$ denotes the compact operators. Since the natural map $\eta : \mathcal{V}(R) \to \mathcal{V}(B(\mathcal{H})/K(\mathcal{H}))$ is injective, and $\mathcal{V}(B(\mathcal{H})/K(\mathcal{H})) = \{0\} \cup \{[1_R]\}$, it follows that η is indeed an isomorphism. Thus R is a non-simple ring for which $\mathcal{V}(R) \setminus \{0\}$ is a group.

6.2 The Whitehead Group $K_1(L_K(E))$

Having established an explicit description of the Grothendieck group $K_0(L_K(E))$ of a Leavitt path algebra in the previous section, we now turn our attention to the Whitehead group $K_1(L_K(E))$.

Definition 6.2.1 For each unital ring R and positive integer n we consider $\mathrm{GL}_n(R)$, the group of invertible $n \times n$ matrices over R. Clearly $\mathrm{GL}_n(R)$ embeds in $\mathrm{GL}_{n+1}(R)$, via the assignment $M \mapsto \begin{pmatrix} M & 0 \\ 0 & 1 \end{pmatrix}$. In this way we may form the group $\varinjlim_{n\in\mathbb{N}} \mathrm{GL}_n(R)$, which is denoted by $\mathrm{GL}(R)$. For any group G (written multiplicatively), the commutator subgroup $[G, G]$ is the (necessarily normal) subgroup of G generated by elements of the form $xyx^{-1}y^{-1}$ for $x, y \in G$.

We define $K_1(R)$ to be the abelian group

$$K_1(R) = GL(R)/[GL(R), GL(R)].$$

$K_1(R)$ is often called the *Whitehead group of* R. If R is non-unital, we let $R^1 = R \oplus \mathbb{Z}$ be the standard unitization of R, and define $K_1(R)$ to be the kernel of the map $K_1(\pi)\colon K_1(R^1) \to K_1(\mathbb{Z})$ induced by the canonical projection $\pi\colon R^1 \to \mathbb{Z}$.

Recall that for any unital ring R we denote the group of units of R by $R^\times$.

Examples 6.2.2 Although the indicated definition of $K_1(R)$ is relatively straightforward, computing $K_1(R)$ in specific situations is typically a highly nontrivial task.

(i) If K is a field, then $K_1(K) \cong K^\times$. There are a number of ways to establish this result (none of which is immediate), including the utilization of an old linear algebra result of Dickson which shows that (except for two specific exceptions) we have, for each n, $[GL_n(K), GL_n(K)] = SL_n(K)$ (where $SL_n(K)$ denotes the $n \times n$ matrices over K of determinant 1). The generalization of this result to division rings D was established by Dieudonné: $K_1(D) = D^\times/[D^\times, D^\times]$.

(ii) If R is a purely infinite simple unital ring, then by Ara et al. [29, Theorem 2.3] we have $K_1(R) \cong R^\times/[R^\times, R^\times]$. We will show below, in the case when $R = L_K(E)$ is a purely infinite simple Leavitt path algebra, how to describe this group explicitly in terms of E and K. Recall (Remark 3.8.4) that if R is a unital ring having the property that for each $0 \neq r \in R$ there exist $x, y \in R$ with $xry = 1$, then R is either a division ring or R is purely infinite simple. So the result [29, Theorem 2.3] can in a sense be viewed as an extension of Dieudonné's result for division rings mentioned in the previous item.

(iii) Of clear interest in the current context is $K_1(K[x, x^{-1}])$. As shown originally by Bass et al. [44, Corollary 3 to Theorem 2], $K_1(K[x, x^{-1}]) \cong K^\times \oplus \mathbb{Z}$. A generalized version of this result will be utilized to achieve Theorem 6.2.4.

If $M = (m_{i,j})$ is an $m \times n$ integer-valued matrix and R is any unital ring, then M induces a homomorphism of groups

$$M : \prod_{i=1}^{n} R^\times \to \prod_{i=1}^{m} R^\times$$

given by exponentiation. Specifically, if $M = (m_{i,j}) \in \mathrm{M}_{m\times n}(\mathbb{Z})$ and $\rho = (r_t) \in \prod_{i=1}^{n} R^\times$, then for each $1 \le i \le m$ the ith entry in $M \cdot \rho$ is given by

$$(M \cdot \rho)_i = \prod_{j=1}^{n} r_j^{m_{i,j}}.$$

This group homomorphism will play an important role in the description of $K_1(L_K(E))$, where M will be the matrix $(I_{ns} - A_{ns})^t$ described in the previous section.

Remark 6.2.3

(i) There is an alternate definition of $K_1(R)$ which starts by considering a category with objects equal to the elements of the monoid $\mathcal{V}(R)$, and with appropriately defined morphisms. It follows almost immediately from this alternate definition that if R and S are Morita equivalent rings, then $K_1(R) \cong K_1(S)$. (See e.g., [159, Proposition III.1.6.4]. This isomorphism may also be shown using Definition 6.2.1 as a starting point, but the required argument is more intricate.) In particular, by Examples 6.2.2(1), if K is a field then $K_1(\mathrm{M}_m(K)) \cong K^\times$.

(ii) It is not hard to see that K_1 preserves direct sums: for rings $\{R_i | i \in I\}$, $K_1(\oplus_{i\in I} R_i) \cong \oplus_{i\in I} K_1(R_i)$. In particular, $K_1(\oplus_{i=1}^n \mathrm{M}_{m_i}(K)) \cong \prod_{i=1}^n K^\times$.

We describe now the steps which will allow us to achieve a description of $K_1(L_K(E))$ in the situation where E is a finite graph having no sources. (We will subsequently comment on the situation for more general graphs.) For a Leavitt path algebra $L_K(E)$, the structure of the zero-component $L_K(E)_0$ was explicitly given in Corollary 2.1.16. To wit, $L_K(E)_0$ is built as a direct limit of K-algebras, each of which is a direct sum of full matrix rings over K. By Remark 6.2.3(ii), we would therefore anticipate that achieving an explicit description of $K_1(L_K(E)_0)$ is plausible, and that the group $K^\times$ should play a key role. We then show that $K_1(L_K(E))$ can be built from $K_1(L_K(E)_0)$ and $K_0(L_K(E)_0)$ by viewing $L_K(E)$ as a skew Laurent polynomial ring over $L_K(E)_0$ (see [25]).

Specifically, we have from [25, Lemma 2.4] that, if E has no sources, then $L_K(E)$ is a skew Laurent polynomial ring over $L_K(E)_0$, as follows. For each vertex v_i ($1 \le i \le d$) of E let e_i denote an edge for which $r(e_i) = v_i$ (that such e_i exist requires the no-source hypothesis). Let t_+ denote $\sum_{i=1}^d e_i$, and let t_- denote $t_+^* = \sum_{i=1}^d e_i^*$. It follows easily that $t_- t_+ = \sum_{i=1}^d v_i = 1_{L_K(E)}$, and that $p = t_+ t_-$ is an idempotent in $L_K(E)$. Then $L_K(E) = L_K(E)_0[t_+, t_-, \phi]$, where $\phi : L_K(E)_0 \to pL_K(E)_0 p$ is the corner isomorphism given by $\phi(b) = t_+ b t_-$ for all $b \in L_K(E)_0$.

Let A be a unital K-algebra with automorphism α. There is an elegant result of Siebenmann [141] which connects various K-theoretic information of A to K-theoretic information of the skew-ring (A, α):

$$K_1(A) \xrightarrow{1-\alpha_*} K_1(A) \xrightarrow{j} K_1(A, \alpha) \xrightarrow{p} K_0(A) \xrightarrow{1-\alpha_*} K_0(A).$$

(The group homomorphism α_* is induced by the ring automorphism α in an easily described way.) The group $K_1(A, \alpha)$ is the *class-torsion group* of the pair (A, α), defined in [141]; see also [124, Definition 2.15].

As presented in [20, Corollary 4.5], if A has some additional properties (in particular, if A is von Neumann regular), this result may be modified to yield the following exact sequence:

$$K_1(A) \xrightarrow{1-\alpha_*} K_1(A) \xrightarrow{j} K_1(A[t_+, t_-, \alpha]) \xrightarrow{p} K_0(A) \xrightarrow{1-\alpha_*} K_0(A),$$

where α is assumed to be a corner isomorphism $\alpha : A \to pAp$, and $\alpha_* \colon K_i(A) \to K_i(A)$ is the map induced by the composition $A \xrightarrow{\alpha} pAp \hookrightarrow A$. Specifically, since $L_K(E)_0$ is locally matricial it satisfies the aforementioned hypotheses, so that we get an exact sequence

$$K_1(L_K(E)_0) \xrightarrow{1-\phi_*} K_1(L_K(E)_0) \to K_1(L_K(E)_0[t_+, t_-, \phi]) \to K_0(L_K(E)_0) \xrightarrow{1-\phi_*} K_0(L_K(E)_0). \qquad (\dagger)$$

Using the description of the connecting homomorphisms of the directed union $L_K(E)_0 = \bigcup_{n\in\mathbb{N}} L_{0,n}$ (Corollary 2.1.16) and the arguments in the proof of [23, Theorem 5.10], one gets that the cokernel of the map $K_1(L_K(E)_0) \xrightarrow{1-\phi_*} K_1(L_K(E)_0)$ and the kernel of the map $K_0(L_K(E)_0) \xrightarrow{1-\phi_*} K_0(L_K(E)_0)$ are

$$\mathrm{Coker}(I_{ns} - A_{ns})^t : \prod_{v\in E^0\setminus \mathrm{Sink}(E)} K^\times \to \prod_{v\in E^0} K^\times$$

and

$$\mathrm{Ker}(I_{ns} - A_{ns})^t : \bigoplus_{v\in E^0\setminus \mathrm{Sink}(E)} \mathbb{Z} \to \bigoplus_{v\in E^0} \mathbb{Z}$$

respectively.

Using all these facts, we obtain the following description of the Whitehead group of $L_K(E)$.

Theorem 6.2.4 *Let E be a finite graph without sources, and K any field. Then $K_1(L_K(E))$ is isomorphic to the direct sum of abelian groups*

$$\Big(\mathrm{Coker}(I_{ns} - A_{ns})^t : \prod_{v\in \mathrm{Reg}(E)} K^\times \to \prod_{v\in E^0} K^\times\Big) \bigoplus \Big(\mathrm{Ker}(I_{ns} - A_{ns})^t : \bigoplus_{v\in \mathrm{Reg}(E)} \mathbb{Z} \to \bigoplus_{v\in E^0} \mathbb{Z}\Big).$$

Proof By the displayed sequence (†) and the above remarks, we have an exact sequence of abelian groups

$$0 \longrightarrow \mathscr{C} \longrightarrow K_1(L_K(E)) \longrightarrow \mathscr{K} \longrightarrow 0,$$

where

$$\mathscr{C} = \mathrm{Coker}(I_{ns} - A_{ns})^t : \prod_{v\in \mathrm{Reg}(E)} K^\times \to \prod_{v\in E^0} K^\times$$

and

$$\mathscr{K} = \mathrm{Ker}(I_{ns} - A_{ns})^t : \bigoplus_{v\in \mathrm{Reg}(E)} \mathbb{Z} \to \bigoplus_{v\in E^0} \mathbb{Z}.$$

Since $\mathscr{K}$ is a free abelian group, the sequence splits, and the result follows. □

We note that when $A \in \mathrm{M}_n(\mathbb{Z})$, then $(I_n - A)^t = I_n - A^t$, a fact we will often use (for notational clarity) throughout the sequel.

Example 6.2.5 Let $n \geq 2$, and let R_n be the rose with n petals graph. Then the adjacency matrix A_E of E is (n), so the matrix $I_1 - A_E^t$ is $(1-n)$. In particular, $\mathrm{Ker}((1-n) : \mathbb{Z}^1 \to \mathbb{Z}^1) = \{0\}$, while $\mathrm{Coker}((1-n) : (K^\times)^1 \to (K^\times)^1) \cong K^\times/(K^\times)^{(n-1)}$, where $(K^\times)^{(n-1)}$ denotes the nonzero elements of K which can be written as $(n-1)^{st}$ powers. Thus by Theorem 6.2.4 we get

$$K_1(L_K(R_n)) \cong K^\times/(K^\times)^{(n-1)}.$$

From this, we note that, unlike the situation for K_0, in general the structure of the field K plays a role in the description of $K_1(L_K(E))$. In particular, we see that $K_1(L_K(R_2))$ is trivial for any field K. Furthermore, $K_1(L_K(R_n))$ is trivial for all $n \geq 2$ whenever K is algebraically closed.

Example 6.2.6 Let $E = R_1$ be the graph having one vertex and one loop, as usual. Thus the adjacency matrix A_E of E is (1), so the matrix $I_1 - A_E$ is (0). In particular, $\mathrm{Ker}((0) : \mathbb{Z}^1 \to \mathbb{Z}^1) = \mathbb{Z}$, while $\mathrm{Coker}((0) : (K^\times)^1 \to (K^\times)^1) = K^\times$. So by Theorem 6.2.4 we get

$$K_1(L_K(R_1)) \cong \big(\mathrm{Coker}((0) : (K^\times)^1 \to (K^\times)^1)\big) \oplus \big(\mathrm{Ker}((0) : \mathbb{Z}^1 \to \mathbb{Z}^1)\big) = K^\times \oplus \mathbb{Z}.$$

Since $L_K(R_1) \cong K[x, x^{-1}]$, we have recovered the Bass–Heller–Swan result mentioned in Examples 6.2.2(iii).

With some minor adjustment, we can use Theorem 6.2.4 to obtain the description alluded to in Examples 6.2.2(ii). Let E be a finite graph for which $L_K(E)$ is purely infinite simple. Let v be a source in E, and let E_v denote the subgraph of E obtained by eliminating v and all edges in $s^{-1}(v)$. (See Definition 6.3.26 below.)

Corollary 6.2.7 *Let E be a finite graph for which $L_K(E)$ is purely infinite simple. Let F be a source-free graph obtained from E by repeated applications of the source elimination process. Let A_F denote the adjacency matrix of F, and let $m = |F^0|$. Then*

$$K_1(L_K(E)) \cong \Big(\mathrm{Coker}((I_m - A_F)^t : \prod_{v\in F^0} K^\times \to \prod_{v\in F^0} K^\times)\Big) \bigoplus \Big(\mathrm{Ker}((I_m - A_F)^t : \bigoplus_{v\in F^0} \mathbb{Z} \to \bigoplus_{v\in F^0} \mathbb{Z})\Big).$$

Proof As we will show below in Proposition 6.3.28, when $L_K(E)$ is simple then the source elimination process preserves Morita equivalence. In particular, if F is obtained from E by repeated applications of the source elimination process, then $L_K(F)$ is Morita equivalent to $L_K(E)$. So by Remark 6.2.3(i) we get that $K_1(L_K(F)) \cong K_1(L_K(E))$. In addition, since $L_K(F)$ is then purely infinite simple, by Theorem 3.1.10 we see that F has no sinks. Now apply Theorem 6.2.4. □

As is evident from Theorems 6.1.9 and 6.2.4, the matrix $(I_{ns} - A_{ns})^t$ plays a pivotal role in the description of both $K_0(L_K(E))$ and $K_1(L_K(E))$. Indeed, information about the structure of $K_0(L_K(E))$ is often sufficient to understand the structure of $K_1(L_K(E))$, as shown here.

Proposition 6.2.8 *Let E and F be finite graphs having neither sinks nor sources, for which $|E^0| = |F^0|$. If $K_0(L_K(E)) \cong K_0(L_K(F))$, then $K_1(L_K(E)) \cong K_1(L_K(F))$.*

Proof Let m denote $|E^0| = |F^0|$. Since neither graph has sinks, the matrices I_{ns} and A_{ns} are the square matrices I_m and A_E (resp., A_F). By Theorem 6.1.9, $\mathrm{Coker}(I_m - A_E^t) \cong \mathrm{Coker}(I_m - A_F^t)$, where these matrices are viewed as linear transformations from $\mathbb{Z}^m$ to $\mathbb{Z}^m$. This in turn implies (by the Fundamental Theorem of Finitely Generated Abelian Groups) the existence of invertible matrices $P, Q \in \mathrm{M}_m(\mathbb{Z})$ such that $I_m - A_F^t = P(I_m - A_E^t)Q$. Thus $\mathrm{Ker}(I_m - A_F^t) \cong \mathrm{Ker}(I_m - A_E^t)$, as these are thereby subgroups of $\mathbb{Z}^m$ having equal rank. Moreover, the *PAQ*-equivalence of $I_m - A_E^t$ and $I_m - A_F^t$ also yields by a standard argument that the abelian groups $\mathrm{Coker}(I_m - A_E^t : \prod_{i=1}^m K^\times \to \prod_{i=1}^m K^\times)$ and $\mathrm{Coker}(I_m - A_F^t : \prod_{i=1}^m K^\times \to \prod_{i=1}^m K^\times)$ are isomorphic as well. Now use Theorem 6.2.4. □

Remark 6.2.9 The result of Theorem 6.2.4 holds verbatim for all row-finite graphs, even those graphs with sources, where we interpret n as ∞ whenever appropriate. (See e.g., [23, Theorem 7.7]. In particular, a general "source elimination process" is described in [23, Lemma 6.1].)

Remark 6.2.10 Having given in Sects. 6.1 and 6.2 a detailed description of the K-theoretic groups $K_0(L_K(E))$ and $K_1(L_K(E))$, one might be led to inquire about a description of the higher K-theoretic groups for Leavitt path algebras. Indeed, such a description of all of the algebraic K-theoretic groups $K_i(L_K(E))$ $(i \geq 2)$ is achieved in [23] for any row-finite graph E, to which we refer the interested reader. We note that, in general, one cannot determine $K_n(L_K(E))$ from the groups $K_i(L_K(E))$ $(0 \leq i \leq n-1)$. In particular, unlike in the situation for graph C^*-algebras, Bott periodicity does not hold for Leavitt path algebras.

6.3 The Algebraic Kirchberg–Phillips Question

We start this section with the following basic question: If two Leavitt path algebras $L_K(E)$ and $L_K(F)$ are ring-theoretically related (e.g., if they are isomorphic, or if they are Morita equivalent), is there some connection between the graphs E and F? On one level the answer must of course be *yes*: for instance, using results from previous chapters, ring-theoretic information such as simplicity, chain conditions, etc., is encoded in the graph, so that if E has a germane property, then so must F.

But one may ask for a tighter connection between E and F; for instance, if $L_K(E)$ and $L_K(F)$ are isomorphic, is it possible to realize the graph F as some sort of "transformed" version of the graph E? That is, does there exist some sequence of "graph transformations" which starts at E and ends at F?

There is no clear understanding of whether this is necessarily the case for an arbitrarily chosen pair of graphs E and F. However, there is one very important context in which many isomorphisms or Morita equivalences between Leavitt path algebras can in fact be realized as arising from such a sequence of graph transformations, specifically, when the Leavitt path algebras are purely infinite simple. Moreover, in this case, the existence of isomorphisms and Morita equivalences is guaranteed by a coincidence of elementary information about the adjacency matrices of the two graphs, including information about the K_0 groups of the algebras.

One of the major lines of investigation in the theory of C^*-algebras, ongoing since the 1970s, is known as the "Elliott program", which refers to the search for user-friendly invariants for various classes of C^*-algebras. More to the point, suppose A and B are C^*-algebras in a specified class. If certain K-theoretic information about A and B matches up, can we conclude that A and B are related in some essential way? Of interest here is the following important result of this type.

Theorem 6.3.1 (The Kirchberg–Phillips Theorem in the Context of Graph C^*-Algebras) *Suppose E and F are countable row-finite graphs for which $C^*(E)$ and $C^*(F)$ are purely infinite simple. Suppose also that $K_0(C^*(E)) \cong K_0(C^*(F))$; if E and F are finite, assume furthermore that this isomorphism takes $[1_{C^*(E)}]$ to $[1_{C^*(F)}]$. Assume in addition that $K_1(C^*(E)) \cong K_1(C^*(F))$. Then $C^*(E) \cong C^*(F)$ homeomorphically as C^*-algebras.*

Remark 6.3.2 The result we have presented as Theorem 6.3.1 is a specific consequence of a much more general result about C^*-algebras, proved independently by both Phillips and Kirchberg in 2000; see e.g., [103, 128]. The hypotheses required to apply [128, Theorem 4.2.4] include not only information about the purely infinite simplicity and K-theory of the algebras, but additional structural information as well. However, these additional requirements are always satisfied for the graph C^*-algebras of countable row-finite graphs (see [146, Remark A.11.13] for a discussion). We note that only the existence of an isomorphism between C^*-algebras is ensured by Phillips [128, Theorem 4.2.4]; the isomorphism is not explicitly constructed. Also, Rørdam had previously established a related version of Theorem 6.3.1 in [135, Theorem 6.5], where it is shown that, using the same hypotheses, a homeomorphism $C^*(E) \cong C^*(F)$ necessarily follows if E and F are finite graphs having neither sinks nor sources; and, indeed, in this situation, the existence of an isomorphism between the K_1 groups is not required as part of the hypotheses.

With the above discussion in mind, and given the close connection between purely infinite simple unital graph C^*-algebras and purely infinite simple unital Leavitt path algebras described in Chap. 5, it is reasonable to ask whether there might be an algebraic result analogous to Theorem 6.3.1.

Question 6.3.3 (The Algebraic Kirchberg–Phillips Question for Leavitt Path Algebras of Finite Graphs) Let K be any field. Suppose E and F are finite graphs for which $L_K(E)$ and $L_K(F)$ are purely infinite simple, and suppose that $K_0(L_K(E)) \cong K_0(L_K(F))$ via an isomorphism which takes $[1_{L_K(E)}]$ to $[1_{L_K(F)}]$. Are the Leavitt path K-algebras $L_K(E)$ and $L_K(F)$ necessarily isomorphic?

We remind the reader that for a purely infinite simple ring R, we have chosen to denote the (stable equivalence classes $[\ ]_0$ of) elements in $K_0(R)$ using the notation of (isomorphism classes $[\ ]$ of) elements in $\mathscr{V}(R) \setminus \{[0]\}$; see Remark 6.1.13.

Throughout the section K denotes an arbitrary field, and all indicated ring isomorphisms are in fact K-algebra isomorphisms. For the remainder of this section we describe the current (as of 2017) state of affairs regarding the resolution of the Algebraic Kirchberg–Phillips Question 6.3.3.

We start by presenting a computational tool which proves to be quite useful in this discussion. Let $M \in \mathrm{M}_n(\mathbb{Z})$, and view M as a linear transformation $M : \mathbb{Z}^n \to \mathbb{Z}^n$ via left multiplication on columns. As indicated earlier, if P, Q are invertible in $\mathrm{M}_n(\mathbb{Z})$, then $\mathrm{Coker}(M) \cong \mathrm{Coker}(PMQ)$. Consequently, if $N \in \mathrm{M}_n(\mathbb{Z})$ is a matrix which is constructed by performing any sequence of $\mathbb{Z}$-elementary row and/or column operations starting with M, then $\mathrm{Coker}(M) \cong \mathrm{Coker}(N)$ as abelian groups. (A $\mathbb{Z}$-elementary row operation is one of: switch two rows; multiply a row by -1; add an integer-multiple of a row to another row. An analogous description holds for $\mathbb{Z}$-elementary column operations.)

Definition 6.3.4 Let $M \in \mathrm{M}_n(\mathbb{Z})$. The *Smith normal form of M* is the diagonal matrix $S \in \mathrm{M}_n(\mathbb{Z})$ having the following two properties.

(i) The diagonal entries of S consist of non-negative integers $s_1, s_2, \ldots, s_n$ for which: (1) if t denotes the number of entries on this list which equal 0, then the list is written so that $s_1, s_2, \ldots, s_t = 0$; and (2) s_i is a divisor of s_{i+1} for all $t+1 \le i \le n-1$.
(ii) There is a sequence of $\mathbb{Z}$-elementary row and/or column operations which starts at M and ends at S.

It can easily be shown that for any $M \in \mathrm{M}_n(\mathbb{Z})$, the Smith normal form of M exists and is unique. If $D \in \mathrm{M}_n(\mathbb{Z})$ is diagonal, with diagonal entries $d_1, d_2, \ldots, d_n$, then, viewing D as a linear transformation from $\mathbb{Z}^n$ to $\mathbb{Z}^n$, we obviously have $\mathrm{Coker}(D) \cong \mathbb{Z}/d_1\mathbb{Z} \oplus \mathbb{Z}/d_2\mathbb{Z} \oplus \cdots \oplus \mathbb{Z}/d_n\mathbb{Z}$. (In this context we interpret $\mathbb{Z}/1\mathbb{Z}$ as the trivial group $\{0\}$.) This observation, with the previous discussion, immediately gives the following.

Proposition 6.3.5 *Let $M \in \mathrm{M}_n(\mathbb{Z})$, and let S denote the Smith normal form of M. Suppose the diagonal entries of S are $s_1, s_2, \ldots, s_n$. Then*

$$\mathrm{Coker}(M) \cong \mathbb{Z}/s_1\mathbb{Z} \oplus \mathbb{Z}/s_2\mathbb{Z} \oplus \cdots \oplus \mathbb{Z}/s_n\mathbb{Z}.$$

As a result, by the Fundamental Theorem of Finitely Generated Abelian Groups, if M and M' are square matrices (not necessarily of the same size) for which $\mathrm{Coker}(M) \cong \mathrm{Coker}(M')$, *then the sequence of "not equal to 1" entries in the Smith normal form S of M equals the sequence of "not equal to 1" entries in the Smith normal form S' of M'.*

As an example, if $M = \begin{pmatrix} 0 & -3 \\ -1 & -1 \end{pmatrix}$, then it is straightforward to show that the Smith normal form of M is $S = \begin{pmatrix} 1 & 0 \\ 0 & 3 \end{pmatrix}$, so that by Proposition 6.3.5 we conclude that $\mathrm{Coker}(M) \cong \mathbb{Z}/3\mathbb{Z}$.

Combining Corollary 6.1.10 with Proposition 6.3.5, we get the following useful result.

Corollary 6.3.6 *Suppose E is a finite graph having no sinks, and let $|E^0| = n$. Let S be the Smith normal form of the matrix $I_n - A_E^t$, with diagonal entries $s_1, s_2, \ldots, s_n$. Then*

$$K_0(L_K(E)) \cong \mathbb{Z}/s_1\mathbb{Z} \oplus \mathbb{Z}/s_2\mathbb{Z} \oplus \cdots \oplus \mathbb{Z}/s_n\mathbb{Z}.$$

Examples 6.3.7 We refer to the graphs E, F, G, and H presented in Examples 3.2.7. By the Purely Infinite Simplicity Theorem 3.1.10, the corresponding Leavitt path algebra of each of these is readily seen to be purely infinite simple. So by Proposition 6.1.12, we have that the nonzero elements of the $\mathcal{V}$-monoid form a group, isomorphic to the Grothendieck group of the algebra. Now apply Corollary 6.3.6 to establish the following previously-mentioned isomorphisms.

$$I_3 - A_E^t = \begin{pmatrix} 1 & -1 & -1 \\ -1 & 0 & -1 \\ 0 & -1 & 1 \end{pmatrix}, \text{ whose Smith Normal Form is } \begin{pmatrix} 1 & 0 & 0 \\ 0 & 1 & 0 \\ 0 & 0 & 3 \end{pmatrix}, \text{ so that } K_0(L_K(E)) \cong \mathbb{Z}/3\mathbb{Z}.$$

$$I_3 - A_F^t = \begin{pmatrix} 0 & -1 & 0 \\ -1 & 1 & -1 \\ 0 & -1 & 0 \end{pmatrix}, \text{ whose Smith Normal Form is } \begin{pmatrix} 0 & 0 & 0 \\ 0 & 1 & 0 \\ 0 & 0 & 1 \end{pmatrix}, \text{ so that } K_0(L_K(F)) \cong \mathbb{Z}.$$

$$I_3 - A_G^t = \begin{pmatrix} 1 & -1 & -1 \\ -1 & 1 & -1 \\ -1 & -1 & 1 \end{pmatrix}, \text{ whose Smith Normal Form is } \begin{pmatrix} 1 & 0 & 0 \\ 0 & 2 & 0 \\ 0 & 0 & 2 \end{pmatrix}, \text{ so that } K_0(L_K(G)) \cong \mathbb{Z}/2\mathbb{Z} \oplus \mathbb{Z}/2\mathbb{Z}.$$

$$I_2 - A_H^t = \begin{pmatrix} -4 & -4 \\ -2 & -2 \end{pmatrix}, \text{ whose Smith Normal Form is } \begin{pmatrix} 0 & 0 \\ 0 & 2 \end{pmatrix}, \text{ so that } K_0(L_K(H)) \cong \mathbb{Z} \oplus (\mathbb{Z}/2\mathbb{Z}).$$

Remark 6.3.8

(i) It is easy to see that if M' is a matrix obtained from $M \in \mathrm{M}_n(\mathbb{Z})$ by applying any of the three $\mathbb{Z}$-elementary row (resp., column) operations, then $\det(M') = \det(M)$ or $\det(M') = -\det(M)$. Consequently, if S is the Smith normal form of M, then either $\det(S) = \det(M)$ or $\det(S) = -\det(M)$.

(ii) Proposition 6.3.5 yields that $\mathrm{Coker}(M)$ is infinite if and only if $s_i = 0$ for some i, which clearly happens if and only if $\det(S) = 0$.

Much of the following discussion is taken from [11]. The key results which have been utilized in the investigation of the Algebraic Kirchberg–Phillips Question are provided by deep work in the theory of symbolic dynamics. We assemble some of the relevant facts in the next few results, then state as Proposition 6.3.15 the conclusion appropriate for our needs. In the following discussion, if A is any non-negative integer-valued matrix, then E_A denotes the directed graph whose adjacency matrix is A.

Definition 6.3.9 We call a graph transformation *standard* if it is one of these six types: in-splitting, in-amalgamation, out-splitting, out-amalgamation, expansion, or contraction. (These six types of graph transformations will be defined below.) Analogously, we call a function which transforms a non-negative integer matrix A to a non-negative integer-valued matrix B *standard* if the corresponding graph operation from E_A to E_B is standard.

Definition 6.3.10 If E and F are graphs having no sources and no sinks, a *flow equivalence from E to F* is a sequence $E = E_0 \to E_1 \to \cdots \to E_n = F$ of graphs and standard graph transformations which starts at E and ends at F. We say that E and F are *flow equivalent* if there is a flow equivalence from E to F. Analogously, a flow equivalence between matrices A and B is defined to be a flow equivalence between the graphs E_A and E_B.

The notion of flow equivalence can be described in topological terms (see e.g., [114]). The definition given in Definition 6.3.10 agrees with the topologically-based definition for source-free, sink-free graphs by an application of [126, Theorem], [160, Corollary 4.4.1], and [114, Corollary 7.15]. Although the graphs which appear in our main result will be allowed to have sources (but not sinks), this particular definition of flow equivalence will serve us most efficiently.

Definition 6.3.11 A graph E is called

(i) *irreducible* if, given any two vertices v, w of E, there exists a path μ with $s(\mu) = v$ and $r(\mu) = w$,
(ii) *nontrivial* if E does not consist solely of a single cycle, and
(iii) *essential* if E contains no sources and no sinks.

An irreducible (resp., nontrivial, essential) non-negative integer-valued matrix A is one whose corresponding graph E_A is irreducible (resp., nontrivial, essential).

For a finite graph E (excepting the graph $E = R_0 = \bullet$), it is not hard to see that if E is irreducible, then E is essential. Consequently, in a number of the results below, one may replace the pair of hypotheses "E is irreducible" and "E is essential" with the single hypothesis "$E \neq R_0$ is irreducible". However, from the point of view of symbolic dynamics, the concepts "irreducible" and "essential" have broader interpretations; in these broader contexts, the two ideas are quite distinct one from the other. The results we will utilize here from symbolic dynamics were developed in this broader framework. Thus in order to more clearly focus on the connection between Leavitt path algebras and symbolic dynamics, we choose to use the two hypotheses "E is irreducible" and "E is essential" in various results, rather than the seemingly more efficient "$E \neq R_0$ is irreducible".

The following deep, powerful theorem of Franks provides most of the heavy lifting in the context of the current discussion.

Theorem 6.3.12 (Franks' Theorem [81, Theorem]) *Suppose that A and B are nontrivial irreducible essential square non-negative integer matrices, of sizes $n \times n$ and $m \times m$, respectively. Then A and B are flow equivalent if and only if*

$$\det(I_n - A) = \det(I_m - B) \text{ and } \operatorname{Coker}(I_n - A) \cong \operatorname{Coker}(I_m - B),$$

where I_n and I_m denote the identity matrices of sizes $n \times n$ and $m \times m$, respectively.

Recasting Franks' Theorem in the context of graphs, we get

Corollary 6.3.13 *Suppose E and F are finite, irreducible, nontrivial, essential graphs with $|E^0| = n$ and $|F^0| = m$. Then there exists a sequence of standard* *graph transformations which starts with E and ends with F if and only if*

$$\det(I_n - A_E) = \det(I_m - A_F) \quad \text{and} \quad \text{Coker}(I_n - A_E) \cong \text{Coker}(I_n - A_F).$$

Clearly there is a relationship between the notions which appear in Franks' Theorem, and notions which play a role in the theory of Leavitt path algebras. Specifically, it is a straightforward exercise in graph theory to establish the equivalence of the first pair of statements of this next result. The equivalence of the second pair constitutes the heart of the Purely Infinite Simplicity Theorem 3.1.10.

Lemma 6.3.14 *Let E be a finite graph and K any field. The following are equivalent.*

(1) E is irreducible, essential, and nontrivial.
(2) E contains no sources, E is cofinal, E satisfies Condition (L), and E contains at least one cycle.
(3) E contains no sources, and $L_K(E)$ is purely infinite simple.

By examining the Smith normal form of each matrix (recall Definition 6.3.4), it is easy to show that $\text{Coker}(I_n - A) \cong \text{Coker}(I_n - A^t)$ for any square matrix A. Furthermore, it is clear that $\det(I_n - A) = \det(I_n - A^t)$. So, by Corollary 6.1.10, we see that Corollary 6.3.13 and Lemma 6.3.14 combine to yield

Proposition 6.3.15 *Let E and F be finite graphs having no sources, and for which $L_K(E)$ and $L_K(F)$ are purely infinite simple. Suppose* $\det(I_n - A_E^t) = \det(I_m - A_F^t)$ *and* $K_0(L_K(E)) \cong K_0(L_K(F))$. *Then there is a sequence of standard graph transformations which starts at E and ends at F.*

In summary, Franks' Theorem yields that when $L_K(E)$ and $L_K(F)$ are purely infinite simple, if the K_0 groups of these Leavitt path algebras are isomorphic, and the determinants of the appropriate matrices are equal, and E and F are source-free, then in fact there is a connection between the graphs E and F.

We now establish that, perhaps remarkably, the connection between the graphs ensured by Franks' Theorem 6.3.12 produces a Morita equivalence between the corresponding Leavitt path algebras. (In fact, we will also be able to drop the "source-free" requirement in this context.) We now explicitly define the six aforementioned standard graph transformations, and show that each preserves Morita equivalence of the corresponding Leavitt path algebras. The key tool is the following lemma, which is straightforward to establish using standard ring-theoretic techniques.

Lemma 6.3.16 ([11, Lemma 1.1]) *Suppose R and S are simple unital rings. Let $\pi : R \to S$ be a nonzero, not-necessarily-identity-preserving ring homomorphism, and let g denote the idempotent $\pi(1_R)$ of S. If $gSg = \pi(R)$, then there exists a Morita equivalence $\Phi : R-Mod \to S-Mod$.*

Moreover, the equivalence Φ restricts to a monoid isomorphism $\Phi_{\mathcal{V}} : \mathcal{V}(R) \to \mathcal{V}(S)$ with the property that for any idempotent $e \in R$, $\Phi_{\mathcal{V}}([Re]) = [S\pi(e)]$.

Definition 6.3.17 Let $E = (E^0, E^1, r, s)$ be a directed graph, and let $v \in E^0$. Let v^* and f be symbols not in $E^0 \cup E^1$. We form the *expansion graph E_v from E at v* as follows:

$$E_v^0 = E^0 \cup \{v^*\}\,, \; E_v^1 = E^1 \cup \{f\}\,,$$

$$s_{E_v}(e) = \begin{cases} v & \text{if } e = f \\ v^* & \text{if } s_E(e) = v \\ s_E(e) & \text{otherwise} \end{cases}\,, \quad \text{and} \quad r_{E_v}(e) = \begin{cases} v^* & \text{if } e = f \\ r_E(e) & \text{otherwise} \end{cases}\,.$$

Conversely, if E and G are graphs, and there exists a vertex v of E for which $E_v = G$, then E is called a *contraction* of G.

Example 6.3.18

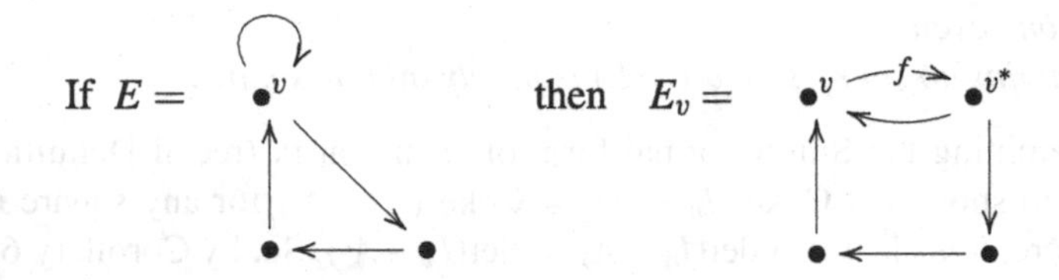

Proposition 6.3.19 *Let E be a row-finite graph such that $L_K(E)$ is simple and unital, and let $v \in E^0$. Then $L_K(E)$ is Morita equivalent to $L_K(E_v)$, via a Morita equivalence*

$$\Phi^{\exp} : L_K(E) - Mod \to L_K(E_v) - Mod$$

for which $\Phi^{\exp}_{\mathcal{V}}([L_K(E)w]) = [L_K(E_v)w]$ for all vertices w of E.

Proof We begin by noting that, as an easy application of the Simplicity Theorem 2.9.1, $L_K(E)$ is simple and unital if and only if $L_K(E_v)$ is simple and unital.

For each $w \in E^0$ we set $Q_w = w$; for each $e \in s^{-1}(v)$ we set $T_e = fe$ and $T_e^* = e^*f^*$; and for each $e \in E^1 \setminus s^{-1}(v)$ we set $T_e = e$ and $T_e^* = e^*$. We claim that $\{Q_w, T_e, T_e^* \mid w \in E^0, e \in E^1\}$ is an E-family in $L_K(E_v)$. The Q_w's are mutually orthogonal idempotents because the w's are. The elements T_e for $e \in E^1$ clearly satisfy $T_e^* T_f = 0$ whenever $e \neq f$. For $e \in E^1$, it is easy to check that $T_e^* T_e = Q_{r(e)}$. Note that $\sum_{e \in s_{E_v}^{-1}(v)} T_e T_e^* = f\left(\sum_{e \in s_{E_v}^{-1}(v^*)} ee^*\right) f^* = ff^* = v = Q_v$ (we utilize here that, as $s_{E_v}^{-1}(v) = \{f\}$, we have $ff^* = v$ by the CK2 relation at v). The same property holds immediately for all $w \in E^0$ with $w \neq v$, thereby establishing the claim.

Therefore, by the Universal Property of $L_K(E)$ 1.2.5, there is a K-algebra homomorphism $\pi : L_K(E) \to L_K(E_v)$ that maps $w \mapsto Q_w$, $e \mapsto T_e$, and $e^* \mapsto T_e^*$. Note that π maps w to $Q_w \neq 0$, so π is nonzero. We now claim that $\pi(L_K(E)) = \pi(1_{L_K(E)})\, L_K(E_v)\, \pi(1_{L_K(E)})$, where $\pi(1_{L_K(E)}) = \sum_{w \in E^0} w$, viewed as an element of $L_K(E_v)$. The inclusion $\pi(L_K(E)) \subseteq \pi(1_{L_K(E)})\, L_K(E_v)\, \pi(1_{L_K(E)})$ is immediate. For the other direction, it suffices to consider arbitrary nonzero terms in $\pi(1_{L_K(E)})\, L_K(E_v)\, \pi(1_{L_K(E)})$ of the form $\mu_1\mu_2^*$, where μ_1 and μ_2 are paths in E_v, $s(\mu_1), s(\mu_2) \neq v^*$, and $r(\mu_1) = r(\mu_2)$.

Let α be the path in E obtained by removing the edge f from μ_1 any place that it occurs, and similarly let β be the path obtained by removing f from μ_2. We claim that $\pi(\alpha\beta^*) = \mu_1\mu_2^*$. There are two cases. If $r(\mu_1) \neq v^* \neq r(\mu_2)$, then $\mu_1 = \pi(\alpha)$ and $\mu_2 = \pi(\beta)$, and the result follows. Otherwise, $r(\mu_1) = v^* = r(\mu_2)$. But because μ_1 and μ_2 both begin at a vertex other than v^*, and the only edge entering v^* is f, we must have $\mu_1 = \nu_1 f$ and $\mu_2 = \nu_2 f$, for paths ν_1, ν_2 in E_v, where $r(\nu_1) = v = r(\nu_2)$. But then $\mu_1\mu_2^* = \nu_1 ff^*\nu_2^* = \nu_1 v \nu_2^* = \nu_1\nu_2^*$, and thus we are back in the first case, so $\pi(\alpha\beta^*) = \mu_1\mu_2^*$, completing the argument.

Applying Lemma 6.3.16, we conclude that $L_K(E)$ is Morita equivalent to $L_K(E_v)$, and that the Morita equivalence restricts to the map on the corresponding $\mathscr{V}$-monoids given above. □

If F is a contraction of E (i.e., if there exists a vertex v of F for which $E = F_v$), then we denote by $\Phi^{\mathrm{cont}} = (\Phi^{\mathrm{exp}})^{-1}$ the Morita equivalence $L_K(F)$–Mod $\to$ $L_K(E)$–Mod.

The remaining four standard graph operations require somewhat more cumbersome machinery to build than the expansion and contraction operations. The following definition is presented in [46, Sect. 5].

Definition 6.3.20 Let $E = (E^0, E^1, r, s)$ be a directed graph. For each $v \in E^0$ with $r^{-1}(v) \neq \emptyset$, partition the set $r^{-1}(v)$ into disjoint nonempty subsets $\mathscr{E}_1^v, \dots, \mathscr{E}_{m(v)}^v$ where $m(v) \geq 1$. (If v is a source then set $m(v) = 0$.) Let $\mathscr{P}$ denote the resulting partition of E^1. We form the *in-split graph $E_r(\mathscr{P})$ from E using the partition $\mathscr{P}$* as follows:

$$E_r(\mathscr{P})^0 = \{v_i \mid v \in E^0, 1 \leq i \leq m(v)\} \cup \{v \mid m(v) = 0\},$$
$$E_r(\mathscr{P})^1 = \{e_j \mid e \in E^1, 1 \leq j \leq m(s(e))\} \cup \{e \mid m(s(e)) = 0\},$$

and define $r_{E_r(\mathscr{P})}, s_{E_r(\mathscr{P})} : E_r(\mathscr{P})^1 \to E_r(\mathscr{P})^0$ by

$$s_{E_r(\mathscr{P})}(e_j) = s(e)_j \text{ and } s_{E_r(\mathscr{P})}(e) = s(e)$$
$$r_{E_r(\mathscr{P})}(e_j) = r(e)_i \text{ and } r_{E_r(\mathscr{P})}(e) = r(e)_i \text{ where } e \in \mathscr{E}_i^{r(e)}.$$

Conversely, if E and G are graphs, and there exists a partition $\mathscr{P}$ of E^1 for which $E_r(\mathscr{P}) = G$, then E is called an *in-amalgamation* of G.

Example 6.3.21 Let E be the graph $e \circlearrowright \bullet^v \underset{g}{\overset{f}{\rightleftarrows}} \bullet^w$. Denote by $\mathscr{P}$ the partition of E^1 that places each edge in its own singleton partition class. Then

$$E_r(\mathscr{P}) = \quad e_1 \circlearrowright \bullet^{v_1} \xrightarrow{f_1} \bullet^{w_1}, \quad \bullet^{v_2} \xrightarrow{e_2} \bullet^{v_1}, \quad \bullet^{v_2} \xrightarrow{f_2} \bullet^{w_1}, \quad \bullet^{w_1} \xrightarrow{g_1} \bullet^{v_2}$$

Using tools similar to those used in the proof of Proposition 6.3.19, one may establish the following.

Proposition 6.3.22 ([11, Proposition 1.11]) *Let E be a directed graph with no sources or sinks, such that $L_K(E)$ is simple and unital. Let $\mathscr{P}$ be a partition of E^1 as in Definition 6.3.20, and $E_r(\mathscr{P})$ the in-split graph from E using $\mathscr{P}$. Then $L_K(E)$ is Morita equivalent to $L_K(E_r(\mathscr{P}))$, via a Morita equivalence*

$$\Phi^{\text{ins}} : L_K(E) - Mod \to L_K(E_r(\mathscr{P})) - Mod$$

for which $\Phi^{\text{ins}}_{\mathscr{V}}([L_K(E)v]) = [L_K(E_r(\mathscr{P}))v_1]$ for all vertices v of E.

If F is an in-amalgamation of E (i.e., if there exists a vertex partition $\mathscr{P}$ of F for which $E = F_r(\mathscr{P})$), then we denote by $\Phi^{\text{inam}} = (\Phi^{\text{ins}})^{-1}$ the Morita equivalence $L_K(F) - Mod \to L_K(E) - Mod$.

We now utilize a definition from [46, Sect. 3].

Definition 6.3.23 Let $E = (E^0, E^1, r, s)$ be a directed graph. For each $v \in E^0$ with $s^{-1}(v) \neq \emptyset$, partition the set $s^{-1}(v)$ into disjoint nonempty subsets $\mathscr{E}_v^1, \dots, \mathscr{E}_v^{m(v)}$ where $m(v) \geq 1$. (If v is a sink then set $m(v) = 0$.) Let $\mathscr{P}$ denote the resulting partition of E^1. We form the *out-split graph $E_s(\mathscr{P})$ from E using the partition $\mathscr{P}$* as follows:

$$E_s(\mathscr{P})^0 = \{v^i \mid v \in E^0, 1 \leq i \leq m(v)\} \cup \{v \mid m(v) = 0\},$$

$$E_s(\mathscr{P})^1 = \{e^j \mid e \in E^1, 1 \leq j \leq m(r(e))\} \cup \{e \mid m(r(e)) = 0\},$$

and define $r_{E_s(\mathscr{P})}, s_{E_s(\mathscr{P})} : E_s(\mathscr{P})^1 \to E_s(\mathscr{P})^0$ for each $e \in \mathscr{E}^i_{s(e)}$ by

$$s_{E_s(\mathscr{P})}(e^j) = s(e)^i \text{ and } s_{E_s(\mathscr{P})}(e) = s(e)^i$$

$$r_{E_s(\mathscr{P})}(e^j) = r(e)^j \text{ and } r_{E_s(\mathscr{P})}(e) = r(e).$$

Conversely, if E and G are graphs, and there exists a partition $\mathscr{P}$ of E^1 for which $E_s(\mathscr{P}) = G$, then E is called an *out-amalgamation* of G.

Example 6.3.24 Let E again be the graph $e \circlearrowright \bullet^v \underset{\xleftarrow{g}}{\xrightarrow{f}} \bullet^w$ given in Example 6.3.21, and again denote by $\mathscr{P}$ the partition of E^1 that places each edge in its own singleton partition class. Then

$$E_s(\mathscr{P}) = \quad e^1 \circlearrowright \bullet^{v^1} \xleftarrow{g^1} \bullet^{w^1}, \quad \bullet^{v^1} \xrightarrow{e^2} \bullet^{v^2}, \quad \bullet^{v^2} \xrightarrow{f^1} \bullet^{w^1}, \quad \bullet^{w^1} \xrightarrow{g^2} \bullet^{v^2}$$

The following result may be established by (again) using tools similar to those used in the proof of Proposition 6.3.19. (We note that for the "out-split" and "out-amalgamation" operations we in fact get an isomorphism of the corresponding Leavitt path algebras; this property does not hold in general for the other four standard graph operations.)

Proposition 6.3.25 ([11, Proposition 1.14]) *Let E be a row-finite graph, $\mathscr{P}$ a partition of E^1 as in Definition 6.3.23, and $E_s(\mathscr{P})$ the out-split graph from E using $\mathscr{P}$. Then $L_K(E)$ is K-algebra isomorphic to $L_K(E_s(\mathscr{P}))$. This isomorphism yields a Morita equivalence*

$$\Phi^{\rm outs} : L_K(E) - Mod \to L_K(E_s(\mathscr{P})) - Mod$$

for which $\Phi^{\rm outs}_{\mathscr{P}}([L_K(E)v]) = [L_K(E_s(\mathscr{P}))\sum_{i=1}^{m(v)} v^i]$ for every vertex v of E.

If F is an out-amalgamation of E (i.e., if there exists a vertex partition $\mathscr{P}$ of F for which $E = F_s(\mathscr{P})$), then we denote by $\Phi^{\rm outam} = (\Phi^{\rm outs})^{-1}$ the Morita equivalence $L_K(F) - Mod \to L_K(E) - Mod$.

We note that the three Propositions 6.3.19, 6.3.22, and 6.3.25 may be extended to wider classes of graphs, see [11, Sect. 3].

The six standard graph transformations have now been presented; each preserves Morita equivalence classes of the corresponding Leavitt path algebras (at least in the case where the algebras are purely infinite simple). We now show that we may remove the "source-free" hypothesis in this context.

Definition 6.3.26 Let $E = (E^0, E^1, r, s)$ be a directed graph with at least two vertices, and let $v \in E^0$ be a source. We form the *source elimination graph* $E_{\backslash v}$ of E as follows:

$$E^0_{\backslash v} = E^0 \backslash \{v\}\,, \qquad E^1_{\backslash v} = E^1 \backslash s^{-1}(v)\,, \qquad s_{E_{\backslash v}} = s|_{E^1_{\backslash v}}\,, \qquad \text{and} \qquad r_{E_{\backslash v}} = r|_{E^1_{\backslash v}}\,.$$

Example 6.3.27 Let E be the graph $\bullet \rightleftarrows \bullet \longleftarrow \bullet^v$. Then $E_{\backslash v} = \bullet \rightleftarrows \bullet$.

It is easy to see that as long as the graph E contains a cycle, repeated source elimination can be used to convert E into a graph with no sources. The following result may be established by (yet again) using tools similar to those used in the proof of Proposition 6.3.19.

Proposition 6.3.28 ([11, Proposition 1.4]) *Let E be a finite graph containing at least two vertices such that $L_K(E)$ is simple, and let $v \in E^0$ be a source. Then $L_K(E_{\setminus v})$ is Morita equivalent to $L_K(E)$, via a Morita equivalence*

$$\Phi^{\text{elim}} : L_K(E_{\setminus v}) - Mod \to L_K(E) - Mod$$

for which $\Phi^{\text{elim}}_{\mathcal{V}}([L_K(E_{\setminus v})w]) = [L_K(E)w]$ for all vertices w of $E_{\setminus v}$.

Consequently, let E be a finite graph for which $L_K(E)$ is purely infinite simple. Then there exists a graph E' which contains no sources, with the property that $L_K(E)$ is Morita equivalent to $L_K(E')$ via a Morita equivalence

$$\Phi^{\text{ELIM}} : L_K(E') - Mod \to L_K(E) - Mod$$

for which $\Phi^{\text{ELIM}}_{\mathcal{V}}([L_K(E')w]) = [L_K(E)w]$ for all vertices w of E'.

Since purely infinite simplicity is a Morita invariant, one consequence of the previous discussion is the following.

Corollary 6.3.29 *Let E and F be finite graphs. If there is a sequence of standard graph transformations and/or source eliminations which starts at E and ends at F, then $L_K(E)$ is purely infinite simple if and only if $L_K(F)$ is purely infinite simple.*

Since our interest here will be in graphs E and F for which $\det(I_n - A_E^t) = \det(I_m - A_F^t)$ and $K_0(L_K(E)) \cong K_0(L_K(F))$, the following notation will prove convenient.

Definition 6.3.30 Let E and G be finite graphs with $|E^0| = n$ and $|G^0| = m$. We write

$$E \equiv_{\det} G$$

if there is an abelian group isomorphism

$$K_0(L_K(E)) \cong K_0(L_K(G)), \text{ and } \det(I_n - A_E^t) = \det(I_m - A_G^t).$$

Lemma 6.3.31 *Let E be a finite graph for which $L_K(E)$ is purely infinite simple, and let v be a source in E. Then $E \equiv_{\det} E_{\setminus v}$.*

Proof Let $n = |E^0|$. Since v is a source, A_E contains a column of zeros. Then a straightforward determinant computation by cofactors along this column gives $\det(I_n - A_E^t) = \det(I_{n-1} - A_{E_{\setminus v}}^t)$. But $L_K(E)$ and $L_K(E_{\setminus v})$ are Morita equivalent by Proposition 6.3.28, so that their K_0 groups are necessarily isomorphic. □

Now we are ready to prove the first of two main results of this section.

Theorem 6.3.32 ([11, Theorem 1.25]) *Let E and F be finite graphs such that $L_K(E)$ and $L_K(F)$ are purely infinite simple. Let $|E^0| = n$ and $|F^0| = m$. Suppose that $E \equiv_{\det} F$; that is, suppose*

$$K_0(L_K(E)) \cong K_0(L_K(F)), \quad \textit{and} \quad \det(I_n - A_E^t) = \det(I_m - A_F^t).$$

Then $L_K(E)$ is Morita equivalent to $L_K(F)$.

Proof By Proposition 6.3.28 there exist graphs E' and F' such that E' and F' contain no sources, for which $L_K(E)$ is Morita equivalent to $L_K(E')$, and for which $L_K(F)$ is Morita equivalent to $L_K(F')$. By hypothesis, and by applying Lemma 6.3.31 at each stage of the source elimination process, we have that

$$\det(I - A^t_{E'}) = \det(I - A^t_E) = \det(I - A^t_F) = \det(I - A^t_{F'}),$$

and that

$$K_0(L_K(E')) \cong K_0(L_K(E)) \cong K_0(L_K(F)) \cong K_0(L_K(F')).$$

Furthermore, $L_K(E')$ and $L_K(F')$ are each purely infinite simple unital by Corollary 6.3.29. So Proposition 6.3.15 applies, and we conclude that there exists a finite sequence of standard graph transformations which starts at E' and ends at F'. By Corollary 6.3.29, since $L_K(E')$ is purely infinite simple unital with no sources, each time such an operation is applied the resulting graph has no sources, and has a corresponding Leavitt path algebra which is purely infinite simple unital. Thus, at each step of the sequence, we may apply the appropriate result from among Propositions 6.3.19, 6.3.22, and 6.3.25, from which we conclude that each step in the sequence preserves the Morita equivalence of the corresponding Leavitt path algebras. Combining these Morita equivalences at each step then yields the Morita equivalence of $L_K(E')$ and $L_K(F')$.

As a result, we have a sequence of Morita equivalences from $L_K(E)$ to $L_K(E')$ to $L_K(F')$ to $L_K(F)$, and the theorem follows. □

Having now established a result which yields Morita equivalence between various Leavitt path algebras, we now turn to the main task of the section, namely, answering in the affirmative the Algebraic Kirchberg–Phillips Question for a large collection of various pairs of purely infinite simple unital Leavitt path algebras. We introduce some additional notation.

Definition 6.3.33 For finite graphs E and G we write

$$E \equiv_{[1]} G$$

if there exists an abelian group isomorphism $\varphi : K_0(L_K(E)) \to K_0(L_K(G))$ for which $\varphi([1_{L_K(E)}]) = [1_{L_K(G)}]$.

We will show that, in the case of Morita equivalent purely infinite simple Leavitt path algebras $L_K(E)$ and $L_K(G)$ of finite graphs E and G, if $E \equiv_{[1]} G$, then $L_K(E) \cong L_K(G)$. (That is, we will answer the Algebraic Kirchberg–Phillips Question in the affirmative in the situation where we have added the hypothesis that the algebras are Morita equivalent.) The argument relies on the adaptation to this context of a deep result of Huang [97, Theorem 1.1].

Definition 6.3.34 Let $E = (E^0, E^1, s_E, r_E)$ be a directed graph. The *transpose graph of* E, denoted E^t, is the graph F for which $F^0 = E^0$, $F^1 = E^1$, and, for each $e \in F^1$, $s_F(e) = r_E(e)$ and $r_F(e) = s_E(e)$.

So E^t is simply the graph E, but with the orientation of the edges reversed.

Suppose E is such that $L_K(E)$ is purely infinite simple unital, and has no sources. Then using Lemma 6.3.14 (and recalling that in this case E has no sinks), it is straightforward to see that E^t has these same properties. Let E^t be denoted both by H_0 and H_n, and let

$$H_0 \xrightarrow{m_1} H_1 \xrightarrow{m_2} H_2 \cdots\cdots \xrightarrow{m_n} H_n$$

be a finite sequence of standard graph transformations which starts and ends with E^t. We write $H_i = G_i^t$ (where $G_i = H_i^t$), and so we have a finite sequence of graph transformations

$$G_0^t \xrightarrow{m_1} G_1^t \xrightarrow{m_2} G_2^t \cdots\cdots \xrightarrow{m_n} G_n^t$$

from E^t to E^t.

For any graph G let $\tau_G : G \to G^t$ be the graph function which is the identity on vertices, but switches the direction of each of the edges. (This is simply the transpose operation on the corresponding adjacency matrices.) In particular, any one of the standard graph transformations $m : G_i^t \to G_{i+1}^t$ yields a graph transformation

$$m' = \tau_{G_{i+1}}^{-1} \circ m \circ \tau_{G_i} : G_i \to G_{i+1}.$$

Lemma 6.3.35 *If* $m : G_i^t \to G_{i+1}^t$ *is a standard graph transformation, then* $m' = \tau_{G_{i+1}}^{-1} \circ m \circ \tau_{G_i} : G_i \to G_{i+1}$ *is also standard.*

Proof It is tedious but straightforward to check each of the following.

(i) If m is an expansion (resp. contraction), then m' is an expansion (resp. contraction).
(ii) If m is an in-splitting (resp. out-splitting), then m' is an out-splitting (resp. in-splitting).
(iii) If m is an in-amalgamation (resp. out-amalgamation), then m' is an out-amalgamation (resp. in-amalgamation). □

As a consequence of Lemma 6.3.35, if we start with any finite sequence of standard graph transformations

$$H_0 \xrightarrow{m_1} H_1 \xrightarrow{m_2} H_2 \cdots\cdots \xrightarrow{m_n} H_n$$

which starts and ends with E^t, then we get a corresponding finite sequence of standard graph transformations

$$G_0 \xrightarrow{m_1'} G_1 \xrightarrow{m_2'} G_2 \cdots\cdots \xrightarrow{m_n'} G_n$$

which starts and ends with E.

When the size of the $m \times m$ identity matrix is clear from context, we will often write I rather than I_m.

By Huang [97, Lemma 3.7], for any graphs E and F, any standard graph transformation $m : E \to F$ yields the so-called *induced* isomorphism φ_m : $\mathrm{Coker}(I - A_E) \to \mathrm{Coker}(I - A_F)$. For each of the six types of standard graph transformations, the corresponding induced isomorphism is explicitly described in [97, Lemma 3.7]. (See [11, Sect. 2] for an explicitly presented example.) Here now is the connection between the Morita equivalences described above and the induced isomorphisms given by Huang.

Proposition 6.3.36 *Let G_i and G_{i+1} be graphs, and K any field. Suppose G_i is such that $L_K(G_i)$ is purely infinite simple unital, and has no sources. Suppose $m_i : G_i^t \to G_{i+1}^t$ is a standard graph transformation, and let $\varphi_{m_i} : \mathrm{Coker}(I - A_{G_i^t}) \to \mathrm{Coker}(I - A_{G_{i+1}^t})$ be the induced isomorphism. Let $m_i' : G_i \to G_{i+1}$ be the corresponding graph transformation, which, by Lemma 6.3.35, is also a standard transformation. Let $\Phi^{m_i'} : L_K(G_i) - Mod \to L_K(G_{i+1}) - Mod$ be the Morita equivalence induced by m_i' as described in Propositions 6.3.19, 6.3.22, and 6.3.25. Then, using the previously described identification between $K_0(L_K(G_i))$ and $\mathrm{Coker}(I - A_{G_i}^t)$ (resp., between $K_0(L_K(G_{i+1}))$ and $\mathrm{Coker}(I - A_{G_{i+1}}^t)$), we have $\Phi_{\mathscr{V}}^{m_i'} = \varphi_{m_i}$.*

Proof Each of the six types of isomorphisms $\Phi_{\mathscr{V}}^{m_i'} : K_0(L_K(G_i)) \to K_0(L_K(G_{i+1}))$ have been explicitly described previously, and each of the six types of induced isomorphisms $\varphi_{m_i} : \mathrm{Coker}(I - A_{G_i^t}) \to \mathrm{Coker}(I - A_{G_{i+1}^t})$ have been explicitly described in [97, Lemma 3.7]. By definition we have $A_{G_i^t} = A_{G_i}^t$ (resp. $A_{G_{i+1}^t} = A_{G_{i+1}}^t$). It is now a tedious but completely straightforward check to verify that, in all six cases, these isomorphisms agree. □

We are finally in position to adapt the result of Huang to this context. For a purely infinite simple ring R, and an automorphism α of $K_0(R) = \mathscr{V}(R) \setminus \{0\}$, we say a Morita equivalence $\Phi : R - Mod \to R - Mod$ *restricts to* α if $\Phi_{\mathscr{V}} = \alpha$.

Proposition 6.3.37 *Suppose $L_K(E)$ is purely infinite simple unital, and let α be any automorphism of $K_0(L_K(E))$. Then there exists a Morita equivalence $\Phi : L_K(E) - Mod \to L_K(E) - Mod$ which restricts to α.*

Proof If E contains sources, then Proposition 6.3.28 guarantees the existence of a Morita equivalence $\Phi^{\mathrm{ELIM}} : L_K(E') - Mod \to L_K(E) - Mod$, where E' has no sources. If $\Psi : L_K(E') - Mod \to L_K(E') - Mod$ is a Morita equivalence which restricts to the automorphism $(\Phi_{\mathscr{V}}^{\mathrm{ELIM}})^{-1} \circ \alpha \circ \Phi_{\mathscr{V}}^{\mathrm{ELIM}}$ of $K_0(L_K(E'))$, then $\Phi^{\mathrm{ELIM}} \circ \Psi \circ (\Phi^{\mathrm{ELIM}})^{-1}$ is a Morita equivalence from $L_K(E) - Mod$ to $L_K(E) - Mod$ which restricts to α. Therefore, it suffices to consider graphs E with no sources.

Since $L_K(E)$ is purely infinite simple, and E has no sources, then E is essential, irreducible, and non-trivial by the Purely Infinite Simplicity Theorem 3.1.10, and hence so is E^t. Since $K_0(L_K(E))$ is identified with $\mathrm{Coker}(I - A_E^t)$, we may view α

as an automorphism of $\mathrm{Coker}(I - A_E^t) = \mathrm{Coker}(I - A_{E^t})$. Therefore, by Huang [97, Theorem 1.1] (details in [96, Theorem 2.15]), there exists a flow equivalence $\mathscr{F}$ from E^t to itself which induces α. Such a flow equivalence can be written as a finite sequence

$$H_0 \xrightarrow{m_1} H_1 \xrightarrow{m_2} H_2 \cdots\cdots \xrightarrow{m_n} H_n$$

which starts and ends with E^t. But this then yields a corresponding finite sequence of standard graph transformations

$$G_0 \xrightarrow{m_1'} G_1 \xrightarrow{m_2'} G_2 \cdots\cdots \xrightarrow{m_n'} G_n$$

which starts and ends with E, as described in Lemma 6.3.35. This sequence of standard graph transformations in turn yields a sequence of Morita equivalences (using Propositions 6.3.19, 6.3.22, and 6.3.25) which starts and ends at $L_K(E)-Mod$. But by Proposition 6.3.36, at each stage of the sequence the restriction of the Morita equivalence to the appropriate K_0 group agrees with the induced map coming from the standard graph transformation. If we denote by $\Phi : L_K(E) - Mod \to L_K(E) - Mod$ the composition of these Morita equivalences, then Φ restricts to the same automorphism of $K_0(L_K(E))$ as does $\mathscr{F}$, namely, the prescribed automorphism α. □

Here now is the second main result of this section.

Theorem 6.3.38 *Let E and G be finite graphs such that $L_K(E)$ and $L_K(G)$ are purely infinite simple unital Leavitt path algebras. Suppose $E \equiv_{[1]} G$; in other words, suppose*

$$K_0(L_K(E)) \cong K_0(L_K(G))$$

via an isomorphism which sends $[1_{L_K(E)}]$ to $[1_{L_K(G)}]$. Suppose also that $L_K(E)$ is Morita equivalent to $L_K(G)$. Then $L_K(E) \cong L_K(G)$.

Proof Let $\varphi : K_0(L_K(E)) \to K_0(L_K(G))$ be an isomorphism with $\varphi([1_{L_K(E)}]) = [1_{L_K(G)}]$. Since $L_K(E)$ and $L_K(G)$ are Morita equivalent by hypothesis, there exists a Morita equivalence $\Gamma : L_K(E) - Mod \to L_K(G) - Mod$. Thus we have the induced isomorphism $\Gamma_{\mathscr{V}} : K_0(L_K(E)) \to K_0(L_K(G))$.

Now consider the group automorphism $\varphi \circ \Gamma_{\mathscr{V}}^{-1} : K_0(L_K(G)) \to K_0(L_K(G))$. By Proposition 6.3.37, there exists a Morita equivalence $\Psi : L_K(G) - Mod \to L_K(G) - Mod$ such that $\Psi_{\mathscr{V}} = \varphi \circ \Gamma_{\mathscr{V}}^{-1}$. Thus, we get a Morita equivalence

$$H := \Psi \circ \Gamma : L_K(E) - Mod \to L_K(G) - Mod$$

with

$$H_{\mathscr{V}} = (\Psi \circ \Gamma)_{\mathscr{V}} = \Psi_{\mathscr{V}} \circ \Gamma_{\mathscr{V}} = \varphi \circ \Gamma_{\mathscr{V}}^{-1} \circ \Gamma_{\mathscr{V}} = \varphi.$$

In particular, $H_{\mathcal{V}}([1_{L_K(E)}]) = \varphi([1_{L_K(E)}]) = [1_{L_K(G)}]$.

(To paraphrase: Huang's result allows us to establish that if there is *some* Morita equivalence between $L_K(E)$ and $L_K(G)$, and there is *some* given isomorphism between $K_0(L_K(E))$ and $K_0(L_K(G))$, then in fact there is a (perhaps different) Morita equivalence between $L_K(E)$ and $L_K(G)$ which restricts to the given isomorphism between the K_0 groups.)

Since $L_K(E)$ and $L_K(G)$ are purely infinite simple rings, Remark 6.1.13 gives that $[1_{L_K(E)}] \in K_0(L_K(E))$ consists of the finitely generated projective left $L_K(E)$-modules isomorphic (as left $L_K(E)$-modules) to the progenerator ${}_{L_K(E)}L_K(E)$, and analogously $[1_{L_K(G)}] \in K_0(L_K(G))$ consists of the finitely generated projective left $L_K(G)$-modules isomorphic (as left $L_K(G)$-modules) to the progenerator ${}_{L_K(G)}L_K(G)$. Thus the equation $H_{\mathcal{V}}([1_{L_K(E)}]) = [1_{L_K(G)}]$ yields that $H({}_{L_K(E)}L_K(E)) \cong {}_{L_K(G)}L_K(G)$. Since Morita equivalences preserve endomorphism rings, we get ring isomorphisms

$$L_K(E) \cong \mathrm{End}_{L_K(E)}(L_K(E)) \cong \mathrm{End}_{L_K(G)}(H(L_K(E))) \cong \mathrm{End}_{L_K(G)}(L_K(G)) \cong L_K(G),$$

and the theorem is established. □

Definition 6.3.39 For finite graphs E and G with $|E^0| = n$ and $|G^0| = m$, we write

$$E \equiv_{\text{triple}} G$$

if there exists an abelian group isomorphism $\varphi : K_0(L_K(E)) \to K_0(L_K(G))$ for which $\varphi([1_{L_K(E)}]) = [1_{L_K(G)}]$, and $\det(I_n - A_E^t) = \det(I_m - A_G^t)$.

Theorem 6.3.38 now yields the following important consequence, which provides an affirmative answer to the Algebraic Kirchberg–Phillips Question in the presence of additional hypotheses.

Theorem 6.3.40 (The Restricted Algebraic Kirchberg–Phillips Theorem) *Let E and G be finite graphs such that $L_K(E)$ and $L_K(G)$ are purely infinite simple Leavitt path algebras. Let $|E^0| = n$ and $|G^0| = m$. Suppose $E \equiv_{\text{triple}} G$; in other words, suppose $K_0(L_K(E)) \cong K_0(L_K(G))$ via an isomorphism which sends $[1_{L_K(E)}]$ to $[1_{L_K(G)}]$, and that* $\det(I_n - A_E^t) = \det(I_m - A_G^t)$. *Then $L_K(E) \cong L_K(G)$.*

Proof Since $E \equiv_{\text{triple}} G$, we have in particular that $E \equiv_{\det} G$, so that $L_K(E)$ and $L_K(G)$ are Morita equivalent by Theorem 6.3.32. At the same time we also have $E \equiv_{[1]} G$, which together with Theorem 6.3.38 gives the isomorphism we seek. □

Indeed, we may draw the same conclusion as in Theorem 6.3.40, using (seemingly) weaker hypotheses.

Corollary 6.3.41 *Let E and G be finite graphs such that $L_K(E)$ and $L_K(G)$ are purely infinite simple Leavitt path algebras. Let $|E^0| = n$ and $|G^0| = m$. Suppose $E \equiv_{[1]} G$; that is, suppose $K_0(L_K(E)) \cong K_0(L_K(G))$ via an isomorphism which sends $[1_{L_K(E)}]$ to $[1_{L_K(G)}]$. Suppose also that the integers* $\det(I_n - A_E^t)$ *and* $\det(I_m - A_G^t)$ *have the same sign (i.e., are either both non-negative or both non-positive). Then $L_K(E) \cong L_K(G)$.*

Proof Since $K_0(L_K(E)) \cong K_0(L_K(G))$ we have by Corollary 6.1.10 that $\mathrm{Coker}(I_n - A_E^t) \cong \mathrm{Coker}(I_m - A_G^t)$, whence by Proposition 6.3.5 the sequences of "not equal to 1" entries in the Smith normal forms of the two matrices $I_n - A_E^t$ and $I_m - A_G^t$ are the same. By Remark 6.3.8, this yields $|\det(I_n - A_E^t)| = |\det(I_m - A_G^t)|$. So $\det(I_n - A_E^t)$ and $\det(I_m - A_G^t)$ having the same sign implies equality of these two integers, whence the result follows from Theorem 6.3.40. □

So we have answered the Algebraic Kirchberg–Phillips Question in the affirmative, under the additional hypotheses that the determinants of the two germane matrices have the same sign. In particular,

Corollary 6.3.42 *Let E and G be finite graphs such that $L_K(E)$ and $L_K(G)$ are purely infinite simple Leavitt path algebras having infinite K_0 groups. Suppose $K_0(L_K(E)) \cong K_0(L_K(G))$ via an isomorphism which sends $[1_{L_K(E)}]$ to $[1_{L_K(G)}]$. Then $L_K(E) \cong L_K(G)$.*

Proof Let $|E^0| = n$ and $|G^0| = m$. By Remark 6.3.8, the condition that $L_K(E)$ and $L_K(G)$ have infinite K_0 groups yields that $\det(I_n - A_E^t) = 0 = \det(I_m - A_G^t)$, and Theorem 6.3.40 then applies. □

Although this next result follows as a consequence of Theorem 6.3.40, the result was first established in [4] (in which the isomorphism is described explicitly), and then re-established in [3] using different techniques. The current approach to establishing the next result hinges on the following observation.

Remark 6.3.43 It is well known (and not hard to prove) that for positive integers d, d', and $n \geq 2$, there is an automorphism of $\mathbb{Z}/(n-1)\mathbb{Z}$ which takes $\overline{d}$ to $\overline{d'}$ if and only if $\mathrm{g.c.d.}(d, n-1) = \mathrm{g.c.d.}(d', n-1)$.

Corollary 6.3.44 *Let n, n', d, d' be positive integers, and K any field. Then there is an isomorphism of K-algebras*

$$\mathrm{M}_d(L_K(1,n)) \cong \mathrm{M}_{d'}(L_K(1,n'))$$

if and only if $n = n'$ and $\mathrm{g.c.d.}(d, n-1) = \mathrm{g.c.d.}(d', n-1)$. *In particular, $L_K(1,n) \cong \mathrm{M}_d(L_K(1,n))$ if and only if* $\mathrm{g.c.d.}(d, n-1) = 1$.

Proof Throughout the proof we utilize various properties mentioned in Remark 6.1.7.

($\Rightarrow$) Since $L_K(1,n)$ and $L_K(1,n')$ are not Morita equivalent for $n \neq n'$ (as their K_0-groups are the nonisomorphic groups $\mathbb{Z}/(n-1)\mathbb{Z}$ and $\mathbb{Z}/(n'-1)\mathbb{Z}$, respectively), an isomorphism of the indicated algebras necessarily implies $n = n'$. Furthermore, the element $[1_{\mathrm{M}_d(L_K(1,n))}]$ of $K_0(\mathrm{M}_d(L_K(1,n))) \cong \mathbb{Z}/(n-1)\mathbb{Z}$ corresponds to $\overline{d}$ in $\mathbb{Z}/(n-1)\mathbb{Z}$; analogously, the element $[1_{\mathrm{M}_{d'}(L_K(1,n))}]$ of $K_0(\mathrm{M}_{d'}(L_K(1,n))) \cong \mathbb{Z}/(n-1)\mathbb{Z}$ corresponds to $\overline{d'}$ in $\mathbb{Z}/(n-1)\mathbb{Z}$. Since any isomorphism of K-algebras induces an isomorphism of K_0 groups which preserves the position of the regular module, we utilize Remark 6.3.43 to conclude that $\mathrm{g.c.d.}(d, n-1) = \mathrm{g.c.d.}(d', n-1)$.

($\Leftarrow$) The hypotheses together with Remark 6.3.43 yield the existence of an automorphism of $\mathbb{Z}/(n-1)\mathbb{Z}$ which takes $\overline{d}$ to $\overline{d'}$. For integers $d, n \geq 2$, we define the graph R^d_n pictured here

$$R^d_n = \quad \bullet^v \xrightarrow{(d-1)} \bullet^w \circlearrowright (n) \quad .$$

(So there are $d-1$ edges from v to w, and n loops at w.) It is easily seen by the Purely Infinite Simplicity Theorem 3.1.10 that $L_K(R^d_n)$ is purely infinite simple. Furthermore, it is established in [3, Lemma 5.1] that $L_K(R^d_n) \cong \mathrm{M}_d(L_K(1,n))$. (This isomorphism can also be verified by utilizing a proof similar to that given in Proposition 2.2.19.) So we have a sequence of group isomorphisms which preserve the indicated elements:

$$\begin{aligned}
(K_0(L_K(R^d_n)), [1_{L_K(R^d_n)}]) &\cong (K_0(\mathrm{M}_d(L_K(1,n))), [1_{\mathrm{M}_d(L_K(1,n))}]) \cong (\mathbb{Z}/(n-1)\mathbb{Z}, \overline{d}) \\
&\cong (\mathbb{Z}/(n-1)\mathbb{Z}, \overline{d'}) \cong (K_0(\mathrm{M}_{d'}(L_K(1,n))), [1_{\mathrm{M}_{d'}(L_K(1,n))}]) \\
&\cong (K_0(L_K(R^{d'}_n)), [1_{L_K(R^{d'}_n)}]).
\end{aligned}$$

Clearly $A_{R^d_n} = \begin{pmatrix} 0 & d-1 \\ 0 & n \end{pmatrix}$, so that $I_2 - A^t_{R^d_n} = \begin{pmatrix} 1 & 0 \\ 1-d & 1-n \end{pmatrix}$, which gives $\det(I_2 - A^t_{R^d_n}) = 1-n$. But an analogous computation yields $\det(I_2 - A^t_{R^{d'}_n}) = 1-n$ as well.

Thus we have all the ingredients required to invoke the Restricted Algebraic Kirchberg–Phillips Theorem 6.3.40, and thereby conclude that $L_K(R^d_n) \cong L_K(R^{d'}_n)$, which in turn yields the desired isomorphism between matrix rings over $L_K(1,n)$. □

Here is another situation in which the Restricted Algebraic Kirchberg–Phillips Theorem may be invoked.

Example 6.3.45 Let E_6 denote the graph pictured here:

$E_6 =$

v_1
v_2
v_3
v_4
v_5
v_6

We have that $L_K(E_6)$ is purely infinite simple by Theorem 3.1.10. Further,

$$A_{E_6} = \begin{pmatrix} 1&1&0&0&0&0 \\ 1&1&1&0&0&0 \\ 0&1&1&1&0&0 \\ 0&0&1&1&1&0 \\ 0&0&0&1&1&1 \\ 0&0&0&0&1&1 \end{pmatrix}, \text{ so that } I_6 - A^t_{E_6} = \begin{pmatrix} 0&-1&0&0&0&0 \\ -1&0&-1&0&0&0 \\ 0&-1&0&-1&0&0 \\ 0&0&-1&0&1&0 \\ 0&0&0&-1&0&-1 \\ 0&0&0&0&-1&0 \end{pmatrix},$$

which by a tedious computation is seen to have Smith normal form equal to I_6. Thus we get by Corollary 6.3.6 that $K_0(L_K(E_6)) = \{0\}$, which trivially then forces $[1_{K_0(L_K(E_6))}] = 0$ in $K_0(L_K(E_6))$. Furthermore, another tedious computation yields that $\det(I_6 - A^t_{E_6}) = -1$. But the purely infinite simple Leavitt path algebra $L_K(R_2) \cong L_K(1,2)$ has this same data: as $A_{R_2} = (2)$, we have $I_1 - A^t_{R_2} = (-1)$, which gives (as previously established) that $K_0(L_K(R_2)) = \{0\}$ (necessarily then with $[1_{K_0(L_K(R_2))}] = 0$), and $\det(I_1 - A^t_{R_2}) = -1$. We conclude by the Restricted Algebraic Kirchberg–Phillips Theorem 6.3.40 that $L_K(R_2) \cong L_K(E_6)$ as K-algebras; in particular, $L_K(E_6) \cong L_K(1,2)$.

Question 6.3.46 We finish this section by presenting a question which has been the subject of significant investigative effort, but remains unresolved as of 2017. We consider the Leavitt path algebras $L_K(E)$ and $L_K(F)$, where E and F are given by

$$E = \circlearrowleft \bullet^{v_1} \rightleftarrows \bullet^{v_2} \circlearrowright \quad \text{and} \quad F = \circlearrowleft \bullet^{v_1} \rightleftarrows \overset{\circlearrowleft}{\bullet}^{v_2} \rightleftarrows \overset{\circlearrowleft}{\bullet}^{v_3} \rightleftarrows \bullet^{v_4} \circlearrowright .$$

By the Purely Infinite Simplicity Theorem 3.1.10 we see immediately that both $L_K(E)$ and $L_K(F)$ are purely infinite simple Leavitt path algebras. It is easy to show that the Smith normal form of $I_2 - A^t_E$ is I_2. Since F is the graph E_4 of Examples 6.1.11, we have already shown that $K_0(F) = \{0\}$. (This can also be established by a tedious computation which yields that the Smith normal form of $I_4 - A^t_F$ is I_4.) Consequently, $(K_0(L_K(E)), [1_{L_K(E)}]) = (\{0\}, 0) = (K_0(L_K(F)), [1_{L_K(F)}])$. But $\det(I_2 - A^t_E) = -1$, while $\det(I_4 - A^t_F) = +1$. So the two Leavitt path algebras $L_K(E)$ and $L_K(F)$ share the same K_0-data, but the signs of the germane determinants are different, so that Theorem 6.3.40 does not apply here.

It is not known whether the Leavitt path algebras $L_K(E)$ and $L_K(F)$ are isomorphic. (Further discussion of this question is presented below in Sect. 7.3.)

6.4 Tensor Products and Hochschild Homology

In this section, we compute the graded structure of the Hochschild homology groups $HH_n(L_k(E))$ of a Leavitt path algebra of a finite graph E. We use this computation to get some non-isomorphism results. In particular, we show that $L_k(1,2) \otimes L_k(1,2)$ is not isomorphic to $L_k(1,2)$. This contrasts with the isomorphism $\mathcal{O}_2 \otimes \mathcal{O}_2 \cong \mathcal{O}_2$ of C^*-algebras, first established by George Elliott (see e.g., [134]).

Throughout this section we fix a field k. (We use k here rather than K for notational clarity, since many of the results herein will involve the letter K in the context of K-theoretic data.) All vector spaces, tensor products and algebras are over k. If R and S are unital k-algebras, then by an (R,S)-bimodule we understand a left module over $R \otimes S^{op}$. By an R-bimodule we shall mean an (R,R)-bimodule, that is, a left module over the enveloping algebra $R^e = R \otimes R^{op}$. Hochschild homology of k-algebras is always taken over k. If M is an R-bimodule, we write

$$HH_n(R,M) = \mathrm{Tor}_n^{R^e}(R,M)$$

for the Hochschild homology of R with coefficients in M; we abbreviate $HH_n(R) = HH_n(R,R)$.

Definitions 6.4.1 Let R be a k-algebra and M an R-bimodule. The Hochschild homology $HH_*(R,M)$ of R with coefficients in M is computed by the *Hochschild complex* $HH(R,M)$ which is given in degree n by

$$HH(R,M)_n = M \otimes R^{\otimes n} .$$

It is equipped with the Hochschild boundary map b defined by

$$b(a_0 \otimes a_1 \otimes \cdots \otimes a_n) = \sum_{i=0}^{n-1} (-1)^i a_0 \otimes \cdots \otimes a_i a_{i+1} \otimes \cdots \otimes a_n + (-1)^n a_n a_0 \otimes \cdots \otimes a_{n-1} .$$

If R and M happen to be $\mathbb{Z}$-graded, then $HH(R,M)$ splits into a direct sum of subcomplexes

$$HH(R,M) = \bigoplus_{m \in \mathbb{Z}} {}_m HH(R,M) .$$

The homogeneous component of degree m of $HH(R,M)_n$ is the linear subspace of $HH(R,M)_n$ generated by all elementary tensors $a_0 \otimes \cdots \otimes a_n$ with a_i homogeneous and $\sum_{i=1}^n |a_i| = m$. One of the first basic properties of the Hochschild complex is that it commutes with filtering colimits. Thus we have

Lemma 6.4.2 *Let I be a filtered ordered set and let $\{(R_i, M_i) : i \in I\}$ be a directed system of pairs (R_i, M_i) consisting of an algebra R_i and an R_i-bimodule M_i, with algebra maps $R_i \to R_j$ and R_i-bimodule maps $M_i \to M_j$ for each $i \leq j$. Let $(R,M) =$ $\mathrm{colim}_i(R_i, M_i)$. Then $HH_n(R,M) = \mathrm{colim}_i HH_n(R_i, M_i)$ $(n \geq 0)$.*

Let R_i be a k-algebra and M_i an R_i-bimodule $(i = 1, 2)$. The Künneth formula [158, 9.4.1] establishes a natural isomorphism

$$HH_n(R_1 \otimes R_2, M_1 \otimes M_2) \cong \bigoplus_{p=0}^{n} HH_p(R_1, M_1) \otimes HH_{n-p}(R_2, M_2) .$$

Another fundamental fact about Hochschild homology which we shall need is Morita invariance. Let R and S be Morita equivalent algebras, and let $P \in R \otimes S^{op} - Mod$ and $Q \in S \otimes R^{op} - Mod$ implement the Morita equivalence. Then [158, Theorem 9.5.6]

$$HH_n(R,M) = HH_n(S, Q \otimes_R M \otimes_R P) . \tag{6.1}$$

Lemma 6.4.3 *Let $R_1, \dots, R_n$ and $S_1, \dots, S_m, \dots$ be a finite and an infinite sequence of algebras, and let $R = \bigotimes_{i=1}^n R_i$, $S_{\leq m} = \bigotimes_{j=1}^m S_j$, and $S = \bigotimes_{j=1}^\infty S_j$. Assume that*

(i) $HH_q(R_i) \neq 0 \neq HH_q(S_j)$ for $q = 0, 1$, $1 \leq i \leq n$, $1 \leq j$,
(ii) $HH_p(R_i) = HH_p(S_j) = 0$ for $p \geq 2$, $1 \leq i \leq n$, $1 \leq j$, and
(iii) $n \neq m$.

Then no two of R, $S_{\leq m}$ and S are Morita equivalent.

Proof By the Künneth formula, we have

$$HH_n(R) = \bigotimes_{i=1}^n HH_1(R_i) \neq 0, \quad \text{while } HH_p(R) = 0 \text{ for } p > n.$$

By the same argument, $HH_p(S_{\leq m})$ is nonzero for $p = m$, and zero for $p > m$. Hence if $n \neq m$, R and $S_{\leq m}$ do not have the same Hochschild homology and therefore they cannot be Morita equivalent, by (6.1). Similarly, by Lemma 6.4.2, we have

$$HH_n(S) = \bigoplus_{J \subset \mathbb{N}, |J|=n} \Big(\bigotimes_{j \in J} HH_1(S_j)\Big) \otimes \Big(\bigotimes_{j \notin J} HH_0(S_j)\Big),$$

so that $HH_n(S)$ is nonzero for all $n \geq 1$, and thus it cannot be Morita equivalent to either R or $S_{\leq m}$. □

Let R be a unital algebra and G a group acting on R by algebra automorphisms. Form the crossed-product algebra $S = R \rtimes G$, and consider the Hochschild complex $HH(S)$. For each conjugacy class ξ of G, the graded submodule $HH^\xi(S) \subset HH(S)$ generated in degree n by the elementary tensors $a_0 \rtimes g_0 \otimes \cdots \otimes a_n \rtimes g_n$ with $g_0 \cdots g_n \in \xi$ is a subcomplex, and we have a direct sum decomposition $HH(S) = \bigoplus_\xi HH^\xi(S)$. The following theorem of Lorenz describes the complex $HH^\xi(S)$ corresponding to the conjugacy class $\xi = [g]$ of an element $g \in G$ as hyperhomology over the centralizer subgroup $Z_g \subset G$.

Theorem 6.4.4 ([115]) *Let R be a unital k-algebra, G a group acting on R by automorphisms, $g \in G$ and $Z_g \subset G$ the centralizer subgroup. Let $S = R \rtimes G$ be the crossed product algebra, and $HH^{\langle g \rangle}(S) \subset HH(S)$ the subcomplex described above. Consider the R-submodule $S_g = R \rtimes g \subset S$. Then there is a quasi-isomorphism*

$$HH^{[g]}(S) \xrightarrow{\sim} \mathbb{H}(Z_g, HH(R, S_g)).$$

In particular, we have a spectral sequence

$$E^2_{p,q} = H_p(Z_g, HH_q(R, S_g)) \Rightarrow HH^{[g]}_{p+q}(S).$$

Remark 6.4.5 Lorenz formulates his result in terms of the spectral sequence alone, but his proof shows that there is a quasi-isomorphism as stated above; an explicit formula is given, for example, in the proof of [66, Lemma 7.2].

Let A be a not-necessarily-unital k-algebra, and denote by $\tilde{A}$ its unitization. Recall from [161] that A is called *H-unital* if the groups $\mathrm{Tor}_n^{\tilde{A}}(k,A)$ vanish for all $n \geq 0$. Wodzicki proved in [161] that A is H-unital if and only if for every embedding $A \lhd R$ of A as a two-sided ideal of a unital ring R, the map

$$HH(A) \to HH(R:A) = \ker(HH(R) \to HH(R/A))$$

is a quasi-isomorphism.

Lemma 6.4.6 *Theorem 6.4.4 still holds if the condition that R be unital is replaced by the condition that it be H-unital.*

Proof This follows from Theorem 6.4.4 and the fact, proved in [66, Prop. A.6.5], that $R \rtimes G$ is H-unital if R is. □

Let R be a unital algebra, and $\phi : R \to pRp$ a corner isomorphism. As in Sect. 6.2 (or see [25]), we consider the skew Laurent polynomial algebra $R[t_+, t_-, \phi]$; recall that this is the R-algebra generated by elements t_+ and t_-, subject to the relations: $t_+a = \phi(a)t_+$; $at_- = t_-\phi(a)$; $t_-t_+ = 1$; and $t_+t_- = p$. The algebra $S = R[t_+, t_-, \phi]$ is $\mathbb{Z}$-graded by setting $\deg(r) = 0$, $\deg(t_\pm) = \pm 1$. The homogeneous component of $R[t_+, t_-, \phi]$ of degree n is given by

$$R[t_+, t_-, \phi]_n = \begin{cases} t_-^{-n}R & \text{if } n < 0 \\ R & n = 0 \\ Rt_+^n & \text{if } n > 0 \,. \end{cases}$$

Proposition 6.4.7 *Let R be a unital ring, $\phi : R \to pRp$ a corner isomorphism, and $S = R[t_+, t_-, \phi]$. Consider the weight decomposition $HH(S) = \bigoplus_{m\in\mathbb{Z}} {}_mHH(S)$. There is a quasi-isomorphism*

$$_mHH(S) \xrightarrow{\sim} \mathrm{Cone}(1 - \phi : HH(R, S_m) \to HH(R, S_m)) \,. \tag{6.2}$$

Proof If ϕ is an automorphism, then for $S = R \rtimes_\phi \mathbb{Z}$, the right-hand side of (6.2) computes $\mathbb{H}(\mathbb{Z}, HH(R, S_m))$, and the proposition becomes the particular case $G = \mathbb{Z}$ of Theorem 6.4.4. In the general case, let A be the colimit of the inductive system

$$R \xrightarrow{\phi} R \xrightarrow{\phi} R \xrightarrow{\phi} \cdots \,.$$

Note that ϕ induces an automorphism $\hat{\phi} : A \to A$. Now A is H-unital, since it is a filtering colimit of unital algebras, and thus the assertion of the proposition is true for the pair $(A, \hat{\phi})$, by Lemma 6.4.6. Hence it suffices to show that for $B = A \rtimes_{\hat{\phi}} \mathbb{Z}$ the maps $HH(S) \to HH(B)$ and $\mathrm{Cone}(1 - \phi : HH(R, S_m) \to HH(R, S_m)) \to \mathrm{Cone}(1 - \phi : HH(A, B_m) \to HH(A, B_m))$ $(m \in \mathbb{Z})$ are quasi-isomorphisms. The analogous property for K-theory is shown in the course of the third step of the proof of [23, Theorem 3.6]. Since the proof in *loc. cit.* uses only that K-theory commutes with filtering colimits and is matrix invariant on those rings for which it satisfies excision, it applies verbatim to Hochschild homology. This concludes the proof. □

Let E be a finite graph. Recall that the algebra $L = L_k(E)$ is equipped with a $\mathbb{Z}$-grading, see Sect. 2.1. The grading is determined by $|v| = 0$ for $v \in E^0$, and $|\alpha| = 1$, $|\alpha^*| = -1$, for $\alpha \in E^1$. Recall that $L_{0,n}$ denotes the linear span of all the elements of the form $\gamma\nu^*$, where γ and ν are paths with $r(\gamma) = r(\nu)$ and $|\gamma| = |\nu| = n$. Recall from Corollary 2.1.16 the description of the algebras $L_{0,n}$ and of the transition homomorphisms $L_{0,n} \to L_{0,n+1}$, for $n \geq 0$.

Assume E has no sources. For each $i \in E^0$, choose an edge α_i such that $r(\alpha_i) = i$. Consider the elements of $L_k(E)$

$$t_+ = \sum_{i \in E_0} \alpha_i \quad \text{and} \quad t_- = t_+^* .$$

As in Sect. 6.2, we have that $t_-t_+ = 1$. Thus, since $|t_\pm| = \pm 1$, the endomorphism

$$\phi : L \to L, \qquad \phi(x) = t_+ x t_- \tag{6.3}$$

is homogeneous of degree 0 with respect to the $\mathbb{Z}$-grading. In particular, it restricts to an endomorphism of L_0. By Ara et al. [25, Lemma 2.4], we have

$$L = L_0[t_+, t_-, \phi]. \tag{6.4}$$

As in the previous sections of this chapter, the adjacency matrix A_E of the finite graph E plays a major role.

We list the vertex set $E^0 = v_1, v_2, \ldots, v_n$ in such a way that the first e_0' vertices are the sinks of E. Accordingly, the first e_0' rows of the matrix A_E are 0. We let N_E denote the matrix obtained from A_E by deleting these first e_0' rows. The matrix that enters the computation of the Hochschild homology of the Leavitt path algebra is

$$\begin{pmatrix} 0 \\ 1_{e_0 - e_0'} \end{pmatrix} - N_E^t \colon \mathbb{Z}^{e_0 - e_0'} \longrightarrow \mathbb{Z}^{e_0}.$$

By a slight abuse of notation, we will write $1 - N_E^t$ for this matrix. Note that $I - N_E^t \in M_{e_0 \times (e_0 - e_0')}(\mathbb{Z})$. Of course $N_E = A_E$ if E has no sinks.

Theorem 6.4.8 *Let E be a finite graph without sources, and k any field. For each $i \in \mathrm{Reg}(E)$, and $m \geq 1$, let $V_{i,m}$ be the vector space generated by all closed paths c of length m with $s(c) = r(c) = i$. Let $\mathbb{Z} = < \sigma >$ act on*

$$V_m = \bigoplus_{i \in \mathrm{Reg}(E)} V_{i,m}$$

by rotation of closed paths. We have:

$$ {}_mHH_n(L_k(E)) = \begin{cases} \mathrm{Coker}(1-\sigma : V_{|m|} \to V_{|m|}) & n = 0, m \neq 0 \\ \mathrm{Coker}(1 - N_E^t) & n = m = 0 \\ \ker(1-\sigma : V_{|m|} \to V_{|m|}) & n = 1, m \neq 0 \\ \ker(1 - N_E^t) & n = 1, m = 0 \\ 0 & n \notin \{0, 1\} . \end{cases} $$

Proof Let $P = KE \subseteq L_k(E)$ be the path algebra of E, and let $W_m \subset P$ be the subspace generated by all paths of length m. For each fixed $n \geq 1$, and $m \in \mathbb{Z}$, consider the following $L_{0,n}$-bimodule

$$ L_{m,n} = \begin{cases} L_{0,n} W_m L_{0,n} & \text{if } m > 0 \\ L_{0,n} W_{-m}^* L_{0,n} & \text{if } m < 0. \end{cases} $$

For notational simplicity we denote $L_k(E)$ by L. The homogeneous part L_m of L of degree m is then

$$ L_m = \bigcup_{n \geq 1} L_{m,n}. $$

If m is positive, then there is a basis of $L_{m,n}$ consisting of the products $\alpha\theta\beta^*$ where each of α, β and θ is a path in E, $r(\alpha) = s(\theta)$, $r(\beta) = r(\theta)$, $|\alpha| = |\beta| = n$ and $|\theta| = m$. Hence the formula

$$ \pi(\alpha\theta\beta^*) = \begin{cases} \theta & \text{if } \alpha = \beta \\ 0 & \text{else} \end{cases} $$

defines a surjective linear map $L_{m,n} \to V_m$. One checks that π induces an isomorphism

$$ HH_0(L_{0,n}, L_{m,n}) \cong V_m \quad (\text{for } m > 0). $$

Similarly

$$ HH_0(L_{0,n}, L_{m,n}) = V_{|m|}^* \cong V_{-m} \quad (\text{for } m < 0). $$

Next, by Corollary 2.1.16, we have

$$ HH_0(L_{0,n}) = k[\mathrm{Reg}(E)] \oplus \bigoplus_{i \in \mathrm{Sink}(E)} k^{r(i,n)}, $$

where

$$r(i,n) = \max\{r \leq n \mid P(r,i) \neq \emptyset\}.$$

Now note that, because $L_{0,n}$ is a product of matrix algebras, it is separable, and thus $HH_1(L_{0,n}, M) = 0$ for any bimodule M. As observed in (6.4), for the automorphism (6.3) we have $L = L_0[t_+, t_-, \phi]$. Hence in view of Proposition 6.4.7 and Lemma 6.4.2, it only remains to identify the maps $HH_0(L_{0,n}, L_{m,n}) \rightarrow HH_0(L_{0,n+1}, L_{m,n+1})$ induced by inclusion and by the homomorphism ϕ. One checks that for $m \neq 0$, these are respectively the cyclic permutation and the identity $V_{|m|} \rightarrow V_{|m|}$. The case $m = 0$ is dealt with in the same way as in [23, Proof of Theorem 5.10]. □

Corollary 6.4.9 *Let E be a finite graph containing at least one nontrivial closed path, and k any field.*

(i) $HH_n(L_k(E)) = \{0\}$ *for* $n \notin \{0, 1\}$.
(ii) ${}_mHH_*(L_k(E)) \cong {}_{-m}HH_*(L(E))$ *for all* $m \in \mathbb{Z}$.
(iii) There exists an $m \in \mathbb{N}$ *such that* ${}_mHH_0(L_k(E))$ *and* ${}_mHH_1(L_k(E))$ *are both nonzero.*

Proof We first reduce to the case where the graph does not have sources. By the proof of [23, Theorem 6.3], there is a finite complete subgraph F of E such that F has no sources, F contains all the non-trivial closed paths of E, $\mathrm{Sink}(F) = \mathrm{Sink}(E)$, and $L_k(F)$ is a full corner in $L_k(E)$ with respect to the homogeneous idempotent $\sum_{v\in F^0} v$. It follows that $HH_*(L_k(E))$ and $HH_*(L_k(F))$ are graded-isomorphic. Therefore we can assume that E has no sources.

The first two assertions are already part of Theorem 6.4.8. For the last assertion, let α be a cycle in E, and let $m = |\alpha|$. Let σ be the cyclic permutation; then $\{\sigma^i\alpha \mid i = 0, \ldots, m-1\}$ is a linearly independent set in $L_K(E)$. Hence $N(\alpha) = \sum_{i=0}^{m-1} \sigma^i\alpha$ is a nonzero element of $V_m^\sigma = {}_mHH_1(L_k(E))$. Since on the other hand N vanishes on the image of $1 - \sigma : V_m \rightarrow V_m$, it also follows that the class of α in ${}_mHH_0(L_k(E))$ is nonzero. □

Theorem 6.4.10 *Let* $E_1, \ldots, E_n$ *and* $F_1, \ldots, F_m$ *be finite graphs, and k any field. Assume that* $n \neq m$ *and that each of the* E_i *and the* F_j *has at least one non-trivial closed path. Then the k-algebras* $L_k(E_1) \otimes \cdots \otimes L_k(E_n)$ *and* $L_k(F_1) \otimes \cdots \otimes L_k(F_m)$ *are not Morita equivalent.*

Proof Immediate from Lemma 6.4.3 and Corollary 6.4.9(iii). □

Example 6.4.11 It follows in particular from Theorem 6.4.10 that the algebras $L_k(1,2) \cong L_k(R_2)$ and $L_k(1,2) \otimes L_k(1,2)$ are not Morita equivalent for any field k. In particular, these two k-algebras are not isomorphic.

Here is another way of proving that these two algebras are not Morita equivalent, due to Jason Bell and George Bergman [48]. Since the weak global dimension of a tensor product of algebras over a field is at least the sum of their global dimensions, it suffices to show that $L_k(1,2)$ has weak dimension 1. Since $L_k(1,2)$ has global dimension 1 (i.e., is hereditary, see Theorem 3.2.5), it suffices to show it is not von Neumann regular. But this follows immediately from Theorem 3.4.1.

Remark 6.4.12 The observation made in Example 6.4.11 provides another situation in which analogous statements about Leavitt path algebras and graph C^*-algebras need not yield identical outcomes. Recall (Example 5.2.4) that for $n \geq 2$, the Cuntz algebra $\mathcal{O}_n$ is defined as $C^*(R_n)$. It is well known that the tensor product $\mathcal{O}_2 \otimes \mathcal{O}_2$ is isomorphic to $\mathcal{O}_2$ as C^*-algebras (see e.g. [134]). (In the C^*-algebra setting, in general the notion of tensor product is not uniquely determined; however, in this case, as $\mathcal{O}_2$ is nuclear, all notions of tensor product here coincide.)

We denote by L_∞ the unital algebra

$$L_\infty = L_k(R_{\mathbb{N}})$$

presented in Example 1.6.13. So L_∞ is generated by elements $x_1, x_1^*, x_2, x_2^*, \dots$, subject to the relations $x_i^* x_j = \delta_{i,j} 1$ for all $i, j \in \mathbb{N}$.

Proposition 6.4.13 *Let E be any finite graph having at least one non-trivial closed path, and k any field. Then $L_\infty \otimes L_k(E)$ and $L_k(E)$ are not Morita equivalent. Similarly $L_\infty \otimes L_\infty$ and L_∞ are not Morita equivalent.*

Proof We have

$$L_\infty = \varinjlim_{n \in \mathbb{N}} C_k(1, n)\,, \tag{6.5}$$

where $C_k(1,n)$ is the Cohn algebra of Sect. 1.5. But $C_k(1,n) \cong L_k(R_n(\emptyset))$ as described in Example 1.5.20. It follows from Theorem 6.4.8 and (6.5) that the formulas in Theorem 6.4.8 for ${}_mHH_n(L_\infty)$, $m \neq 0$, hold, taking as $V_{i,m}$ the vector space generated by all the words in $x_1, x_2, \dots$ of length m, and that ${}_0HH_0(L_\infty) = k$ and ${}_0HH_n(L_\infty) = 0$ for $n \geq 1$. As before, Lemma 6.4.3 gives the result. □

Theorem 6.4.14 *Let $E_1, \dots, E_n$ and $F_1, \dots, F_m, \dots$ be a finite and an infinite sequence of finite graphs, and k any field. Assume that the number of indices i such that F_i has at least one non-trivial closed path is infinite. Then the algebras $L_k(E_1) \otimes \cdots \otimes L_k(E_n)$ and $\bigotimes_{i=1}^{\infty} L_k(F_i)$ are not Morita equivalent.*

Proof Immediate from Lemma 6.4.3 and Corollary 6.4.9(iii). □

Example 6.4.15 Let $L^{(\infty)}$ denote the infinite tensor product $\bigotimes_{i=1}^{\infty} L_k(1,2)$, and let E be any graph having at least one nontrivial closed path. Then $L^{(\infty)} \otimes L_k(E)$ and $L_k(E)$ are not Morita equivalent.

To conclude the section we note that algebraic K-theory cannot distinguish between $L_k(1,2)$ and $L_k(1,2) \otimes L_k(1,2)$, nor between L_∞ and $L_\infty \otimes L_\infty$. For this we need a lemma, which may be of independent interest. A unital ring R is said to be *regular supercoherent* if all the polynomial rings $R[t_1, \dots, t_n]$ are regular coherent in the sense of [83].

Lemma 6.4.16 *Let E be a finite graph and k any field. Then $L_k(E)$ is regular supercoherent.*

Proof Let kE be the usual path algebra of E. It was observed in the proof of [22, Lemma 7.4] that the algebra $kE[t]$ is regular coherent. The same proof gives that all the polynomial algebras $kE[t_1, \dots, t_n]$ are regular coherent. This shows that kE is regular supercoherent. By Ara and Brustenga [22, Proposition 4.1], the universal localization $kE \to L_k(E) = \Sigma^{-1}kE$ is flat on the left. It follows that $L_k(E)$ is left regular supercoherent (see [23, page 23]). Since $L_k(E) \otimes k[t_1, \dots, t_n]$ admits an involution, it follows that $L_k(E)$ is regular supercoherent. □

Proposition 6.4.17 *Let R be a regular supercoherent k-algebra. Then the algebraic K-theories of $L_k(1,2)$ and of $L_k(1,2) \otimes R$ are both trivial.*

Proof Let $E = R_2$ be the graph with one vertex and two loops, as usual. Then $L_k(1,2) \cong L_k(E)$, and by Corollary 1.5.14 we have

$$L_k(1,2) \otimes R \cong L_R(E).$$

Applying [23, Theorem 7.6] we obtain that $K_*(L_R(E)) = K_*(L_k(E)) = \{0\}$. The result follows. □

In our final result of this section, we obtain a K-absorbing result for Leavitt path algebras of finite graphs, indeed for any regular supercoherent algebra.

Proposition 6.4.18 *Let R be a regular supercoherent k-algebra. Then the natural inclusion $R \to R \otimes L_\infty$ induces an isomorphism $K_i(R) \to K_i(R \otimes L_\infty)$ for all $i \in \mathbb{Z}$.*

Proof Using the notation of the proof of Proposition 6.4.13, we see that it is enough to show that the natural map $R \to R \otimes L_k(R_n(\emptyset))$ induces isomorphisms $K_i(R) \to K_i(R \otimes L_k(R_n(\emptyset))$ for all $i \in \mathbb{Z}$ and all $n \geq 1$. Since R is regular supercoherent the K-theory of $R \otimes L_k(R_n(\emptyset)) \cong L_R(R_n(\emptyset))$ can be computed by using [23, Theorem 7.6]. By the explicit form of the graph $E = R_n(\emptyset)$, we see that $A_E = \begin{pmatrix} 0 & 0 \\ n & n \end{pmatrix}$, so that $N_E = (n \ n)$, and $I - N_E^t = \begin{pmatrix} -n \\ 1-n \end{pmatrix}$. We thus obtain that

$$K_i(R \otimes L(R_n(X))) \cong (K_i(R) \oplus K_i(R))/(-n, 1-n)K_i(R).$$

The natural map $R \to L_R(R_n(X))$ factors as

$$R \to Rv \oplus Rw \to L_R(R_n(X)).$$

The first map induces the diagonal homomorphism $K_i(R) \to K_i(R) \oplus K_i(R)$ sending x to (x, x). The second map induces the natural surjection

$$K_i(R) \oplus K_i(R) \to (K_i(R) \oplus K_i(R))/(-n, 1-n)K_i(R).$$

Therefore the natural homomorphism $R \to L_R(R_n(X))$ induces an isomorphism

$$K_i(R) \xrightarrow{\cong} K_i(L_R(R_n(X))).$$

This concludes the proof. □

Corollary 6.4.19 *The natural maps $k \to L_\infty \to L_\infty \otimes L_\infty$ induce K-theory isomorphisms $K_*(k) = K_*(L_\infty) = K_*(L_\infty \otimes L_\infty)$.*

Proof A first application of Proposition 6.4.18 gives $K_*(k) = K_*(L_\infty)$. A second application shows that for $R_n(X)$ as above, the inclusion $L_k(R_n(X)) \to L_k(R_n(X)) \otimes L_\infty$ induces a K-theory isomorphism; passing to the limit, we obtain the result. □

Chapter 7
Generalizations, Applications, and Current Lines of Research

In the first six chapters of this book we have introduced and subsequently described various properties of Leavitt path algebras. Our goal in this final chapter is to round out the presentation by providing the reader with a sense of how the subject fits into the broader mathematical landscape. In Sect. 7.1 we present the descriptions of various constructions which have grown out of, or were motivated by, Leavitt path algebras. In Sect. 7.2 we describe a few longstanding questions (including questions in seemingly unrelated fields) which were resolved (wholly or partially) by using Leavitt path algebras as a tool. These include a question of Higman about infinite, finitely presented simple groups; a question of Kaplansky about prime, non-primitive von Neumann regular algebras; a question about the realization of various monoids as the $\mathscr{V}$-monoid of a von Neumann regular ring; and others. We then conclude the book with Sect. 7.3, in which we sketch some of the open problems which are, at the time of the book's completion, driving much of the research energy in the subject. For additional information, see [1].

7.1 Generalizations of Leavitt Path Algebras

Leavitt path algebras were first defined and investigated in the setting of row-finite graphs in [5, 31]. With this observation as historical context, it is fair to say that two concepts which may be viewed as generalizations of this original notion have already been discussed herein: namely, the Leavitt path algebras for arbitrary graphs (i.e., relax restrictions on the graph E), and relative Cohn path algebras (i.e., relax the restriction that the (CK2) relation be imposed at all elements of $\text{Reg}(E)$).

The goal of the current section is to briefly present a number of additional generalizations of the notion of a Leavitt path algebra which have been taken up in the literature.

G. Abrams et al., *Leavitt Path Algebras*, Lecture Notes in Mathematics 2191,
DOI 10.1007/978-1-4471-7344-1_7

7.1.1 Leavitt Path Algebras of Separated Graphs

The (CK2) condition imposed at any regular vertex in a Leavitt path algebra may be modified in various ways. Such is the motivation for the discussion in this subsection. All of these ideas appear in [27].

In the (CK2) condition, the edges emanating from a given regular vertex v are treated as a single entity, and the single relation $v = \sum_{e \in s^{-1}(v)} ee^*$ is imposed. More generally, one may partition the set $s^{-1}(v)$ into disjoint nonempty subsets, and then impose a (CK2)-type relation corresponding exactly to those subsets. More formally, a *separated graph* is a pair (E, C), where E is a graph, $C = \sqcup_{v \in E^0} C_v$, and, for each $v \in E^0 \setminus \mathrm{Sink}(E)$, C_v is a partition of $s^{-1}(v)$ into pairwise disjoint nonempty subsets. If $v \in \mathrm{Sink}(E)$, then C_v is taken to be the empty family of subsets of $s^{-1}(v)$.

Definition 7.1.1 Let E be any graph and K any field. $C = \sqcup_{v \in E^0} C_v$, as above. Let $\widehat{E}$ denote the extended graph of E, and $K\widehat{E}$ the path K-algebra of $\widehat{E}$. The *Leavitt path algebra of the separated graph* (E, C) *with coefficients in* K is the quotient of $K\widehat{E}$ by the ideal generated by these two types of relations:

(SCK1) for each $X \in C$, $e^*f = \delta_{e,f} r(e)$ for all $e, f \in X$, and
(SCK2) for each non-sink $v \in E^0$, $v = \sum_{e \in X} ee^*$ for every finite $X \in C_v$.

So the usual Leavitt path algebra $L_K(E)$ is exactly $L_K(E, C)$, where each C_v is defined to be the subset $\{s^{-1}(v)\}$ if v is not a sink, and $\emptyset$ otherwise. Leavitt path algebras of separated graphs include a much wider class of algebras than those which arise as Leavitt path algebras in the standard construction. For instance, the algebras of the form $L_K(m, n)$ for $m \geq 2$ originally studied by Leavitt in [112] do not arise as $L_K(E)$ for any graph E. On the other hand, as shown in [27, Proposition 2.12], $L_K(m, n)$ ($m \geq 2$) appears as a full corner of the Leavitt path algebra of an explicitly described separated graph (having two vertices and $m + n$ edges). In particular, $L_K(m, n)$ is Morita equivalent to the Leavitt path algebra of a separated graph.

Of significantly more importance is the following Bergman-like realization result, which shows that the collection of Leavitt path algebras of separated graphs is extremely broad.

Theorem 7.1.2 ([27, Sect. 4]) *Let M be any conical abelian monoid. Then there exists a graph E, and partition $C = \sqcup_{v \in E^0} C_v$, for which $\mathcal{V}(L_K(E, C)) \cong M$.*

Consequently, $\mathcal{V}(L_K(E, C))$ need not share the separativity nor the refinement properties of the standard Leavitt path algebras $L_K(E)$ (see Sect. 3.6). Furthermore, the ideal structure of $L_K(E, C)$ is in general significantly more complex than that of $L_K(E)$. Nonetheless, a description of the idempotent-generated ideals of $L_K(E, C)$ can be achieved (solely in terms of graph-theoretic information).

7.1.2 Kumjian–Pask Algebras

Any directed graph $E = (E^0, E^1, s, r)$ may be viewed as a category Γ_E; the objects of Γ_E are the vertices E^0, and, for each pair $v, w \in E^0$, the morphism set $\mathrm{Hom}_{\Gamma_E}(v, w)$ consists of those elements of Path(E) having source v and range w. Composition is concatenation. In addition, the set $\mathbb{Z}^+$ may be viewed as the category $\Gamma_{\mathbb{Z}^+}$ having one object, and morphisms given by the elements of $\mathbb{Z}^+$, where composition is addition. At this level of abstraction, the length map ℓ : Path(E) $\to \mathbb{Z}^+$ yields a functor $\Phi_\ell : \Gamma_E \to \Gamma_{\mathbb{Z}^+}$, which satisfies the following *factorization* property on morphisms: if $\lambda \in$ Path(E) and $\ell(\lambda) = m+n$, then there exist unique $\mu, \nu \in$ Path(E) such that $\ell(\mu) = m, \ell(\nu) = n$, and $\lambda = \mu\nu$. Conversely, we may view a category as the morphisms of the category, where the objects are identified with the identity morphisms. Then any category Λ which admits a functor $d : \Lambda \to \Gamma_{\mathbb{Z}^+}$ having the factorization property can be viewed as a directed graph E_Λ in the expected way.

With these observations as motivation, one defines a *higher rank* graph, as follows.

Definition 7.1.3 Let k be a positive integer. View the additive monoid $(\mathbb{Z}^+)^k$ as a category with one object, and view a category as the morphisms of the category, where the objects are identified with the identity morphisms. A *graph of rank k* (or simply a *k-graph*) is a countable category Λ, together with a functor $d : \Lambda \to (\mathbb{Z}^+)^k$, which satisfies the factorization property: if $\lambda \in \Lambda$ and $d(\lambda) = \overline{m} + \overline{n}$ for some $\overline{m}, \overline{n} \in (\mathbb{Z}^+)^k$, then there exist unique $\mu, \nu \in \Lambda$ such that $d(\mu) = \overline{m}, d(\nu) = \overline{n}$, and $\lambda = \mu\nu$. (So the usual notion of a graph is a 1-graph in this more general context.)

Given any k-graph (Λ, d) and field K, one may define the *Kumjian–Pask K-algebra* $KP_K(\Lambda, d)$. (We omit the somewhat lengthy details of the construction; see [37] for the complete description.)

If $k = 1$, and d is the usual length function, then the Kumjian–Pask algebra $KP_K(\Lambda, d)$ is precisely the Leavitt path algebra $L_K(E_\Lambda)$.

7.1.3 Steinberg Algebras: The Groupoid Approach

A *groupoid* $\mathcal{G}$ is a small category in which every morphism has an inverse. Notationally, if f is a morphism in $\mathcal{G}$ with domain x and codomain y, then we write $x = s(f)$ and $y = r(f)$; so a groupoid $\mathcal{G}$ has the property that for each morphism $f : s(f) \to r(f)$ there exists a $g : r(f) \to s(f)$ for which $f \circ g = 1_{r(f)}$ and $g \circ f = 1_{s(f)}$. A *topological groupoid* is a groupoid in which the underlying set is equipped with a topology, in which both the product (i.e., composition) and inversion functions are continuous (where the set of pairs of composable morphisms is given the induced topology from the product topology).

In [144], Steinberg introduced, for any topological groupoid $\mathcal{G}$ satisfying various additional topological conditions (Hausdorff and ample), and any commutative unital ring K, the *K-algebra of the groupoid* $\mathcal{G}$, denoted $K\mathcal{G}$. Formally, $K\mathcal{G}$ is the

K-module spanned by the functions from $\mathscr{G}$ to K which have compact open support and which are continuous on their support (where K has the discrete topology). The algebra $K\mathscr{G}$ is now known as the *Steinberg K-algebra* of the groupoid; in addition, the more common notation for $K\mathscr{G}$ has become $A_K(\mathscr{G})$.

In [104], Kumjian and Pask build a groupoid G_Λ corresponding to any given k-graph Λ. In particular, for a directed graph (i.e., 1-graph) E, a groupoid G_E is associated to E. The construction of G_E is explicitly described in [63, Sect. 2]; we refer the reader to that article for the details. In [61, Proposition 4.3], Clark, Farthing, Sims and Tomforde show that for a row-finite k-graph (Λ, d) with no sources, then $A_{\mathbb{C}}(G_\Lambda) \cong KP_{\mathbb{C}}(\Lambda, d)$, the Kumjian–Pask algebra described above. In particular [61, Remark 4.4], if Λ is a row-finite directed graph with no sources, then $A_{\mathbb{C}}(G_E) \cong L_{\mathbb{C}}(E)$. Subsequently, in [63, Example 3.2], Clark and Sims establish that this isomorphism indeed holds for arbitrary directed graphs. While the two conditions (CK1) and (CK2) are of course not explicitly included in the general definition of a Steinberg algebra, it turns out that these are natural consequences of the Steinberg algebra construction when $\mathscr{G} = G_E$; for instance, the (CK2) condition follows from the trivial observation that for a regular vertex v of E, the set $\{p \in \text{Path}(E) \mid s(p) = v\}$ equals the finite disjoint union $\bigsqcup_{e \in s^{-1}(v)} Q_e$, where $Q_e = \{\alpha \in \text{Path}(E) \mid \alpha = e\beta \text{ for some } \beta \in \text{Path}(E)\}$.

The importance of being able to interpret Leavitt path algebras as Steinberg algebras is twofold. First, the notion of a *groupoid* C^**-algebra* has been investigated by a number of authors; in the specific case of a graph-groupoid G_Λ, the groupoid C^*-algebra is isomorphic to the graph C^*-algebra $C^*(\Lambda)$. For any groupoid $\mathscr{G}$, it has been shown that $A_{\mathbb{C}}(\mathscr{G})$ is dense in the groupoid C^*-algebra $C^*(\mathscr{G})$ (see e.g., [61, Proposition 4.2]). Consequently, the groupoid approach provides a context in which both Leavitt path algebras over $\mathbb{C}$ and graph C^*-algebras live, and thus provides a more general set of tools which are helping to begin to explain the compelling connections between $L_{\mathbb{C}}(E)$ and $C^*(E)$ (see Sect. 5.6).

Secondly, a number of results have been established for various types of Steinberg algebras (i.e., those associated to various types of groupoids). Many of these results have thereby been used to re-establish known results about Leavitt path algebras, and provide some new results as well. For example:

- The groupoids for which the corresponding Steinberg algebra is simple are described in [54]. This in turn yields the Simplicity Theorem 2.9.1 as a direct consequence.
- Results established in [61] (for coefficients in $\mathbb{C}$) and [60] (for coefficients in an arbitrary commutative ring) give both the Graded and Cuntz–Krieger Uniqueness Theorems 2.2.15 and 2.2.16 as consequences.
- Steinberg establishes in [145, Theorem 4.10] necessary and sufficient conditions which give the primitivity of $A_K(\mathscr{G})$ in terms of the structure of $\mathscr{G}$. In the case where $\mathscr{G} = G_E$ is the graph groupoid of the graph E, then the *effectiveness* of G_E corresponds to Condition (L) in E, while the *existence of a dense orbit* in E corresponds to the downward directedness of E. In fact, the dense orbit condition turns out to automatically yield the (CSP) condition on E if E is not

row-finite (see Definition 7.2.3 below), so that Theorem 7.2.5 below can also be re-established from the groupoid perspective.

- Using the tools provided by the Steinberg algebra model of Leavitt path algebras, in [62, Theorems 3.6 and 3.11] the authors describe completely the center of $L_R(E)$ for an arbitrary graph E and commutative ring R. The groupoid model is quite powerful here, as the results of [62] simultaneously yield some previously established descriptions of the center in specific situations; see e.g., [38] and [65].

7.1.4 Non-Field Coefficients

We finish this section by noting that while a great deal of the energy expended on understanding $L_K(E)$ has focused on the graph E, one may also relax the requirement that the coefficients be taken from a field K. For a commutative unital ring R and graph E one may form the *path ring RE of E with coefficients in R* in the expected way; it is then easy to see how to subsequently define the *Leavitt path ring $L_R(E)$ of E with coefficients in R*. (This idea was utilized in Sect. 6.4 without explicit mention.) While some of the results given herein when R is a field do not hold verbatim in the more general setting (e.g., the Simplicity Theorem), one can still understand much of the structure of $L_R(E)$ in terms of the properties of E and R; see e.g., [148]. A situation in which there are results about Leavitt path algebras $L_{\mathbb{Z}}(E)$, but for which there are no (currently) known corresponding results about Leavitt path algebras $L_K(E)$ for K a field, is discussed below, see Theorem 7.3.3.

7.2 Applications of Leavitt Path Algebras

In this section we present a number of instances in which Leavitt path algebras, or their close cousins, have been used to answer various general ring-theoretic questions, questions which on the surface might seem to have little to do with Leavitt path algebras.

7.2.1 Isomorphisms Between Matrix Rings over Leavitt Algebras: Applications to Higman–Thompson Groups

We reconsider the Leavitt algebras $L_K(1, n)$ for $n \geq 2$, the motivating examples of Leavitt path algebras. Fix $n \in \mathbb{N}$ and K any field, and let R denote $L_K(1, n)$. By construction we have ${}_RR \cong {}_RR^n$ as left R-modules; so by taking endomorphism rings and using the standard representation of these as matrix rings, we get $R \cong M_n(R)$ as K-algebras. Furthermore, by repeatedly invoking the module isomorphism

${}_R R \cong {}_R R^n$, we get ${}_R R \cong {}_R R^s$ for any $s = 1 + j(n-1)$ (for all $j \in \mathbb{N}$), which similarly yields $R \cong \mathrm{M}_s(R)$ as K-algebras for all such s. By standard matrix computations, this then also gives $R \cong \mathrm{M}_{s^t}(R)$ for all $t \in \mathbb{N}$, where s is of the indicated form.

It is not difficult to show that isomorphisms of this form do not represent all possible isomorphisms between $L_K(1,n)$ and a matrix ring over itself. For instance, one can show (by explicitly writing down matrices which multiply correctly) that $R = L_K(1,4)$ has $R \cong \mathrm{M}_2(R)$, and 2 is clearly not of the form $1 + j(4-1)$ for $j \in \mathbb{N}$. But an analysis of this particular case leads easily to the general observation that if $d \mid s^t$ for some $t \in \mathbb{N}$, then $R \cong \mathrm{M}_d(R)$ (by an explicitly described isomorphism).

An upshot of the previous remarks is the natural question: Given $n \in \mathbb{N}$, for which $d \in \mathbb{N}$ is $L_K(1,n) \cong \mathrm{M}_d(L_K(1,n))$ as K-algebras? The analogous question was posed by Paschke and Salinas for matrix rings over the Cuntz algebras $\mathcal{O}_n$ in [127]: given $n \in \mathbb{N}$, for which $d \in \mathbb{N}$ is $\mathcal{O}_n \cong \mathrm{M}_d(\mathcal{O}_n)$ as C^*-algebras? The resolution of this analogous question required many years of effort. In the end, the solution may be obtained as a consequence of the Kirchberg–Phillips Theorem 6.3.1: $\mathcal{O}_n \cong \mathrm{M}_d(\mathcal{O}_n)$ if and only if g.c.d.$(d, n-1) = 1$. So while the C^*-algebra question was resolved for matrices over the Cuntz algebras, the solution did not shed any light on the analogous Leavitt algebra question, both because the C^*-algebra solution required analytic tools, and because it did not produce an explicit isomorphism between the germane algebras.

An easy consequence of a result of Leavitt [112, Theorem 5] is that, when g.c.d.$(d, n-1) > 1$, then $L_K(1,n) \not\cong \mathrm{M}_d(L_K(1,n))$. With this and the Cuntz algebra result in hand, it is reasonable to conjecture that $L_K(1,n) \cong \mathrm{M}_d(L_K(1,n))$ if and only if g.c.d.$(d, n-1) = 1$. Clearly if $d \mid n^t$ for some $t \in \mathbb{N}$ then g.c.d.$(d, n-1) = 1$, so that by a previous remark the conjecture is validated in this situation. The key idea which led to Theorem 7.2.1 below was to explicitly produce an isomorphism in situations more general than this. The method of attack was clear: one reaches the desired conclusion by finding a subset of $\mathrm{M}_d(L_K(1,n))$ of size $2n$ which both behaves as in the appropriate $L_K(1,n)$ relations (1.1), and also generates $\mathrm{M}_d(L_K(1,n))$ as a K-algebra.

The smallest pair d, n for which g.c.d.$(d, n-1) = 1$ but $d \nmid n^t$ for any $t \in \mathbb{N}$ is the case $d = 3, n = 5$. Finding subsets of $\mathrm{M}_3(L_K(1,5))$ of size $2 \cdot 5 = 10$ which behave as in (1.1) is not hard. However, the sets of matrices one is led to by slightly modifying the process used in the aforementioned $d \mid n^t$ case yields sets of matrices in $\mathrm{M}_3(L_K(1,5))$ which do not generate $\mathrm{M}_3(L_K(1,5))$ as a K-algebra. Nonetheless, an alternate and eventually successful approach arose from a process which involves viewing matrices over Leavitt algebras as Leavitt path algebras for various graphs, and then manipulating the underlying graphs appropriately. Specifically, various graph operations as described in Sect. 6.3 were used to produce a sequence of explicitly-described isomorphisms which starts with $L_K(R_5)$ and ends with $L_K(M_3R_5)$ (see Definition 2.2.17). By explicitly tracing through this sequence, and then using the isomorphisms $L_K(R_5) \cong L_K(1,5)$ and $L_K(M_3R_5) \cong \mathrm{M}_3(L_K(1,5))$, an appropriate specific set of ten generating matrices in $\mathrm{M}_3(L_K(1,5))$ was identified. This in turn led in a relatively natural way to a method for generalizing the same process and corresponding result to arbitrary d, n.

Theorem 7.2.1 ([4, Theorems 4.14 and 5.12]) *Let $2 \leq n \in \mathbb{N}$, and let K be any field. Then*

$$L_K(1,n) \cong \mathrm{M}_d(L_K(1,n)) \iff \text{g.c.d.}(d, n-1) = 1.$$

More generally,

$$\mathrm{M}_d(L_K(1,n)) \cong \mathrm{M}_{d'}(L_K(1,n)) \iff \text{g.c.d.}(d, n-1) = \text{g.c.d.}(d', n-1).$$

Moreover, when g.c.d.$(d, n-1) =$ g.c.d.$(d', n-1)$, *an isomorphism (indeed, many isomorphisms)* $\mathrm{M}_d(L_K(1,n)) \to \mathrm{M}_{d'}(L_K(1,n))$ *can be explicitly described.*

There are two historically important consequences of the explicit construction of the isomorphisms which yield Theorem 7.2.1. First, when $K = \mathbb{C}$ and g.c.d.$(d, n-1) = 1$, the explicit nature of an isomorphism $L_{\mathbb{C}}(1,n) \cong \mathrm{M}_d(L_{\mathbb{C}}(1,n))$ constructed in the proof of the theorem allows (by a straightforward completion process) for the explicit construction of an isomorphism $\mathcal{O}_n \cong \mathrm{M}_d(\mathcal{O}_n)$; such an explicit isomorphism between these C^*-algebras was unknown at the time. Second, the explicit construction led to the resolution of a longstanding question in group theory. In the mid 1970s, G. Higman produced, for each pair $r, n \in \mathbb{N}$ with $n \geq 2$, an infinite, finitely presented simple group, the now-so-called *Higman–Thompson group* $G^+_{n,r}$. A complete classification up to isomorphism of these groups eluded Higman and others for over four decades. However, E. Pardo was able to use the construction given in the proof of Theorem 7.2.1 to settle the question.

Theorem 7.2.2 ([125, Theorem 3.6]) $G^+_{n,r} \cong G^+_{m,s}$ *if and only if* $m = n$ *and* g.c.d.$(r, n-1) =$ g.c.d.$(s, n-1)$.

Sketch of Proof The forward implication was already known by Higman. Conversely, one first shows that $G^+_{n,\ell}$ can be realized as an appropriate subgroup of the invertible elements of $\mathrm{M}_\ell(L_{\mathbb{C}}(1,n))$ for any $\ell \in \mathbb{N}$. Then one verifies that the explicit isomorphism from $\mathrm{M}_r(L_{\mathbb{C}}(1,n))$ to $\mathrm{M}_s(L_{\mathbb{C}}(1,n))$ provided in the proof of Theorem 7.2.1 takes $G^+_{n,r}$ onto $G^+_{n,s}$. □

7.2.2 *Primitive Leavitt Path Algebras: A Systematic Answer to a Question of Kaplansky*

In Theorem 4.1.10 the primitive Leavitt path algebras $L_K(E)$ arising from row-finite graphs are classified as those for which E is downward directed and satisfies Condition (L). The extension of this primitivity result to arbitrary graphs requires an extra condition.

Definition 7.2.3 The graph E has the *Countable Separation Property* (CSP) if there exists a countable set $S \subseteq E^0$ with the property that for every $v \in E^0$ there exists an $s \in S$ for which $v \geq s$.

Remark 7.2.4 For instance, any graph E for which E^0 is at most countable has CSP. On the other hand, let X be any nonempty set, and let $\mathscr{F}(X)$ denote the collection of nonempty finite subsets of X. The graph $E_{\mathscr{F}(X)}$ is defined by setting $E^0_{\mathscr{F}(X)} = \mathscr{F}(X)$, $E^1_{\mathscr{F}(X)} = \{e_{A,A'} \mid A, A' \in \mathscr{F}(X), \text{ and } A \subsetneq A'\}$, $s(e_{A,A'}) = A$, and $r(e_{A,A'}) = A'$ for each $e_{A,A'} \in E^1_{\mathscr{F}(X)}$. Clearly $E_{\mathscr{F}(X)}$ is acyclic. It is a standard exercise to show that $E_{\mathscr{F}(X)}$ has CSP if and only if X is at most countable.

The equivalent ideal-theoretic conditions provided in the previously-cited [107, Lemma 11.28] which ensure the primitivity of an algebra may again be invoked in this more general setting: to wit, a relatively technical argument presented in [10] establishes that if E does not have CSP, then the unitization of $L_K(E)$ cannot admit an ideal of the appropriate form (see Sect. 4.1).

Theorem 7.2.5 ([10, Theorem 5.7]) *Let E be an arbitrary graph and K any field. Then the Leavitt path algebra $L_K(E)$ is primitive if and only if E is downward directed, E satisfies Condition (L), and E has the Countable Separation Property.*

The structure of prime and primitive algebras has long been a focus of attention. The spark for much of the interest in such structures was a question posed in 1970 by Kaplansky [102, p. 2]: "*Is a regular prime ring necessarily primitive?*" Kaplansky continued: "*It seems unlikely that the answer is affirmative, but a counter-example may have to be weird.*" An example of such a ring (a very clever although somewhat ad hoc construction of a specific group algebra) was first given in 1977 by Domanov [75]. But the use of Theorem 7.2.5, together with Theorem 3.4.1 and Remark 7.2.4, allows for the construction of the following infinite class of prime, non-primitive, von Neumann regular algebras.

Corollary 7.2.6 *Let X be any uncountable set and K any field. Then the Leavitt path algebra $L_K(E_{\mathscr{F}(X)})$ is a prime, non-primitive, von Neumann regular K-algebra.*

In a similar manner, infinite classes of graphs other than those of the form $E_{\mathscr{F}(X)}$ which are acyclic (and so vacuously satisfy Condition (L)), are downward directed, and do not have CSP may be constructed, thereby leading to additional examples of algebras which answer Kaplansky's question in the negative.

The analysis which inspired Theorem 7.2.5 led to a similar result about C^*-algebras.

Theorem 7.2.7 ([14, Theorem 3.8]) *Let E be an arbitrary graph. Then the graph C^*-algebra $C^*(E)$ is primitive if and only if E is downward directed, E satisfies Condition (L), and E has the Countable Separation Property.*

Theorem 7.2.7 thereby gave a general approach to producing C^*-algebras of a type whose existence was put into question by Dixmier in the 1960s. (In addition, this result along with Theorem 7.2.5 furthered the tight connection between certain aspects of Leavitt path algebras and their graph C^*-algebra counterparts, see also Sect. 5.6.)

7.2.3 *The Regular Algebra of a Graph: The Realization Problem for von Neumann Regular Rings*

The "Realization Problem for von Neumann Regular Rings" asks whether every countable conical refinement monoid can be realized as the monoid $\mathcal{V}(R)$ for some von Neumann regular ring R. As the only von Neumann regular Leavitt path algebras are those associated to acyclic graphs (see Theorem 3.4.1), it would initially seem that Leavitt path algebras would not be fertile ground in the context of the Realization Problem. Nonetheless, Ara and Brustenga developed an elegant construction which provides the key connection. Using the *algebra of rational power series on E*, and appropriate localization techniques (*inversion*), they showed how to construct a K-algebra $Q_K(E)$ (the *regular algebra of E*), which has the following properties.

Theorem 7.2.8 ([21, Theorem 4.2]) *Let E be a finite graph and K any field. Then there exists a K-algebra $Q_K(E)$ for which:*

(i) there is an embedding of K-algebras $L_K(E) \hookrightarrow Q_K(E)$,
(ii) $Q_K(E)$ is unital von Neumann regular, and
(iii) $\mathcal{V}(L_K(E)) \cong \mathcal{V}(Q_K(E))$.

Consequently, using Bergman's Theorem 1.4.3, Theorem 7.2.8 yields that any monoid which arises as the graph monoid M_E for a finite graph E has a positive solution to the Realization Problem. This result represented (at the time) a significant broadening of the class of monoids for which the Realization Problem had a positive solution. Theorem 7.2.8 extends relatively easily to row-finite graphs (see [21, Theorem 4.3]), with the proviso that $Q_K(E)$ need not be unital in that generality.

Although significant progress has been made in resolving the Realization Problem for von Neumann regular rings, there is not as of 2017 a complete answer. In particular, a characterization of graph monoids amongst finitely generated conical refinement monoids has been achieved in [34]. A survey of the main ideas relevant to this endeavor can be found in [19].

7.3 Current Lines of Research in Leavitt Path Algebras

In this final section of our book we consider some of the important current research problems in the field. For additional information, see "The graph algebra problem page":

www.math.uh.edu/~tomforde/GraphAlgebraProblems/ListOfProblems.html

This website was built and is being maintained by Mark Tomforde of the University of Houston.

7.3.1 *The Classification Question for Purely Infinite Simple Leavitt Path Algebras, a.k.a. "The Algebraic Kirchberg–Phillips Question"*

In Sect. 6.3 we established the Restricted Algebraic Kirchberg–Phillips Theorem 6.3.40, which asserts that if E and F are finite graphs for which $L_K(E)$ and $L_K(F)$ are purely infinite simple, for which there is an isomorphism $\varphi : K_0(L_K(E)) \to K_0(L_K(F))$ having $\varphi([1_{L_K(E)}]) = [1_{L_K(F)}]$, and for which $\det(I - A_E^t) = \det(I - A_F^t)$, then $L_K(E) \cong L_K(F)$ as K-algebras.

What is generally agreed to be the most compelling unresolved question in the subject of Leavitt path algebras (as of 2017) may then be stated concisely as:

Question 7.3.1 (The Algebraic Kirchberg–Phillips (KP) Question) Can the hypothesis on the determinants in the Restricted Algebraic Kirchberg–Phillips Theorem 6.3.40 be dropped?

More formally, the Algebraic KP Question is the following "Classification Question". Let E and F be finite graphs, and K any field. Suppose $L_K(E)$ and $L_K(F)$ are purely infinite simple. If $K_0(L_K(E)) \cong K_0(L_K(F))$ via an isomorphism for which $[L_K(E)] \mapsto [L_K(F)]$, is it necessarily the case that $L_K(E) \cong L_K(F)$?

With the Restricted Algebraic Kirchberg–Phillips Theorem 6.3.40 having been established, there are three possible answers to the Algebraic Kirchberg–Phillips Question:

No. That is, if the two graphs E and F have $\det(I - A_E^t) \neq \det(I - A_F^t)$, then $L_K(E) \ncong L_K(F)$ for any field K.

Yes. That is, the existence of an isomorphism of the indicated type between the K_0 groups is sufficient to yield an isomorphism of the associated Leavitt path algebras, for any field K.

Sometimes. That is, for some pairs of graphs E and F, and/or for some fields K, the answer is *No*, and for other pairs the answer is *Yes*.

One of the elegant aspects of the Algebraic KP Question is that its answer will be interesting, regardless of which of the three possibilities turns out to be correct. If the answer is *No*, then isomorphism classes of purely infinite simple unital Leavitt path algebras will match exactly the flow equivalence classes of the germane set of graphs, which would suggest that there is some deeper, as-of-yet-not-understood connection between the Leavitt path algebras and symbolic dynamics. If the answer is *Yes*, this would yield further compelling evidence for an as-yet-not-discovered direct connection between various Leavitt path algebra results and the corresponding C^*-algebra results. If the answer is *Sometimes*, then this would likely require the development and utilization of a completely new set of tools in the subject. (Indeed, the *Sometimes* answer might be the most interesting of the three.)

The analogous Kirchberg–Phillips Question regarding Morita equivalence asks whether or not the determinant hypothesis in Theorem 6.3.32 can be dropped. But the two questions will have the same answer: if isomorphic K_0 groups yields Morita equivalence of the Leavitt path algebras, then the Morita equivalence together with the previously invoked Huang's Theorem [96, Theorem 1.1] will yield isomorphism of the algebras.

Suppose E is a finite graph for which $L_K(E)$ is purely infinite simple. There is a way to associate with E a new (finite) graph E_-, for which $L_K(E_-)$ is purely infinite simple, for which $K_0(L_K(E)) \cong K_0(L_K(E_-))$, and for which $\det(I - A_E) = -\det(I - A_{E_-})$. This is called the "Cuntz splice" process, which appends to a vertex $V \in E^0$ two additional vertices and six additional edges, as shown here pictorially:

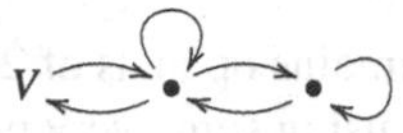

V

Although the indicated isomorphism between $K_0(L_K(E))$ and $K_0(L_K(E_-))$ need not in general send $[1_{L_K(E)}]$ to $[1_{L_K(E_-)}]$, the Cuntz splice process allows for an easy way to produce many specific examples of pairs of Leavitt path algebras to analyze in the context of the Algebraic KP Question. The most basic pair of such algebras arises from the following two graphs:

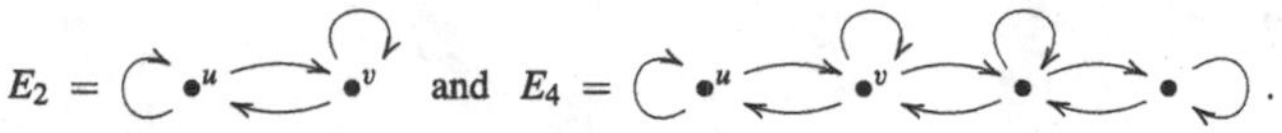

$E_2 = \bullet^u \quad \bullet^v$ and $E_4 = \bullet^u \quad \bullet^v \quad \bullet \quad \bullet$.

We note that $E_4 = (E_2)_-$. These two graphs are precisely those mentioned in Question 6.3.46.

There is an alternate approach to establishing the (analytic) Kirchberg–Phillips Theorem 6.3.1 in the limited context of graph C^*-algebras. Using the same symbolic dynamics techniques as those used to establish Theorem 6.3.40, one can establish the C^*-version of the Restricted Algebraic Kirchberg–Phillips Theorem (i.e., one which involves the determinants). One then "crosses the determinant gap" for a single pair of algebras, by showing that $C^*(E_2) \cong C^*(E_4)$; this is done using a powerful analytic tool (KK-theory). Finally, again using analytic tools, one shows that this one particular crossing of the determinant gap allows for the crossing of the gap for all germane pairs of graph C^*-algebras. But neither KK-theory, nor the tools which yield the extension from one crossing to all crossings, seem to accommodate analogous algebraic techniques.

The pair $\{E_2, E_4\}$ can appropriately be viewed as the smallest pair of graphs of interest in this context, as follows. A graph has *Condition (Sing)* if there are no parallel edges in the graph (i.e., the incidence matrix A_E consists only of 0's and 1's). It can be shown that, up to graph isomorphism, there are 2 (resp., 34) graphs having two (resp., three) vertices, and having Condition (Sing), and for which the corresponding Leavitt path algebras are purely infinite simple; see [3] for an explicit description of these. For each of these 36 graphs E, $\det(I - A_E^t) \leq 0$. So finding an appropriate pair of graphs having Condition (Sing) and having unequal (signs of the) determinants requires at least one of the two graphs to contain at least four vertices.

7.3.2 Tensor Products

We noted in Example 6.4.11 that for any field K, the algebras $L_K(1,2)$ and $L_K(1,2) \otimes L_K(1,2)$ are not Morita equivalent, so of course cannot be isomorphic. But the relationship between these two algebras remains the focus of significant interest. In particular,

Question 7.3.2 For a field K, does there exist a unital homomorphism $\phi: L_K(1,2) \otimes L_K(1,2) \to L_K(1,2)$?

Although Question 7.3.2 remains open as of 2017, there have been some related results achieved. In particular, using some very powerful techniques (an analysis of the Thompson group), Brownlowe and Sørensen in [58] have established:

Theorem 7.3.3 *There is no unital $*$-embedding of $L_{\mathbb{Z}}(1,2) \otimes L_{\mathbb{Z}}(1,2)$ into $L_{\mathbb{Z}}(1,2)$.*

There are a number of additional unresolved questions regarding tensor products of Leavitt path algebras, for example:

Question 7.3.4 Is $L_K(1,2) \otimes_K L_K(1,3)$ isomorphic to $L_K(1,2) \otimes_K L_K(1,2)$ as K-algebras?

7.3.3 The Classification Question for Graphs with Finitely Many Vertices and Infinitely Many Edges

We consider now the collection $\mathscr{S}$ of those graphs E having finitely many vertices, but (countably) infinitely many edges, and for which $L_K(E)$ is (necessarily unital) purely infinite simple. The Purely Infinite Simplicity Theorem 3.1.10 extends to this generality, so we can fairly easily determine whether or not a given graph E is in $\mathscr{S}$. Unlike the case for finite graphs, a description of $K_0(L_K(E))$ for $E \in \mathscr{S}$ cannot be given in terms of the cokernel of an integer-valued matrix transformation from $\mathbb{Z}^{|E^0|}$ to $\mathbb{Z}^{|E^0|}$. Nonetheless, there is still a relatively easy way to determine and describe $K_0(L_K(E))$, so that this group remains a very useful tool in this context.

Recall that $\text{Sing}(E)$ denotes the set of singular vertices of the graph E, i.e., the set of vertices which are either sinks, or infinite emitters. Ruiz and Tomforde in [138] obtained the following.

Theorem 7.3.5 *Let $E, F \in \mathscr{S}$. If $K_0(L_K(E)) \cong K_0(L_K(F))$ and $|\text{Sing}(E)| = |\text{Sing}(F)|$, then $L_K(E)$ is Morita equivalent to $L_K(F)$.*

So, while "the determinant of $I - A_E^t$" is clearly not defined here in the usual sense (because there is at least one pair of vertices v, w in E for which there are infinitely many edges from v to w), the isomorphism class of K_0 together with the number of singular vertices is enough information to determine Morita equivalence. Although this is quite striking, it is not completely satisfying, because it is not clear whether or not $|\text{Sing}(E)|$ is an algebraic property of $L_K(E)$.

Continuing the search for a Classification Theorem which is cast completely in terms of algebraic properties of the underlying algebras, Ruiz and Tomforde were able to show that for a certain type of field (those with *no free quotients*), there is such a result. In a manner similar to the computation of $K_0(L_K(E))$ for $E \in \mathscr{S}$, there is a way to relatively easily compute $K_1(L_K(E))$ as well.

Theorem 7.3.6 ([138, Theorem 7.1]) *Suppose $E, F \in \mathscr{S}$, and suppose that K is a field with no free quotients. Then $L_K(E)$ is Morita equivalent to $L_K(F)$ if and only if $K_0(L_K(E)) \cong K_0(L_K(F))$ and $K_1(L_K(E)) \cong K_1(L_K(F))$.*

The collection of fields having no free quotients includes algebraically closed fields, the field of real numbers, finite fields, perfect fields of positive characteristic, and others. However, the field of rational numbers $\mathbb{Q}$ is not included in this list. Indeed, the authors in [138, Example 10.2] give an example of graphs $E, F \in \mathscr{S}$ for which $K_0(L_\mathbb{Q}(E)) \cong K_0(L_\mathbb{Q}(F))$ and $K_1(L_\mathbb{Q}(E)) \cong K_1(L_\mathbb{Q}(F))$, but $L_\mathbb{Q}(E)$ is not Morita equivalent to $L_\mathbb{Q}(F)$. There are many open questions here. For instance, might there be an integer N for which, if $K_i(L_K(E)) \cong K_i(L_K(F))$ for all $0 \leq i \leq N$, then $L_K(E)$ and $L_K(F)$ are Morita equivalent for all fields K? Of note in this context is that, unlike the situation for graph C^*-algebras (in which the aforementioned Bott periodicity yields that K_0 and K_1 are the only distinct K-groups, see the remark made at the end of Sect. 6.2), there is no analogous result for the K-groups of Leavitt path algebras. Further, although a long exact sequence for the K-groups of $L_K(E)$ has been computed in [23, Theorem 7.6] (as mentioned in Chap. 6), this sequence does not yield easily recognizable information about $K_i(L_K(E))$ for $i \geq 2$.

Finally, we mention an intriguing result presented in [82] demonstrates that, if K is a finite extension of $\mathbb{Q}$, then the pair consisting of $(K_0(L_K(E)), K_6(L_K(E)))$ provides a complete invariant for the Morita equivalence classes of Leavitt path algebras arising from graphs in $\mathscr{S}$, while none of the pairs $(K_0(L_K(E)), K_i(L_K(E)))$ for $1 \leq i \leq 5$ provides such.

7.3.4 Graded Grothendieck Groups, and the Corresponding Graded Classification Conjecture

The Algebraic Kirchberg–Phillips Question, motivated by the corresponding C^*-algebra result, is not the only natural classification-type question to ask in the context of Leavitt path algebras. Having in mind the importance that the $\mathbb{Z}$-grading on $L_K(E)$ has been shown to play in the multiplicative structure, Hazrat in [92] has built the machinery which allows for the casting of an analogous question from the graded point of view.

There is a very well-developed theory of graded modules over group-graded rings, which is especially robust when the group is $\mathbb{Z}$, the case of interest for Leavitt path algebras. (For a general overview of these ideas, see Hazrat's book [94].) If $A = \oplus_{t \in \mathbb{Z}} A_t$ is a $\mathbb{Z}$-graded ring and M is a left A-module, then M is *graded* if $M = \oplus_{i \in \mathbb{Z}} M_i$, and $a_t m_i \in M_{t+i}$ whenever $a_t \in A_t$ and $m_i \in M_i$. If M is a $\mathbb{Z}$-graded

A-module, and $j \in \mathbb{Z}$, then the *suspension module* $M(j)$ is a $\mathbb{Z}$-graded A-module, for which $M(j) = M$ as A-modules, with $\mathbb{Z}$-grading given by setting $M(j)_i = M_{j+i}$ for all $i, j \in \mathbb{Z}$.

In the expected way, one can define the notion of a graded finitely generated projective module, and subsequently build the monoid $\mathcal{V}^{\text{gr}}$ of isomorphism classes of such modules, with $\oplus$ as operation. If $[M] \in \mathcal{V}^{\text{gr}}$, then $[M(j)] \in \mathcal{V}^{\text{gr}}$ for each $j \in \mathbb{Z}$, which yields a $\mathbb{Z}$-action on $\mathcal{V}^{\text{gr}}$. In a manner analogous to the non-graded case, one may define the graded Grothendieck group $K_i^{\text{gr}}(A)$ for each $i \geq 0$. Each of these groups becomes a $\mathbb{Z}[x, x^{-1}]$-module, via the suspension operation.

From this graded-module point of view, one can now ask about structural information of the $\mathbb{Z}$-graded K-algebra $L_K(E)$ which might be gleaned from the K_i^{gr} groups. A reasonable initial question is to ask whether the graded version of the Kirchberg–Phillips Theorem holds. That is, suppose that E and F are finite graphs for which $L_K(E)$ and $L_K(F)$ are purely infinite simple, and suppose $K_0^{\text{gr}}(L_K(E)) \cong K_0^{\text{gr}}(L_K(F))$ as $\mathbb{Z}[x, x^{-1}]$-modules, via an isomorphism which takes $[L_K(E)]$ to $[L_K(F)]$. Is it necessarily the case that $L_K(E) \cong L_K(F)$ as $\mathbb{Z}$-graded K-algebras?

As it turns out, the purely infinite simple hypothesis is not the natural one to start with in the graded context. In fact, Hazrat in [92] makes the following conjecture.

Conjecture 7.3.7 Let E and F be any pair of finite graphs and K any field. Then $L_K(E) \cong L_K(F)$ as $\mathbb{Z}$-graded K-algebras if and only if $K_0^{\text{gr}}(L_K(E)) \cong K_0^{\text{gr}}(L_K(F))$ as $\mathbb{Z}[x, x^{-1}]$-modules, via an order-preserving isomorphism which takes $[L_K(E)]$ to $[L_K(F)]$.

In [92, Theorem 4.8], Hazrat verifies Conjecture 7.3.7 in the case when the graphs E and F are *polycephalic* (essentially, mixtures of acyclic graphs, or graphs which can be described as "multiheaded comets" or "multiheaded roses" in which the cycles and/or roses have no exits.)

As described in Sect. 6.2, in work that predates the introduction of the general definition of Leavitt path algebras, the four authors of [25] investigated the notion of a fractional skew monoid ring, which in particular situations is denoted $A[t_+, t_-, \alpha]$. Recast in the language of Leavitt path algebras, the discussion in [25, Example 2.5] yields that, when E is an essential graph (i.e., has no sinks or sources), then $L_K(E) = L_K(E)_0[t_+, t_-, \alpha]$ for suitable elements $t_+, t_- \in L_K(E)$, and a corner isomorphism α of the zero component $L_K(E)_0$.

When E is a finite graph with no sinks, then $L_K(E)$ is strongly graded [93, Theorem 2], which yields (by a classical theorem of Dade) that the category of graded modules over $L_K(E)$ is equivalent to the category of (all) modules over the zero component $L_K(E)_0$. Adopting this point of view, Ara and Pardo [33, Theorem 4.1] prove the following modified version of Conjecture 7.3.7.

Theorem 7.3.8 *Let E and F be finite essential graphs. Write $L_K(E) = L_K(E)_0[t_+, t_-, \alpha]$ as described above. Then the following are equivalent.*

(1) $K_0(L_K(E)_0) \cong K_0(L_K(F)_0)$ via an order-preserving $K[x, x^{-1}]$-module isomorphism which takes $[1_{L_K(E)_0}]$ to $[1_{L_K(F)_0}]$.

(2) *There exists a locally inner automorphism g of $L_K(E)_0$ for which $L_K(F) \cong L_K(E)_0[t_+, t_-, g \circ \alpha]$ as $\mathbb{Z}$-graded K.*

A complete resolution of Conjecture 7.3.7 currently remains elusive.

7.3.5 Connections to Noncommutative Algebraic Geometry

One of the basic ideas of (standard) algebraic geometry is the correspondence between geometric spaces and commutative algebras. Over the past few decades, significant research energy has been focused on appropriately extending this correspondence to the noncommutative case; the resulting theory is called *noncommutative algebraic geometry.*

Suppose A is a $\mathbb{Z}^+$-graded algebra (i.e., a $\mathbb{Z}$-graded algebra for which $A_n = \{0\}$ for all $n < 0$). Let $\mathrm{Gr}(A)$ denote the category of $\mathbb{Z}$-graded left A-modules (with graded homomorphisms), and let $\mathrm{Fdim}(A)$ denote the full subcategory of $\mathrm{Gr}(A)$ consisting of the graded A-modules which equal the sum of their finite-dimensional submodules. Denote by $\mathrm{QGr}(A)$ the quotient category $\mathrm{Gr}(A)/\mathrm{Fdim}(A)$. The category $\mathrm{QGr}(A)$ turns out to be one of the fundamental constructions in noncommutative algebraic geometry. In particular, if E is a directed graph, then the path algebra KE is $\mathbb{Z}^+$-graded in the usual way (by setting $\deg(v) = 0$ for each vertex v, and $\deg(e) = 1$ for each edge e), and so one may construct the category $\mathrm{QGr}(KE)$.

Let E^{nss} denote the graph obtained by repeatedly removing all sinks and sources (and their incident edges) from E.

Theorem 7.3.9 ([142, Theorem 1.3]) *Let E be a finite graph. Then there is an equivalence of categories*

$$\mathrm{QGr}(KE) \sim \mathrm{Gr}(L_K(E^{\mathrm{nss}})).$$

Moreover, since $L_K(E^{\mathrm{nss}})$ is strongly graded, then these categories are also equivalent to the full category of modules over the zero-component $(L_K(E^{\mathrm{nss}}))_0$.

So the Leavitt path algebra construction arises naturally in the context of noncommutative algebraic geometry.

In general, when the $\mathbb{Z}^+$-graded K-algebra A arises as an appropriate graded deformation of the standard polynomial ring $K[x_0, \ldots, x_n]$, then $\mathrm{QGr}(A)$ shares many similarities with projective n-space $\mathbb{P}^n$; parallels between them have been studied extensively. However, in general, an algebra of the form KE does not arise in this way; and for these, it is much more difficult to connect to the geometric aspects of $\mathrm{QGr}(KE)$. In specific situations there are some geometric perspectives available (see e.g., [143]), but the general case is not well understood.

References

1. Gene Abrams. Leavitt path algebras: the first decade. *Bull. Math. Sci.*, 5(1):59–120, 2015.
2. Gene Abrams and Pham N. Ánh. Some ultramatricial algebras which arise as intersections of Leavitt algebras. *J. Algebra Appl.*, 1(4):357–363, 2002.
3. Gene Abrams, Pham N. Ánh, Adel Louly, and Enrique Pardo. The classification question for Leavitt path algebras. *J. Algebra*, 320(5):1983–2026, 2008.
4. Gene Abrams, Pham N. Ánh, and Enrique Pardo. Isomorphisms between Leavitt algebras and their matrix rings. *J. Reine Angew. Math.*, 624:103–132, 2008.
5. Gene Abrams and Gonzalo Aranda Pino. The Leavitt path algebra of a graph. *J. Algebra*, 293(2):319–334, 2005.
6. Gene Abrams and Gonzalo Aranda Pino. Purely infinite simple Leavitt path algebras. *J. Pure Appl. Algebra*, 207(3):553–563, 2006.
7. Gene Abrams, Gonzalo Aranda Pino, Francesc Perera, and Mercedes Siles Molina. Chain conditions for Leavitt path algebras. *Forum Math.*, 22(1):95–114, 2010.
8. Gene Abrams, Gonzalo Aranda Pino, and Mercedes Siles Molina. Locally finite Leavitt path algebras. *Israel J. Math.*, 165:329–348, 2008.
9. Gene Abrams, Jason P. Bell, Pinar Colak, and Kulumani M. Rangaswamy. Two-sided chain conditions in Leavitt path algebras over arbitrary graphs. *J. Algebra Appl.*, 11(3):1250044, 23, 2012.
10. Gene Abrams, Jason P. Bell, and Kulumani M. Rangaswamy. On prime nonprimitive von Neumann regular algebras. *Trans. Amer. Math. Soc.*, 366(5):2375–2392, 2014.
11. Gene Abrams, Adel Louly, Enrique Pardo, and Christopher Smith. Flow invariants in the classification of Leavitt path algebras. *J. Algebra*, 333:202–231, 2011.
12. Gene Abrams and Kulumani M. Rangaswamy. Regularity conditions for arbitrary Leavitt path algebras. *Algebr. Represent. Theory*, 13(3):319–334, 2010.
13. Gene Abrams and Mark Tomforde. Isomorphism and Morita equivalence of graph algebras. *Trans. Amer. Math. Soc.*, 363(7):3733–3767, 2011.
14. Gene Abrams and Mark Tomforde. A class of C^*-algebras that are prime but not primitive. *Münster J. Math.*, 7(2):489–514, 2014.
15. Pere Ara. Extensions of exchange rings. *J. Algebra*, 197(2):409–423, 1997.
16. Pere Ara. Morita equivalence for rings with involution. *Algebr. Represent. Theory*, 2(3):227–247, 1999.
17. Pere Ara. The exchange property for purely infinite simple rings. *Proc. Amer. Math. Soc.*, 132(9):2543–2547 (electronic), 2004.
18. Pere Ara. Rings without identity which are Morita equivalent to regular rings. *Algebra Colloq.*, 11(4):533–540, 2004.

G. Abrams et al., *Leavitt Path Algebras*, Lecture Notes in Mathematics 2191,
DOI 10.1007/978-1-4471-7344-1

19. Pere Ara. The realization problem for von Neumann regular rings. In *Ring theory 2007*, pages 21–37. World Sci. Publ., Hackensack, NJ, 2009.
20. Pere Ara and Miquel Brustenga. K_1 of corner skew Laurent polynomial rings and applications. *Comm. Algebra*, 33(7):2231–2252, 2005.
21. Pere Ara and Miquel Brustenga. The regular algebra of a quiver. *J. Algebra*, 309(1):207–235, 2007.
22. Pere Ara and Miquel Brustenga. Module theory over Leavitt path algebras and K-theory. *J. Pure Appl. Algebra*, 214(7):1131–1151, 2010.
23. Pere Ara, Miquel Brustenga, and Guillermo Cortiñas. K-theory of Leavitt path algebras. *Münster J. Math.*, 2:5–33, 2009.
24. Pere Ara and Alberto Facchini. Direct sum decompositions of modules, almost trace ideals, and pullbacks of monoids. *Forum Math.*, 18(3):365–389, 2006.
25. Pere Ara, María A. González-Barroso, Kenneth R. Goodearl, and Enrique Pardo. Fractional skew monoid rings. *J. Algebra*, 278(1):104–126, 2004.
26. Pere Ara and Kenneth R. Goodearl. Stable rank of corner rings. *Proc. Amer. Math. Soc.*, 133(2):379–386 (electronic), 2005.
27. Pere Ara and Kenneth R. Goodearl. Leavitt path algebras of separated graphs. *J. Reine Angew. Math.*, 669:165–224, 2012.
28. Pere Ara, Kenneth R. Goodearl, Kevin C. O'Meara, and Enrique Pardo. Separative cancellation for projective modules over exchange rings. *Israel J. Math.*, 105:105–137, 1998.
29. Pere Ara, Kenneth R. Goodearl, and Enrique Pardo. K_0 of purely infinite simple regular rings. *K-Theory*, 26(1):69–100, 2002.
30. Pere Ara and Martin Mathieu. *Local multipliers of C^*-algebras*. Springer Monographs in Mathematics. Springer-Verlag London, Ltd., London, 2003.
31. Pere Ara, María A. Moreno, and Enrique Pardo. Nonstable K-theory for graph algebras. *Algebr. Represent. Theory*, 10(2):157–178, 2007.
32. Pere Ara and Enrique Pardo. Stable rank of Leavitt path algebras. *Proc. Amer. Math. Soc.*, 136(7):2375–2386, 2008.
33. Pere Ara and Enrique Pardo. Towards a K-theoretic characterization of graded isomorphisms between Leavitt path algebras. *J. K-Theory*, 14(2):203–245, 2014.
34. Pere Ara and Enrique Pardo. Representing finitely generated refinement monoids as graph monoids. *J. Algebra*, 480:79–123, 2017.
35. Pere Ara, Gert K. Pedersen, and Francesc Perera. An infinite analogue of rings with stable rank one. *J. Algebra*, 230(2):608–655, 2000.
36. Pere Ara and Francesc Perera. Multipliers of von Neumann regular rings. *Comm. Algebra*, 28(7):3359–3385, 2000.
37. Gonzalo Aranda Pino, John Clark, Astrid an Huef, and Iain Raeburn. Kumjian-Pask algebras of higher-rank graphs. *Trans. Amer. Math. Soc.*, 365(7):3613–3641, 2013.
38. Gonzalo Aranda Pino and Kathi Crow. The center of a Leavitt path algebra. *Rev. Mat. Iberoam.*, 27(2):621–644, 2011.
39. Gonzalo Aranda Pino, Kenneth R. Goodearl, Francesc Perera, and Mercedes Siles Molina. Non-simple purely infinite rings. *Amer. J. Math.*, 132(3):563–610, 2010.
40. Gonzalo Aranda Pino, Enrique Pardo, and Mercedes Siles Molina. Exchange Leavitt path algebras and stable rank. *J. Algebra*, 305(2):912–936, 2006.
41. Gonzalo Aranda Pino, Enrique Pardo, and Mercedes Siles Molina. Prime spectrum and primitive Leavitt path algebras. *Indiana Univ. Math. J.*, 58(2):869–890, 2009.
42. Gonzalo Aranda Pino, Kulumani Rangaswamy, and Lia Vaš. $*$-regular Leavitt path algebras of arbitrary graphs. *Acta Math. Sin. (Engl. Ser.)*, 28(5):957–968, 2012.
43. Hyman Bass. K-theory and stable algebra. *Inst. Hautes Études Sci. Publ. Math.*, 22:5–60, 1964.
44. Hyman Bass, Alex Heller, and Richard G. Swan. The Whitehead group of a polynomial extension. *Inst. Hautes Études Sci. Publ. Math.*, 22:61–79, 1964.
45. Teresa Bates, Jeong Hee Hong, Iain Raeburn, and Wojciech Szymański. The ideal structure of the C^*-algebras of infinite graphs. *Illinois J. Math.*, 46(4):1159–1176, 2002.

46. Teresa Bates and David Pask. Flow equivalence of graph algebras. *Ergodic Theory Dynam. Systems*, 24(2):367–382, 2004.
47. Teresa Bates, David Pask, Iain Raeburn, and Wojciech Szymański. The C^*-algebras of row-finite graphs. *New York J. Math.*, 6:307–324 (electronic), 2000.
48. Jason P. Bell and George M. Bergman. *Personal communication, 2011.*
49. Sterling K. Berberian. *Baer *-rings*. Springer-Verlag, New York, 1972. Die Grundlehren der mathematischen Wissenschaften, Band 195.
50. George M. Bergman. On Jacobson radicals of graded rings. *Unpublished.* http://math.berkeley.edu/~gbergman/papers/unpub/JG.pdf, pages 1–10.
51. George M. Bergman. Coproducts and some universal ring constructions. *Trans. Amer. Math. Soc.*, 200:33–88, 1974.
52. Bruce Blackadar. *K-theory for operator algebras*, volume 5 of *Mathematical Sciences Research Institute Publications*. Cambridge University Press, Cambridge, second edition, 1998.
53. Gary Brookfield. Cancellation in primely generated refinement monoids. *Algebra Universalis*, 46(3):343–371, 2001.
54. Jonathan Brown, Lisa Orloff Clark, Cynthia Farthing, and Aidan Sims. Simplicity of algebras associated to étale groupoids. *Semigroup Forum*, 88(2):433–452, 2014.
55. Lawrence G. Brown. Homotopy of projection in C^*-algebras of stable rank one. *Astérisque*, 232:115–120, 1995. Recent advances in operator algebras (Orléans, 1992).
56. Lawrence G. Brown and Gert K. Pedersen. C^*-algebras of real rank zero. *J. Funct. Anal.*, 99(1):131–149, 1991.
57. Lawrence G. Brown and Gert K. Pedersen. Non-stable K-theory and extremally rich C^*-algebras. *J. Funct. Anal.*, 267(1):262–298, 2014.
58. Nathan Brownlowe and Adam P. W. Sørensen. $L_{2,\mathbb{Z}} \otimes L_{2,\mathbb{Z}}$ does not embed in $L_{2,\mathbb{Z}}$. *J. Algebra*, 456:1–22, 2016.
59. Huanyin Chen. On separative refinement monoids. *Bull. Korean Math. Soc.*, 46(3):489–498, 2009.
60. Lisa Orloff Clark and Cain Edie-Michell. Uniqueness theorems for Steinberg algebras. *Algebr. Represent. Theory*, 18(4):907–916, 2015.
61. Lisa Orloff Clark, Cynthia Farthing, Aidan Sims, and Mark Tomforde. A groupoid generalisation of Leavitt path algebras. *Semigroup Forum*, 89(3):501–517, 2014.
62. Lisa Orloff Clark, Dolores Martin Barquero, Cándido Martin Gonzalez, and Mercedes Siles Molina. Using the Steinberg algebra model to determine the center of any Leavitt path algebra. To appear in Israel Journal of Mathematics. *arXiv*, 1604.01079:1–14, 2016.
63. Lisa Orloff Clark and Aidan Sims. Equivalent groupoids have Morita equivalent Steinberg algebras. *J. Pure Appl. Algebra*, 219(6):2062–2075, 2015.
64. P. M. Cohn. Some remarks on the invariant basis property. *Topology*, 5:215–228, 1966.
65. María G. Corrales García, Dolores Martín Barquero, Cándido Martín González, Mercedes Siles Molina, and José F. Solanilla Hernández. Extreme cycles. The center of a Leavitt path algebra. *Publ. Mat.*, 60(1):235–263, 2016.
66. Guillermo Cortiñas and Eugenia Ellis. Isomorphism conjectures with proper coefficients. *J. Pure Appl. Algebra*, 218(7):1224–1263, 2014.
67. Peter Crawley and Bjarni Jónsson. Refinements for infinite direct decompositions of algebraic systems. *Pacific J. Math.*, 14:797–855, 1964.
68. Joachim Cuntz. Simple C^*-algebras generated by isometries. *Comm. Math. Phys.*, 57(2):173–185, 1977.
69. Joachim Cuntz. The structure of multiplication and addition in simple C^*-algebras. *Math. Scand.*, 40(2):215–233, 1977.
70. Joachim Cuntz. K-theory for certain C^*-algebras. *Ann. of Math. (2)*, 113(1):181–197, 1981.
71. Joachim Cuntz and Wolfgang Krieger. A class of C^*-algebras and topological Markov chains. *Invent. Math.*, 56(3):251–268, 1980.
72. Klaus Deicke, Jeong Hee Hong, and Wojciech Szymański. Stable rank of graph algebras. Type I graph algebras and their limits. *Indiana Univ. Math. J.*, 52(4):963–979, 2003.

73. Jacques Dixmier. Sur les C^*-algèbres. *Bull. Soc. Math. France*, 88:95–112, 1960.
74. Jacques Dixmier. *Les C^*-algèbres et leurs représentations*. Deuxième édition. Cahiers Scientifiques, Fasc. XXIX. Gauthier-Villars Éditeur, Paris, 1969.
75. O.I. Domanov. A prime but not primitive regular ring. *Uspehi Mat. Nauk*, 32(6(198)):219–220, 1977.
76. Søren Eilers, Gunnar Restorff, Efren Ruiz, and Adam P.W. Sørensen. The complete classification of unital graph C^*-algebras: geometric and strong. *arXiv*, 1611.07120v1:1–73, 2016.
77. Ruy Exel. *Partial Dynamical Systems, Fell Bundles and Applications*, volume 224 of *Mathematical Surveys and Monographs Series*. American Mathematical Society, Providence, R.I., 2017.
78. Alberto Facchini and Franz Halter-Koch. Projective modules and divisor homomorphisms. *J. Algebra Appl.*, 2(4):435–449, 2003.
79. Carl Faith. *Lectures on injective modules and quotient rings*. Lecture Notes in Mathematics, No. 49. Springer-Verlag, Berlin-New York, 1967.
80. Antonio Fernández López, Eulalia García Rus, Miguel Gómez Lozano, and Mercedes Siles Molina. Goldie theorems for associative pairs. *Comm. Algebra*, 26(9):2987–3020, 1998.
81. John Franks. Flow equivalence of subshifts of finite type. *Ergodic Theory Dynam. Systems*, 4(1):53–66, 1984.
82. James Gabe, Efren Ruiz, Mark Tomforde, and Tristan Whalen. K-theory for Leavitt path algebras: Computation and classification. *J. Algebra*, 433:35–72, 2015.
83. Stephen M. Gersten. K-theory of free rings. *Comm. Algebra*, 1:39–64, 1974.
84. Miguel Gómez Lozano and Mercedes Siles Molina. Quotient rings and Fountain-Gould left orders by the local approach. *Acta Math. Hungar.*, 97(4):287–301, 2002.
85. Kenneth R. Goodearl. *Notes on real and complex C^*-algebras*, volume 5 of *Shiva Mathematics Series*. Shiva Publishing Ltd., Nantwich, 1982.
86. Kenneth R. Goodearl. *von Neumann regular rings*. Robert E. Krieger Publishing Co. Inc., Malabar, FL, second edition, 1991.
87. Kenneth R. Goodearl. Leavitt path algebras and direct limits. In *Rings, Modules and Representations*, volume 480 of *Contemp. Math.* American Mathematical Society, Providence, R.I., 2009.
88. Kenneth R. Goodearl and Robert B. Warfield, Jr. Algebras over zero-dimensional rings. *Math. Ann.*, 223(2):157–168, 1976.
89. Frederick M. Goodman, Pierre de la Harpe, and Vaughan F. R. Jones. *Coxeter graphs and towers of algebras*, volume 14 of *Mathematical Sciences Research Institute Publications*. Springer-Verlag, New York, 1989.
90. David Handelman. Rings with involution as partially ordered abelian groups. *Rocky Mountain J. Math.*, 11(3):337–381, 1981.
91. Damon Hay, Marissa Loving, Martin Montgomery, Efren Ruiz, and Katherine Todd. Non-stable K-theory for Leavitt path algebras. *Rocky Mountain J. Math.*, 44(6):1817–1850, 2014.
92. Roozbeh Hazrat. The graded Grothendieck group and the classification of Leavitt path algebras. *Math. Ann.*, 355(1):273–325, 2013.
93. Roozbeh Hazrat. A note on the isomorphism conjectures for Leavitt path algebras. *J. Algebra*, 375:33–40, 2013.
94. Roozbeh Hazrat. *Graded rings and graded Grothendieck groups*, volume 435 of *London Mathematical Society Lecture Note Series*. Cambridge University Press, Cambridge, 2016.
95. Richard H. Herman and Leonid N. Vaserstein. The stable range of C^*-algebras. *Invent. Math.*, 77(3):553–555, 1984.
96. Danrun Huang. Flow equivalence of reducible shifts of finite type. *Ergodic Theory Dynam. Systems*, 14(4):695–720, 1994.
97. Danrun Huang. Automorphisms of Bowen–Franks groups of shifts of finite type. *Ergodic Theory Dynam. Systems*, 21(4):1113–1137, 2001.
98. Nathan Jacobson. Some remarks on one-sided inverses. *Proc. Amer. Math. Soc.*, 1:352–355, 1950.

99. Nathan Jacobson. *Structure of rings.* American Mathematical Society, Colloquium Publications, vol. 37. American Mathematical Society, 190 Hope Street, Prov., R. I., 1956.
100. Ja A. Jeong and Gi Hyun Park. Graph C^*-algebras with real rank zero. *J. Funct. Anal.*, 188(1):216–226, 2002.
101. Richard V. Kadison and John R. Ringrose. *Fundamentals of the theory of operator algebras. Vol. I*, volume 15 of *Graduate Studies in Mathematics*. American Mathematical Society, Providence, RI, 1997. Elementary theory, Reprint of the 1983 original.
102. Irving Kaplansky. *Algebraic and analytic aspects of operator algebras.* American Mathematical Society, Providence, R.I., 1970. Conference Board of the Mathematical Sciences Regional Conference Series in Mathematics, No. 1.
103. Eberhard Kirchberg. The classification of purely infinite C^*-algebras using Kasparov theory. *Unpublished, 3rd draft*, pages 1–37, 1994.
104. Alex Kumjian and David Pask. Higher rank graph C^*-algebras. *New York J. Math.*, 6:1–20, 2000.
105. Alex Kumjian, David Pask, and Iain Raeburn. Cuntz–Krieger algebras of directed graphs. *Pacific J. Math.*, 184(1):161–174, 1998.
106. Alex Kumjian, David Pask, Iain Raeburn, and Jean Renault. Graphs, groupoids, and Cuntz–Krieger algebras. *J. Funct. Anal.*, 144(2):505–541, 1997.
107. Tsit-Yen Lam. *A first course in noncommutative rings*, volume 131 of *Graduate Texts in Mathematics*. Springer-Verlag, New York, second edition, 2001.
108. Tsit-Yuen Lam. *Lectures on modules and rings*, volume 189 of *Graduate Texts in Mathematics*. Springer-Verlag, New York, 1999.
109. Joachim Lambek. *Lectures on rings and modules.* Chelsea Publishing Co., New York, second edition, 1976.
110. Charles Lanski, Richard Resco, and Lance Small. On the primitivity of prime rings. *J. Algebra*, 59(2):395–398, 1979.
111. Hossein Larki and Abdolhamid Riazi. Stable rank of Leavitt path algebras of arbitrary graphs. *Bull. Aust. Math. Soc.*, 88(2):206–217, 2013.
112. William G. Leavitt. The module type of a ring. *Trans. Amer. Math. Soc.*, 103:113–130, 1962.
113. William G. Leavitt. The module type of homomorphic images. *Duke Math. J.*, 32:305–311, 1965.
114. Douglas Lind and Brian Marcus. *An introduction to symbolic dynamics and coding.* Cambridge University Press, Cambridge, 1995.
115. Martin Lorenz. On the homology of graded algebras. *Comm. Algebra*, 20(2):489–507, 1992.
116. John C. McConnell and J. Chris Robson. *Noncommutative Noetherian rings*, volume 30 of *Graduate Studies in Mathematics*. American Mathematical Society, Providence, RI, revised edition, 2001. With the cooperation of L.W. Small.
117. Pere Menal and Jaume Moncasi. Lifting units in self-injective rings and an index theory for Rickart C^*-algebras. *Pacific J. Math.*, 126(2):295–329, 1987.
118. Gerard J. Murphy. *C^*-algebras and operator theory.* Academic Press, Inc., Boston, MA, 1990.
119. Constantin Năstăsescu and Freddy van Oystaeyen. *Graded ring theory*, volume 28 of *North-Holland Mathematical Library*. North-Holland Publishing Co., Amsterdam, 1982.
120. Constantin Năstăsescu and Freddy Van Oystaeyen. *Methods of graded rings*, volume 1836 of *Lecture Notes in Mathematics*. Springer-Verlag, Berlin, 2004.
121. W. Keith Nicholson. Lifting idempotents and exchange rings. *Trans. Amer. Math. Soc.*, 229:269–278, 1977.
122. W.K. Nicholson. *I*-rings. *Trans. Amer. Math. Soc.*, 207:361–373, 1975.
123. Narutaka Ozawa. About the Connes embedding conjecture: algebraic approaches. *Jpn. J. Math.*, 8(1):147–183, 2013.
124. Andrei V. Pajitnov and Andrew A. Ranicki. The Whitehead group of the Novikov ring. *K-Theory*, 21(4):325–365, 2000. Special issues dedicated to Daniel Quillen on the occasion of his sixtieth birthday, Part V.

125. Enrique Pardo. The isomorphism problem for Higman–Thompson groups. *J. Algebra*, 344:172–183, 2011.
126. Bill Parry and Dennis Sullivan. A topological invariant of flows on 1-dimensional spaces. *Topology*, 14(4):297–299, 1975.
127. William L. Paschke and Norberto Salinas. Matrix algebras over $\mathscr{O}_n$. *Michigan Math. J.*, 26(1):3–12, 1979.
128. N. Christopher Phillips. A classification theorem for nuclear purely infinite simple C^*-algebras. *Doc. Math.*, 5:49–114 (electronic), 2000.
129. Iain Raeburn. *Graph algebras*, volume 103 of *CBMS Regional Conference Series in Mathematics*. Published for the Conference Board of the Mathematical Sciences, Washington, DC, 2005.
130. Iain Raeburn and Wojciech Szymański. Cuntz–Krieger algebras of infinite graphs and matrices. *Trans. Amer. Math. Soc.*, 356(1):39–59 (electronic), 2004.
131. Kulumani M. Rangaswamy. The theory of prime ideals of Leavitt path algebras over arbitrary graphs. *J. Algebra*, 375:73–96, 2013.
132. Kulumani M. Rangaswamy. On generators of two-sided ideals of Leavitt path algebras over arbitrary graphs. *Comm. Algebra*, 42(7):2859–2868, 2014.
133. Marc A. Rieffel. Dimension and stable rank in the K-theory of C^*-algebras. *Proc. London Math. Soc. (3)*, 46(2):301–333, 1983.
134. Mikael Rørdam. A short proof of Elliott's theorem: $\mathscr{O}_2 \otimes \mathscr{O}_2 \cong \mathscr{O}_2$. *C. R. Math. Rep. Acad. Sci. Canada*, 16(1):31–36, 1994.
135. Mikael Rørdam. Classification of Cuntz–Krieger algebras. *K-Theory*, 9(1):31–58, 1995.
136. Mikael Rørdam, Flemming Larsen, and Niels Laustsen. *An introduction to K-theory for C^*-algebras*, volume 49 of *London Mathematical Society Student Texts*. Cambridge University Press, Cambridge, 2000.
137. Joseph J. Rotman. *An introduction to homological algebra*. Universitext. Springer, New York, second edition, 2009.
138. Efren Ruiz and Mark Tomforde. Classification of unital simple Leavitt path algebras of infinite graphs. *J. Algebra*, 384:45–83, 2013.
139. Konrad Schmüdgen. *Unbounded operator algebras and representation theory*, volume 37 of *Operator Theory: Advances and Applications*. Birkhäuser Verlag, Basel, 1990.
140. Konrad Schmüdgen. Noncommutative real algebraic geometry—some basic concepts and first ideas. In *Emerging applications of algebraic geometry*, volume 149 of *IMA Vol. Math. Appl.*, pages 325–350. Springer, New York, 2009.
141. Laurence C. Siebenmann. A total Whitehead torsion obstruction to fibering over the circle. *Comment. Math. Helv.*, 45:1–48, 1970.
142. S. Paul Smith. Category equivalences involving graded modules over path algebras of quivers. *Adv. Math.*, 230(4–6):1780–1810, 2012.
143. S. Paul Smith. The space of Penrose tilings and the noncommutative curve with homogeneous coordinate ring $k\langle x, y\rangle/(y^2)$. *J. Noncommut. Geom.*, 8(2):541–586, 2014.
144. Benjamin Steinberg. A groupoid approach to discrete inverse semigroup algebras. *Adv. Math.*, 223(2):689–727, 2010.
145. Benjamin Steinberg. Simplicity, primitivity and semiprimitivity of étale groupoid algebras with applications to inverse semigroup algebras. *J. Pure Appl. Algebra*, 220(3):1035–1054, 2016.
146. Mark Tomforde. *Extensions of graph C^*-algebras*. ProQuest LLC, Ann Arbor, MI, 2002. Thesis (Ph.D.)–Dartmouth College.
147. Mark Tomforde. Uniqueness theorems and ideal structure for Leavitt path algebras. *J. Algebra*, 318(1):270–299, 2007.
148. Mark Tomforde. Leavitt path algebras with coefficients in a commutative ring. *J. Pure Appl. Algebra*, 215(4):471–484, 2011.
149. Leonid N. Vaseršteĭn. On the stabilization of the general linear group over a ring. *Math. USSR-Sb.*, 8:383–400, 1969.

150. Leonid N. Vaseršteĭn. The stable range of rings and the dimension of topological spaces. *Funkcional. Anal. i Priložen.*, 5(2):17–27, 1971.
151. Leonid N. Vaseršteĭn. Bass's first stable range condition. In *Proceedings of the Luminy conference on algebraic K-theory (Luminy, 1983), J. Pure Appl. Algebra*, volume 34, pages 319–330, 1984.
152. Ivan Vidav. On some $*$-regular rings. *Acad. Serbe Sci. Publ. Inst. Math.*, 13:73–80, 1959.
153. R.B. Warfield, Jr. Cancellation of modules and groups and stable range of endomorphism rings. *Pacific J. Math.*, 91(2):457–485, 1980.
154. Robert B. Warfield, Jr. Exchange rings and decompositions of modules. *Math. Ann.*, 199:31–36, 1972.
155. Yasuo Watatani. Graph theory for C^*-algebras. In *Operator algebras and applications, Part I (Kingston, Ont., 1980)*, volume 38 of *Proc. Sympos. Pure Math.*, pages 195–197. Amer. Math. Soc., Providence, R.I., 1982.
156. Friedrich Wehrung. Various remarks on separativity and stable rank in refinement monoids. *Unpublished manuscript.*
157. Friedrich Wehrung. The dimension monoid of a lattice. *Algebra Universalis*, 40(3):247–411, 1998.
158. Charles A. Weibel. *An introduction to homological algebra*, volume 38 of *Cambridge Studies in Advanced Mathematics*. Cambridge University Press, Cambridge, 1994.
159. Charles A. Weibel. *The K-book*, volume 145 of *Graduate Studies in Mathematics*. American Mathematical Society, Providence, RI, 2013. An introduction to algebraic K-theory.
160. Robert F. Williams. Classification of subshifts of finite type. *Ann. of Math. (2)*, 98:120–153; errata, ibid. (2) 99 (1974), 380–381, 1973.
161. Mariusz Wodzicki. Excision in cyclic homology and in rational algebraic K-theory. *Ann. of Math. (2)*, 129(3):591–639, 1989.

Index

G. Abrams et al., *Leavitt Path Algebras*, Lecture Notes in Mathematics 2191,
DOI 10.1007/978-1-4471-7344-1

LECTURE NOTES IN MATHEMATICS Springer

Editors in Chief: J.-M. Morel, B. Teissier;

Editorial Policy

1. Lecture Notes aim to report new developments in all areas of mathematics and their applications – quickly, informally and at a high level. Mathematical texts analysing new developments in modelling and numerical simulation are welcome.

   Manuscripts should be reasonably self-contained and rounded off. Thus they may, and often will, present not only results of the author but also related work by other people. They may be based on specialised lecture courses. Furthermore, the manuscripts should provide sufficient motivation, examples and applications. This clearly distinguishes Lecture Notes from journal articles or technical reports which normally are very concise. Articles intended for a journal but too long to be accepted by most journals, usually do not have this "lecture notes" character. For similar reasons it is unusual for doctoral theses to be accepted for the Lecture Notes series, though habilitation theses may be appropriate.

2. Besides monographs, multi-author manuscripts resulting from SUMMER SCHOOLS or similar INTENSIVE COURSES are welcome, provided their objective was held to present an active mathematical topic to an audience at the beginning or intermediate graduate level (a list of participants should be provided).

   The resulting manuscript should not be just a collection of course notes, but should require advance planning and coordination among the main lecturers. The subject matter should dictate the structure of the book. This structure should be motivated and explained in a scientific introduction, and the notation, references, index and formulation of results should be, if possible, unified by the editors. Each contribution should have an abstract and an introduction referring to the other contributions. In other words, more preparatory work must go into a multi-authored volume than simply assembling a disparate collection of papers, communicated at the event.

3. Manuscripts should be submitted either online at www.editorialmanager.com/lnm to Springer's mathematics editorial in Heidelberg, or electronically to one of the series editors. Authors should be aware that incomplete or insufficiently close-to-final manuscripts almost always result in longer refereeing times and nevertheless unclear referees' recommendations, making further refereeing of a final draft necessary. The strict minimum amount of material that will be considered should include a detailed outline describing the planned contents of each chapter, a bibliography and several sample chapters. Parallel submission of a manuscript to another publisher while under consideration for LNM is not acceptable and can lead to rejection.

4. In general, **monographs** will be sent out to at least 2 external referees for evaluation.

   A final decision to publish can be made only on the basis of the complete manuscript, however a refereeing process leading to a preliminary decision can be based on a pre-final or incomplete manuscript.

   Volume Editors of **multi-author works** are expected to arrange for the refereeing, to the usual scientific standards, of the individual contributions. If the resulting reports can be forwarded to the LNM Editorial Board, this is very helpful. If no reports are forwarded or if other questions remain unclear in respect of homogeneity etc, the series editors may wish to consult external referees for an overall evaluation of the volume.

5. Manuscripts should in general be submitted in English. Final manuscripts should contain at least 100 pages of mathematical text and should always include

   - a table of contents;
   - an informative introduction, with adequate motivation and perhaps some historical remarks: it should be accessible to a reader not intimately familiar with the topic treated;
   - a subject index: as a rule this is genuinely helpful for the reader.
   - For evaluation purposes, manuscripts should be submitted as pdf files.

6. Careful preparation of the manuscripts will help keep production time short besides ensuring satisfactory appearance of the finished book in print and online. After acceptance of the manuscript authors will be asked to prepare the final LaTeX source files (see LaTeX templates online: https://www.springer.com/gb/authors-editors/book-authors-editors/manuscriptpreparation/5636) plus the corresponding pdf- or zipped ps-file. The LaTeX source files are essential for producing the full-text online version of the book, see http://link.springer.com/bookseries/304 for the existing online volumes of LNM). The technical production of a Lecture Notes volume takes approximately 12 weeks. Additional instructions, if necessary, are available on request from lnm@springer.com.

7. Authors receive a total of 30 free copies of their volume and free access to their book on SpringerLink, but no royalties. They are entitled to a discount of 33.3 % on the price of Springer books purchased for their personal use, if ordering directly from Springer.

8. Commitment to publish is made by a *Publishing Agreement*; contributing authors of multiauthor books are requested to sign a *Consent to Publish form*. Springer-Verlag registers the copyright for each volume. Authors are free to reuse material contained in their LNM volumes in later publications: a brief written (or e-mail) request for formal permission is sufficient.

Addresses:

Professor Jean-Michel Morel, CMLA, École Normale Supérieure de Cachan, France
E-mail: moreljeanmichel@gmail.com

Professor Bernard Teissier, Equipe Géométrie et Dynamique,
Institut de Mathématiques de Jussieu – Paris Rive Gauche, Paris, France
E-mail: bernard.teissier@imj-prg.fr

Springer: Ute McCrory, Mathematics, Heidelberg, Germany,
E-mail: lnm@springer.com

Printed in the United States
By Bookmasters